WHERE DARWIN MEETS THE BIBLE

WHERE DARWIN MEETS THE BIBLE

CREATIONISTS AND EVOLUTIONISTS IN AMERICA

Larry A. Witham

OXFORD

UNIVERSITY PRESS

OXFORD
UNIVERSITY PRESS

Oxford University Press, Inc., publishes works that further
Oxford University's objective of excellence
in research, scholarship, and education.

Oxford   New York
Auckland   Cape Town   Dar es Salaam   Hong Kong   Karachi
Kuala Lumpur   Madrid   Melbourne   Mexico City   Nairobi
New Delhi   Shanghai   Taipei   Toronto

With offices in
Argentina   Austria   Brazil   Chile   Czech Republic   France   Greece
Guatemala   Hungary   Italy   Japan   Poland   Portugal   Singapore
South Korea   Switzerland   Thailand   Turkey   Ukraine   Vietnam

Copyright © 2002 by Oxford University Press, Inc.

First published in 2002 by Oxford University Press, Inc.
198 Madison Avenue, New York, New York 10016

First issued as an Oxford University Press paperback, 2005

www.oup.com

Oxford is a registered trademark of Oxford University Press

Library of Congress Cataloging-in-Publication Data
Witham, Larry, 1952–
Where Darwin meets the Bible : creationists and evolutionists in America / Larry A. Witham.
p.   cm.
Includes bibliographical references and index.
ISBN-13 978-0-19-518281-1 (pbk.)
ISBN 0-19-515045-7; 0-19-518281-2 (pbk.)
1. Evolution (Biology)—Religious aspects.   2. Religion and science—United States.
3. Creationism—United States.   I. Title
BL263 .W593 2002
231.7'652'0973—dc21   2002022028

1 3 5 7 9 8 6 4 2

Printed in the United States of America
on acid-free paper

PREFACE

When I conducted my first interview for this book in 1995, I was thinking of a short journalistic project. The next five years flew by, of course. Fortunately, they were some of the most eventful and colorful in America's great debate between evolutionists and creationists. A similar interest in science and religion seemed to crest as well. What results in these pages, I believe, is a story of people, places, events, and ideas that is both pro-science and pro-religion. But its predominant theme is how we, a religiously inclined society, try to understand nature, which is mostly the bailiwick of science.

Science can be a hard topic. One may think of it as having three areas to be deciphered, beginning with a giant parts list of the universe (often in Latin). Next are the mathematical formulas that explain the substance of the parts and how they move, and then what scientists call the "metaphors"—tree of life, big bang, double helix, gravity, matter, chaos—by which our minds may grasp the sheer complexity. With simplicity and general readers in mind, I have avoided the mathematics and draw sparingly on the parts and the metaphors. This book talks mostly with scientists and science educators, and as to what science is as a whole, they will all have something to say.

Theologians are in this conversation as well, but in fewer number. Though I have done graduate studies in theology, history of religion, and the Bible, I presume no expertise in my simple treatment of such topics. Whatever craft I may bring to telling or explaining a story was learned in a Washington, D.C., newsroom, where almost daily since 1982 I have covered American society and the current events of religion, and where my colleagues at the *Washington Times* have taught me much. "Journalists are only as good as their sources," says the truism, and in this project I've probably met some of the best sources, in both knowledge and generosity. Appreciation is primarily due to the more than one hundred people who agreed to interviews. Many of them have checked for accuracy in how I used their statements. On occasion, I have had to insert myself into the narrative to give an interview a proper context, since the interviews were conducted over several years.

Neither professionals in science nor those in religion have given journalists high marks for explaining their fields fairly or accurately.[1] During this project, an occasional concern was that a reporter's approach, which sometimes quotes the person on the street alongside the Ph.D. in the academy, might wrongly appear to legitimize one view or another. Fortunately, the journalist's task is not to quibble over "intellectual equals" in a social or academic debate but to give full compass to responsible voices. What is more, this book is a cultural history. Who can doubt that evolutionists and creationists, for example, are equal forces in American culture?

There are several scholars, like the proverbial "shoulders" on which neophytes stand, on whose classic works I have relied for my own initiation. Among them was the historian Edward J. Larson, whom I interviewed for a 1996 news story; we went on to conduct three opinion surveys of scientists and theologians. The findings were published in *Nature, Scientific American,* and *Christian Century* and are further elaborated upon in this book. Stephen J. Gould, one of the most prominent evolutionists of his generation, in 2002 lost his second bout with cancer, and I want to note his kindness in granting an interview for this project. Others to whom I am grateful for having shared original research, sent materials, or commented on chapters include Edward B. Davis, Cynthia McCune, Jonathan Wells, Mark Kalthoff, Roland Hirsch, Doug McNeil, Clyde Wilcox, John Green, and staff at the American Association of University Professors, the National Association of Biology Teachers, the National Center for Science Education, the Institute for Creation Research, and the Discovery Institute.

My warmest thanks to Cynthia Read of Oxford University Press, who backed this manuscript as soon as it flew over the transom and offered sage editorial advice. Thanks also to newsroom colleagues Mary Margaret Green, Stephen Goode, Stacey McCain, Jeremy Redmond, and Cheryl Wetzstein for editorial comments, and to Ken Hanner for supporting news coverage that added to this book. The Library of Congress provided me with a space where, for six weeks, I could hoard books and write during the 1998 Christmas season, and similar thanks are due to staff at the University of Maryland libraries in College Park. During these book-writing years, my son became quite the science student himself. Both he and my wife had the patience of Job as my preoccupation consumed evenings, weekends, and days off. It was a testimony, of course, to their love and support.

February 2002   L. A. W
Burtonsville, Maryland

CONTENTS

WHERE DARWIN MEETS THE BIBLE

INTRODUCTION: WAYS OF KNOWING

"Will it never end?"

This lament may be heard at every new eruption of the evolution-creation debate in America. Begun in 1859 with Charles Darwin's *On the Origin of Species*, the debate has come and gone like a storm. From time to time modern science claims victory in the debate: the scientific "way of knowing" has settled the question. But after a calm, the gale returns.

Darwin meets the Bible just about everywhere in America, at the great intersections where creationism meets evolutionism and where science meets religion. They are daunting crossroads, congested with technical science, sacred theologies, moral concerns, ideological agendas, and political hardball. Most Americans, according to opinion polls, know little of what they might find there. Generally, they have avoided these points of contact like giant traffic jams.

Darwin hinted at what was to come. He called his *Origin of Species* "one long argument." Across its pages, he pressed that argument: nature itself, by the gradual "natural selection" of beneficial traits in organisms, has produced the complexity of the natural world, including the human mind.[1] He explained in naturalistic terms what had hitherto been viewed as a "special creation," a complex and wondrous world put here by the heavenly Ideas believed in by Plato, the aloof Clock Maker of deism, or the personal God of the Bible. When Darwin's argument leapt from the pages of the *Origin* into society, both science and religion became arenas of debate. Still today, science is in polite turmoil over whether Darwin's mechanism of "natural selection" can explain how all things came to be. Religion, too, continues to ask: Does evolution do away with God or refute the Scriptures? What kind of God could coexist with evolution's sweeping claims?

Amid this rollicking debate, some have argued vigorously that the exchange is fruitless and counterproductive. When *Scientific American* reported on religion and scientists, a slew of letters expressed the wish that "the same energy that goes into the science-religion debate could be redirected to improving the world."[2] One popular solution has been simply to separate science and religion entirely. One is about facts, the other about beliefs. The two "ways of knowing" are said to exist side by side—separate and in peace. Not surprisingly, this partition is happily welcomed by most scientists, most theologians, and the general public.

The National Academy of Sciences has promoted this solution, saying science and religion are "mutually exclusive" kinds of knowledge. In his president's address to the American Association for the Advancement of Science in 2000, Harvard paleontologist Stephen J. Gould proposed a "respectful separation." He allowed that the spheres of science and religion should meet in "frequent and searching dialogue," which for strategic reasons is essential in a nation so religious as America. "It's the only way we'll ever talk to the majority of Americans," Gould said. "If they think that science opposes religion intrinsically, how can we ever prevail?"[3]

This simple map for "two ways of knowing," however, does not keep the crossroads flowing smoothly. Science and religion can easily operate in their two separate modes in the laboratory and in the sanctuary, but in society they actively mingle. They meet in culture, education, and politics, and the debate continues: Who draws the line between facts and values? How do we decide what is scientific knowledge and what is religious or philosophical knowledge? Which is more valid—expert opinion or common sense?

When the appeal to separate "ways of knowing" is not enough, those who would end the evolution-creation argument make another case. The debate is an anomaly, they say. It has arisen in only one spot of all humanity, a small enclave of Bible fundamentalism in America.

But clearly, Bible literalism is not the only force rocking the boat of evolution. "The desire to escape Darwinism is a common theme of contemporary thought," naturalist philosopher David Papineau wrote in 1995. "It spreads far beyond creationist circles into the strongholds of secular rationalism. . . . To official Darwinians, this kind of secular skepticism is almost worse than creationism. It is bad enough that people who believe the Bible literally should dismiss Darwin. But members of the scientific community ought to know better."[4]

There is still one more plank in the "argument-is-over" platform. Darwinian evolution states only the obvious—and its triumph is therefore a fait accompli. Evolution rests on the indisputable fact that gene mutations or mixing, called "changes in gene frequency," produce "populations" different from their parents. The beaks of surviving birds have indeed evolved in size; surviving insects and their progeny have without a doubt evolved immunities to insecticides. Since the 1940s, however, even the most literal creationists have granted this power, and even more, to "microevolution," says historian Ronald L. Numbers. To corroborate Genesis, he says, "These people have to get the entire earth populated with all its diversity through microevolution, and they're willing to allow for natural selection to be one of the principal mechanisms."[5]

What the creationists reject are evolution's higher claims on nature—that the human mind, for example, evolved from aimless molecules. They object to the way evolution has defined science, and to the way it influences society.

These grander evolutionary claims have many articulate theorists, among them biologist Ernst Mayr. At his home near Harvard one spring morning, Mayr told me that he was all for separating science from religion. "No scientist would interfere with any believer and what they do with the Bible," said Mayr, whose mother tried unsuccessfully to rear him as a Lutheran. "But believers shouldn't use that to try to undermine or refute scientific statements." With no malice toward religion, Mayr nevertheless says it, too, falls under "the influence of Darwin on modern thought," the topic of his 2000 *Scientific American* article. "Almost every component in modern man's belief system," Mayr said, "is somehow affected by Darwinian principles." He says elsewhere, in fact, that scientists who merely pursue discoveries or "technological innovation" have missed the whole point of Darwin's scientific genius.[6]

That genius has been central to the rise of philosophical materialism, which has aimed to dethrone God and the supernatural. It tells people that only matter in motion is worthy of belief. While this philosophical overthrow turns many religious people against every aspect of Darwinism, other believers will reject the philosophy but take the Darwinian science. A significant group of Christians in the sciences posit that God works through the evolutionary mechanism to "create." They believe that science has improved on ideas about God, yielding a new and improved "theology after Darwin." Evolution is the backdrop for "finding Darwin's God," says biologist Kenneth Miller, who assures fellow theists that there is "no reason for believers to draw a line in the sand between God and Darwin."[7]

Americans like the idea of reconciling God to evolution, according to some surveys. They also like to think that God can intervene in the world he created. Yet here is where Darwin, in his writings and credo "Natura non facit saltum"—nature does not make leaps—presents a challenge.[8] How do biblical faiths live with science's rule that God may not intervene in nature? Prayer is often a request that God do just that. The reconciling gets even more complex, moving from God and nature to mind and matter, fact and revelation, freedom and necessity, morality and determinism.

These sorts of brain twisters have often made Americans want to postpone the evolution-creation debate for another day. In fact, American literacy on the topic is surprisingly low. Many Americans think evolution primarily states that humans evolved from apes. Only 15 percent know what Darwin meant by "natural selection." On the other hand, half of Americans say they have never heard the term "creationism"; just two in ten are "very familiar" with creationist claims. Only four in ten adult citizens, moreover, can name the four Gospels or say who delivered the Sermon on the Mount. It is more than likely that very few Americans know that Genesis has two creation stories.[9]

Despite such ignorance of the issues, nowhere does the debate reach such dizzying heights, and political lows, as in the United States. A general loyalty to the Bible seems to be the catalyst. "The creation story is not going to go away as a political issue, for the obvious cultural reason that the Bible is not going to stop being the central book in our intellectual heritage," says historian Gary Wills.

When Americans are asked to name the most important book in history, they pick the Bible over Darwin's *Origin of Species* by twenty to one.[10]

They also hold the scientific profession in high esteem—more than in any other industrialized nation—and may know that science has fueled half of America's economic prosperity since the 1950s.[11] Naturally, the struggle to balance such fruitful enterprises as science and religious belief has drawn in every other social sector. Under the duress of the evolution-creation debate, judges are pitted against legislators and school boards against courts. The priorities of applied scientists can clash with those of theoretical scientists. In public education, most Americans want both evolution and creation taught evenhandedly, but teachers despair at such juggling. How do we teach academic traditions and academic freedom, and how may religious freedom exist in a bureaucratic society?

So broad is the sweep of the evolution-creation debate that it seems likely to be perennial. But is it helpful? Opinion is obviously divided. But when accepted as inevitable, it certainly can stimulate learning and also test various important social claims. For example, science leaders assert that only evolution education can produce the scientific minds necessary for America to compete in the world economy. Moralists, in turn, argue that mere technical training, or the propagation of a materialist worldview, robs young people of moral values and the nation of moral capital. Ideally, a society should have moral, enlightened people. How to get there seems worth arguing over.

While Americans hotly debate evolution, it is almost unanimously accepted in Western Europe and Japan, the former being the most secular part of the world, and the latter a Buddhist society that has developed scientifically. Still, this does not make America the only nation with beliefs that seem at odds with evolutionary theory. Japan was a secular nation before the Darwinian revolution, and though polls show it is evolutionist today, its people still maintain a moderate mysticism about ancestors and are liberally open to nonmechanistic medicine. Neither has post-Christian Europe bled away all its supernatural and nonrational beliefs under a Darwinian triumph. To be sure, evolution is the view officially held in research universities and science academies the world over. But the argument is unsettled for the masses, from Muslims in Indonesia to Roman Catholics in Latin America.[12]

The United States will not settle this argument for other cultures. Yet what better place is there to keep the argument going? Europe once provided a society where religion gave "presupposition, sanction, and even motivation for science," says historian John Hedley Brooke. But perhaps only the American configuration will allow that interaction to continue. "Neither science nor religion has had a stable and permanent definition in American culture," argues historian James Gilbert. "They continually shift in meaning and in their relation to each other."[13]

The forces for evolutionism and creationism in America have both emerged from the 1990s with powerful new tools and constituencies. For evolution, institutional science has led the way with calls for a new "civic scientist" who can win public confidence. For the first time since its founding in 1946, the Society for the Study of Evolution created a public outreach arm. The National Academy of Sciences established a Web site on evolution, issued the lengthy guidebook *Teaching Evolution and the Nature of Science*, and updated its 1984 criticism of "scientific" creationism. School textbooks, once thin on evolution, since the 1990s have given it "unabashed" coverage. The new science standards movement, which identifies evolution as one of five "unifying concepts and processes in science," was making its mark in all the states; the National Science Teachers Association heard in a 2000 report, "The century-long struggle to have evolution emphasized in the science classrooms of this nation has reached a significant and new stage."[14]

More Americans have entered higher education, and a college education is a significant indicator—though no guarantee—that a person will accept the theory of evolution. From 1971 to 1997, enrollment at a college or university had jumped from 44 to 65 percent of all high school graduates. Evolution has been protected in public education by U.S. Supreme Court rulings in 1968 and 1987, a status confirmed again by the Court in 2000. Eight of the nine high court justices, with Antonin Scalia dissenting, had no interest in reversing a federal court ruling against Louisiana's 1994 disclaimer law. Louisiana had required biology or earth science teachers to say that instruction in the "scientific theory of evolution" was "not intended to influence or dissuade the Biblical version of Creation or any other concept," and to encourage students to "exercise critical thinking" on the subject. Once again, the federal courts put evolution virtually beyond criticism in classrooms.[15]

Federal science is also keen to educate the public about evolution in the wake of the human genome revolution. In 2000, the National Human Genome Research Institute made one of its five-year research goals the question: "As new genetic technologies and information provide additional support for the central role of evolution in shaping the human species, how will society accommodate the challenges that this may pose to traditional religious and cultural views of humanity?"[16]

The popular culture has smiled on evolution as well. News coverage of the antievolution vote of the Kansas school board in 1999 would have pleased H. L. Mencken, who in the 1920s had pilloried Bible-thumping creationists. Public television's Bill Nye "the Science Guy" called the Kansas decision "nutty," and sister program *NOVA* began to publicize its fall 2001 showing of *Evolution*, a seven-part documentary—and mother of all Darwinian telecasts. Evolution has meanwhile expanded its reach into the liberal arts and the world of television talk shows with the eye-catching new field of evolutionary psychology. This new mode of Darwinian interpretation has opened every quirk of humanity—from fashion and sex to sports and Wall Street—to speculation on what "survival value" it had in man's "evolutionary past."[17]

This apparent "triumph of evolution," however, has not hindered similar new strides for the creationists. As early as 1981, the magazine *Science* remarked on the "increasing philosophical skill" of those who attack evolution. Two decades later, science philosopher Robert T. Pennock warned of "the size and renewed power of the movement," now labeled the "new antievolutionists" or the "new creationists." In his book *Tower of Babel* (1999), Pennock launched a highbrow attack on what has become an equally highbrow creationism, which uses mathematics, biochemistry, and philosophy to argue for design in nature.[18]

More than ever, it is clear that Bible fundamentalists hardly exhaust the spectrum of antievolutionists. One of the new critics is law school professor Phillip E. Johnson, a prolific author and speaker who, by putting Darwin "on trial," has demanded a new vigilance on the part of evolutionists. "The basic controversy is the definition of science," Johnson told me in his Berkeley home one summer. "For evolutionists, science explains the world in materialistic terms. If something is outside of science, it is outside of reality."[19]

Biochemist Michael Behe is a tenured professor and Roman Catholic who has no problem with evolution in general. But he angered American science with his *Darwin's Black Box* (1996), a book that said evolution failed to explain the complexity of molecular life. The book received an astounding eighty reviews, most of them in science journals, where its challenge to Darwinism was generally attacked. When his book was still a controversy, Behe gave me a tour of his Lehigh University laboratory, wearing his trademark blue jeans and flannel shirt. "They say to me, 'Well, of course! You're a biochemist. You don't know how to think like an evolutionist,'" he said. "And I say, 'Yes, you're right, because I see these difficulties that nobody has addressed.'"[20]

Behe represents a new criticism of Darwinism, a criticism that sidesteps Genesis and the age of the earth. It uses terms such as "intelligent design" and the "anthropic principle," which states that nature seems to have been fine-tuned for the arrival of human existence. The debate has switched from defending religious scripture to making scientists explain the holes in evolutionary theory. The debate has switched, what is more, to asking why, if Darwinian science is not a philosophy, does it so often lead to disbelief?

These two lines of attack are hardly the invention of creationists alone. Readers of Michael Crichton's novel *The Lost World* (1995) heard the hero Ian Malcolm, a brilliant mathematician enamored of chaos theory, saying, "Everybody agrees evolution occurs, but nobody understands how it works. There are big problems with the theory. And more and more scientists are admitting it." Scientists do not openly advertise the "big problems" with evolution, correctly assuming that creationists will use them for political advantage. Science historian William Provine, meanwhile, is far more candid in spreading his Darwinian gospel that evolution logically leads to atheism. "And that's why the vast majority of working evolutionists are in fact atheists," he says, pacing a University of Tennessee stage.[21]

The bulwark of creationism, of course, is America's religiosity and belief in God. Creationists have long resorted to saying "God did it" wherever science has no answer, and have routinely been criticized for bringing their "God of the gaps" into empirical science.

Yet creationists are finding new metaphors. Just as Bill Gates says, "DNA is like a computer program, but far, far more advanced than any software we've ever created," creationists make a theological argument for "intelligent information" that shapes biological life. In early 2001, the U.S. government and the private company Celera Genomics together released a full DNA sequence of the human genome, the "code of life," and Celera's top computer scientist mused that its complexity suggested "design." He was not thinking of "God or gods," he clarified, but "there's a huge intelligence there. I don't see that as being unscientific. Others may, but not me." Theists have long been inspired by technology, but now they are likening hardware and software to matter and spirit. A generation of computer-literate Americans soon may ask, Is the universe self-running, or functioning on DOS, a divine operating system?[22]

From another direction, moreover, Americans have shown increased reluctance to give science a blank check on every question of the day. A *Science* headline in 1980 told that story: "Public Doubts about Science." It hinted at a troubled love affair; America was becoming disillusioned with scientific progress, an aloof profession, or a "way of knowing" that seems to put more frustrating technical and mathematical demands on life. Doubts about science have also grown as more people are persuaded that knowledge is mere opinion, the cultural relativism commonly termed "postmodern." "The postmodernism movement hasn't been particularly warm and receptive to religions," a humanities professor told a U.S. government commission in 1998. But it did "make clear that the old emphasis on a kind of scientific way of understanding the world is somewhat naive." While the U.S. Senate is anything but antiscience, in 2001 it almost unanimously urged teachers of biological evolution to "prepare students to distinguish the data and testable theories of science from philosophical or religious claims that are made in the name of science." All of these second thoughts about science may add up to a cultural boost for creationism.[23]

Long before DNA, computers, and postmodernism, stories of religion and science have been among the greatest ever told. Genesis narrates how God formed an entire universe. In late Renaissance Italy, Pope Urban VIII brought his friend Galileo Galilei before the Inquisition over how science, philosophy, and theology may view the world. Two centuries later, in Victorian England, Darwin was born, traveled the world as a creationist, and left behind a revolution in science. On American soil, the Scopes "Monkey Trial" of 1925 became a duel of titans: the Bible-believing William Jennings Bryan and the agnostic rationalist Clarence Darrow. Thanks to Scopes, the evolution-creation debate has become America's IQ test. Where you stand can be an instant pass or fail on being modern or backward, faithful or apostate. The snap-quiz approach, of course, is hardly conducive to a healthy conversation.

At a paleontology convention in Washington, I caught up one day with David Raup, a "devout evolutionist" and former senior scientist at the Field Museum of Natural History, and asked him about the debate. "There are fanatics on both sides," he offered. He said that neither evolutionists nor creationists seem willing to learn from an opponent's criticisms. "Unfortunately, since the two sides don't generally talk to each other, there's no decent devil's advocate," he said.[24] Some creationists have closed off the discussion by declaring evolution "the malignant influence of 'that old serpent, called the Devil.'" Some evolutionists have shut it down by warning that the man who doubts evolution "inevitably attracts the speculative psychiatric eye to himself."[25]

The story that follows welcomes devil's advocates on either side. It places where Darwin meets the Bible in the open sunlight. Though a contemporary story, it will frequently reach back to the past. The last chapter will gaze speculatively into the future.

1. DARWIN'S LEGACY IN AMERICA

The Appalachian Mountains run an arching course from Maine to Alabama like a parenthesis on the American East. One of the tallest peaks in that gigantic wrinkle of upturned stone, Virginia's Mount Rogers, is named for a contemporary of Charles Darwin who, like Darwin, was intrigued by nature's beginnings. Geologist William Barton Rogers looked out over his state's folkloric ridges and valleys and asked how they had come to be. His answer, though later proved incorrect, was an early step in a revolution in geology that gave birth to the revolution in biology fathered by Darwin's *On the Origin of Species*.

The formation of the Appalachians was not understood until the 1960s, more than a century after Rogers hypothesized that volcanic eruptions deep in the earth had pushed the mountains into being. Since then, most scientists have agreed that the movement of colossal plates on the ocean floor buckled the land, forming the great mountain range. When the root idea of plate tectonics—that continents move—was proposed early in the century, most scientists reproached it as an "impossible hypothesis" that was "very dangerous" for science, but the public had no emotional stake in the debate.[1] Not so with Darwin's theory that species, including humans, arose from natural selection acting on variations in organisms, now attributed to genetic mutations. Darwin's "descent with modification" proposed that simpler forms evolved into more complex ones, an idea that probed into human origins—our arrival, our nature, and our place in the universe. The resulting cultural debate makes plate tectonics pale by comparison.

"It is curious how nationality influences opinion," Darwin wrote to a friend soon after he learned of the German and French reactions to his *Origin of Species*.[2] But the national character of Americans assured that the tumult over Darwinism would escalate most on this side of the Atlantic. The nation began with a strong religious bent, but treacherous oceans and wilderness added a twist, turning American minds to "nature's God" and making it natural to see the Creator in creation. When it came to science, moreover, the nation has tended to value the practical over the theoretical, which was a more European affection; America's sense of egalitarian social beginnings has also made its society wary of elites, whether clerical or scientific. Finally, the nation was born with an insistence that taxes not be used to spread ideas with which the taxpayer may disagree—ideas of religion, politics, or science, especially in public schools.[3]

These national traits have profoundly influenced the evolution-creation debate. Over the years, this story has played out in many American locations, but it is summarized particularly well by visits to Virginia, Massachusetts, and Tennessee.

■　■　■

The Appalachians in southwest Virginia cross over the Cumberland Plateau, a landmass whose edge rises up twelve hundred feet like a giant doorstep to the city of Blacksburg. Since 1870, the old coal town has been home to Virginia Polytechnic Institute and State University, better known as Virginia Tech, the state's largest research institution. When the plateau bursts with springtime flowers, Virginia Tech botany professor Duncan Porter reaches an apotheosis in his course on plant taxonomy. "We begin working in the lab in January," said Porter, who, when I visited him in 1996, was an avuncular fifty-nine-year-old with a fair beard and gray, thinning hair. "It's only in the last four weeks that we go out in the field, when it starts flowering. And they get very excited about that." Porter hopes his hundreds of students carry away the evolutionary idea that all things are related. "I take evolution as a fact," he said. "Where the theory comes in is not the theory of evolution, it's the theory of evolution by natural selection. Natural selection has some problems, not evolution. All you have to do is look around you, and I don't see how anyone can not accept that evolutionary change takes place."

When Darwin traveled to South America on a sea journey that lasted from 1831 to 1836, he shipped back a cache of plant specimens, on which Porter is the leading authority. Traveling to England to catalogue Darwin's plants, Porter also became an authority on Darwin's writings and the most recent director of the Darwin Correspondence Project. The project will not reach its initial goal—to publish Darwin's fifteen thousand letters in thirty-three volumes by 2009, the bicentenary of Darwin's birth—but Porter realized more than ever the social and philosophical depth of Darwin's revolution. In 1984 he teamed up with Virginia Tech English professor Peter Graham, a tall Connecticut native, to teach a humanities course on Darwin; then they instigated the honors course "Darwin: Myths and Reality."

If the average American student is unfamiliar with Darwin and the Victorian age, said Professor Graham, they are not much better on literature, including the Bible. Only a few of his students know the story of Job, the Old Testament treatise on suffering, natural evil, and justice in a God-made world; it is the very conundrum Darwinism answers by saying the laws of nature are what dictate the suffering. Students have a "cluster of ideas" about Darwin, some sound and others not, Graham said. "That he was a bald man with a big bushy white beard, looking like your stereotypical Victorian patriarch, and that he was an invalid. That he was a recluse. That he had these atheistical scientific ideas that were this enormous challenge to a very rigid and orthodox religious world." The students learn the family history—that the Darwin and Wedgwood families often intermarried, and that the men were freethinkers, but the women were pious.

"Another myth that students have is this idea of the scientist as isolated genius, working alone and coming up with world-changing ideas," said Professor Graham. "The effort for Darwin was collaboration, all his life long. Gathering information from people all over the world, and bouncing his ideas off of his friends." Students are also astonished that Darwin, being independently wealthy, could return from the Galapagos Islands adventure and abruptly retire to a rural home outside London, spending his remaining forty years in quiet research and voluminous writing. "That really surprises them that having to work for a living was an obstacle to someone's scientific interests." In their introduction to a later book, *The Portable Darwin*, Porter and Graham explained quite plainly what questions Darwin had grappled with, questions that he bequeathed to each new generation: "Is there a place for God in a naturally evolving world? If so, what kind of God?"

Nearly a century and a half after publication of Darwin's *Origin of Species*, Porter represents another Darwin legacy: the career evolutionary biologist. A native of the California Central Valley, Porter loved nature, studied at Stanford and Harvard, taught biology at the University of San Francisco, and then became a curator at the Missouri Botanical Garden. His eastward drift finally carried him to Washington, D.C., where for a year at the Smithsonian Institution's National Museum of Natural History he was editor in chief of the Flora North America project. Then he joined the National Science Foundation, where in the early 1970s the panel he served on awarded $5 million in grants each year for research. Most of it went for "getting data on evolutionary changes."

The field of biology itself has changed dramatically since Porter entered the guild. Hands-on scrutiny of specimens has given way to mechanical analysis of the molecular structure of tiny DNA samples or the running of mathematical models on a supercomputer. "When molecular biology arose in the 1960s, natural history sort of became passé," said Porter. In thirty years of classroom experience, he has been struck more profoundly by these scientific changes than by the perennial evolution-creation debate. "I've never had a student come up and argue evolution or creationism with me," he said. When dissent arises—infrequently—it is reflected in teacher evaluations. "It has happened only a couple of times, but a student may write, 'Dr. Porter better look out and give up this evolution and go back to God.'"

He takes no offense. Since his arrival at Virginia Tech in 1975, he and his wife have nurtured their four children in the Episcopal Church, in which he is a communicant. "Well, I believe you can be a Christian and an evolutionist," said Porter, who was reared a Methodist but professed to be an agnostic during most of his career. "In fact, I am a Christian, and I am an evolutionist."

In this, Porter differs from most evolutionary biologists, who generally are agnostics, as Darwin became, or atheists. Porter's greatest affinity is to the 87 percent of Americans who say they are Christian. That affinity narrows down, however, when it comes to evolution, for only about 40 percent of Americans would agree roughly with Porter that God has "guided" evolution over millions of years (which he personally qualifies as "in the sense that natural selection is one of God's laws that regulate the universe"). Another 44 percent of Americans, according to a 1997 Gallup poll, embrace the creationist stumbling block to science: they believe God brought humans, and perhaps the earth itself, into being by a "special creation" only thousands of years ago.[4] At the time that the HMS *Beagle* sailed, every American believer was a creationist, and so was the young Darwin.

. . .

Darwin never visited the United States. If, on his journey, he had docked in America, it would have been during a "golden age" of geology. Despite our association of Darwin with tortoises and finches in the Galapagos Islands, he considered himself a geologist. As he wrote to his sister, "There is nothing like geology; the pleasure of the first days partridge shooting or first days hunting cannot be compared to finding a fine group of fossil bones, which tell their story of former times."[5]

In Virginia at that time, the premier name in geology was William Barton Rogers. A Scotch-Irishman born to a Presbyterian family in Philadelphia, Rogers began his career as a professor of natural philosophy, or science, at the University of Virginia. No naturalist is more famous than Darwin, but a look at the careers of the two contemporaries illustrates the contrast between the European and American ways of doing science.

Rogers's father was also a science teacher, who so admired Erasmus Darwin, grandfather of Charles, for his book on organic evolution that he named his third son Henry Darwin Rogers. William, the second son, had at his father's knee met former president Thomas Jefferson, who was building his university down the hill from his home at the Appalachian-hugging Monticello. In 1835, while Darwin was on the high seas, William Rogers was appointed Geologist of Virginia and began a five-year geological survey of the state, which then was twice its modern size. Contemporaries claimed that he took the "first broad reading of American geology."[6]

In the 1830s, state geologists were the captains of American science. The American Association of Geologists and Naturalists was the first national scientific body. William Rogers was its chairman in 1847, when the group dissolved, to be reborn the next year as the American Association for the Advancement of Science, now the world's largest science federation. Rogers married into a Boston family and, with his wife, took a first European tour to attend the 1849 British Association meeting in Birmingham, England. There he met Charles Darwin. By that time, Darwin had written down his theory of evolution by natural selection, but he was circumspect with both the earliest sketch of 1842 and also the longer *Essay* of 1844, in which he had elaborated on his theories, leaving his wife money to publish the *Essay* after his death.[7]

In the year of the *Essay*, Darwin joked in a letter about what his theory meant: "Species are not (it is like confessing a murder) immutable."[8] Before and immediately after the *Beagle* voyage, Darwin still believed that God had laid down the species by special acts of creation. Seven months after he returned home, he "firmly believed in the gradual origin of new species."[9] By 1838 he believed that nature alone, without divine action, could create new species. The power to create, he said, was found mostly in the mechanism of natural selection, which is "daily and hourly scrutinizing, throughout the world, every variation, even the slightest; rejecting that which is bad, preserving and adding up all that is good."[10] This was not an idea necessarily seen in nature. Darwin was drawing as much on the economic treatise of Thomas Malthus. The Malthusian viewpoint helped Darwin cast nature in terms of hungry, multiplying mouths amid limited food. It was truly a "struggle for existence."

When Rogers met Darwin, however, all this was still in the Englishman's head. Describing the Birmingham meeting in a letter home, Darwin wrote of his boredom and "all the spouting" at the sessions. He also said he had gotten sick. Rogers's letter home was written in exultation. "I made quite a respectable speech, which was often loudly applauded," he told his three brothers. Four days later Rogers presented his "law of flexures" theory, which proposed that the Appalachian ridges and valleys had been pushed up or folded over by volcanic pressures below. "They laid on the compliments so thick that I could hardly stand up under them."[11]

Having observed the Andes Mountains, Darwin also thought that volcanic pressure had raised them. He and Rogers, whose wives were both named Emma, had also shared a view of geology that, unaware of tectonic plates, was wrong. But it mattered little. The greater revolution in geology was taking place around the question of time—and its vastness in natural history.

The new view of time was epitomized in the 1830 work *Principles of Geology* by lawyer-turned-geologist Charles Lyell, a friend of Darwin. Lyell posited that the same laws of erosion and accumulation observed working so gradually in nature during his lifetime had shaped the earth eons earlier. Natural history, said Lyell, was ruled by the "undeviating uniformity of secondary causes," an explanation that came to be described as "uniformitarian." Nature, Lyell said, needed no intervention by the primary cause—except in the divine creation of the human mind. Darwin said the new outlook "altered the whole tone of one's mind." Lyell's idea paved the way for the demise of the preferred belief of Victorian Anglican religion, God as designer.[12]

In triumph, uniformitarian thinking eclipsed the scriptural view of Creation by cataclysm, followed by a global flood. Darwin, a former theology student—and Anglican priest, if not for the *Beagle* voyage—waxed eloquent about the new view of time and nature. It "impresses my mind almost in the same manner as does the vain endeavor to grapple with the idea of eternity," he said. Boundless time could create boundless natural variety, and so Darwin asked rhetorically,

"What may not nature effect?" Time was the Creator. "The belief that species were immutable productions was almost unavoidable as long as the history of the world was thought to be of short duration." A revolution in geology had spawned a revolution in biology.[13]

In the United States of the 1840s, the Darwinian revolution was still on a far horizon. Funding for Rogers's geological survey of Virginia, begun in 1835, had come only with a struggle. The state's economic depression had caused a brain drain, and the tensions that would erupt into the Civil War were building. Rogers looked north. His in-laws lived in Boston, as did his younger brother, Henry Darwin, with whom he had long shared his dream of opening a "polytechnic school" in the city. In Virginia, Rogers had seen riots and murders at the university, and he envisioned a more studious atmosphere. The polytechnic would have a practical focus, the brothers had said, but equally important, it would be free of the kind of political purse strings that thwarted scientific research in Virginia. Their dream was realized in 1862, when William founded the Massachusetts Institute of Technology and became its first president and first professor of physics. Henry by then had moved to Scotland to teach.

The Civil War erupted in 1861, two years after publication of *Origin of Species*, and while the war stalled the wider American debate, the Rogers brothers were privy to Darwinian claims from the start. Henry was in Scotland at the time and was witness to British naturalist Thomas H. Huxley's agitation for a priesthood of scientists to supersede the religious aristocracy. "'Darwin' is the great subject just at present, and everybody is talking about it," Huxley wrote to Henry Rogers. "The thoroughly orthodox hold up their hands and lift up their eyes, but know not how to crush the enemy." Henry agreed with Huxley that though Darwin doubtless was correct, he had not proved his theory by demonstration. "Development of species from species, firmly as I believe in it," Henry wrote his brother, "I think it will never be capable of a strictly scientific proof. No more can the opposite doctrine of supernatural creations, and therefore the main point to insist on now is toleration, and no dogmatizing."[14]

His brother William also believed before 1859 in the transmutation of species by means of either "violent and sudden physical changes" or "the gradual modification of species through external conditions."[15] And in Boston in 1860, William Rogers's opinions of *Origin of Species* were as conciliatory as Henry's. Reviewing the *Origin* in the *Boston Courier*, Rogers said that "probably a large majority" of naturalists would hold to the biblical doctrine of immutable species in the face of Darwin's claims. "It is, however, certain that arguments emanating from so philosophical a thinker, and presented with such fairness and simplicity, will . . . in many cases win, at least, their partial assent."[16]

Boston became the first American center of Darwinian debate. In the year of the *Origin*, Harvard regally opened the Museum of Comparative Zoology with a long procession from the museum to the church, with the governor at the front. The museum was founded by the Swiss naturalist Louis Agassiz, the son of

Calvinist clergy—though now a Unitarian—and perhaps the biggest name in American natural science. Agassiz rejected evolution for its failure to explain the building up of complexity in organisms. As a European he also was wedded to the Continent's philosophical idealism and viewed organisms as ideas in the mind of God.

In Boston's first public debates on the evolution topic, presented in four sessions in the spring of 1860, William Rogers held forth as Darwin's ally and Agassiz as his critic. They packed the Boston Society for Natural History. According to historian Edward J. Pfeifer, "When Agassiz and Rogers clashed, the show was worth seeing. Agassiz was handsome, impetuous, and eloquent, but unguarded in speech. Rogers had sharper features, was always alert, and possessed a keener sense of logic."[17] They argued about "persistent types" in nature, geological layers in North America, the alleged migration of species between continents, and the uplift of rocks.[18] William Rogers recounted these "friendly contests" with Agassiz, but he perceived accurately that besides Asa Gray, the Harvard botanist and friend of Darwin, he was having "to do battle almost unaided."[19] Other American naturalists were mum. "The real issue at stake was whether Agassiz's or Darwin's principles would guide future scientific research," says Pfeifer. "Each provided a coherent view, but both could not be right."[20] Agassiz, who died in 1873, was the last great antievolutionist of the American scientific establishment.

Darwin, like Agassiz, did not live to see how the contest fully played out. Darwin died in 1882 at his home, Down House, where according to the custom of the gentry, he had retired in 1842 at age thirty-six. From there he led a busy life of writing, experimenting, and taking trips with his family. The retching illness he had complained of in his Birmingham letter of 1849 was an early sign of ailments, still mysterious to modern doctors, that increasingly afflicted his life. One April night in 1882 he was overtaken by convulsions; the next day, unconscious, he took his last breath in the arms of his wife, Emma. His cousin Francis Galton urged an entombment in Westminster Abbey. Seeing a boon to the scientific priesthood, Huxley shepherded such a petition through the House of Commons. "Getting a free thinker in the Abbey was not easy," say historians Adrian Desmond and James R. Moore. But it worked. The Unitarians, that freethinking wing of British Christianity in which Darwin felt most at home, predominated as pallbearers. But it was from the Anglican pulpit of Saint Paul's Church in London that the abbey burial was declared a sign of "the reconciliation between faith and science."[21]

News of Darwin's death reached the United States, where William Rogers, a nominal Presbyterian with ties to Unitarians, was spending his last years at his Rhode Island cottage. His final project was a small geological map of Virginia. He, like Darwin, had battled illness for years, but on graduation day at MIT in 1882, Rogers climbed the steps of the institute in the back bay of Boston to speak at the outdoor commencement. It was a cloudless Tuesday, May 30, not five weeks after Darwin had been lowered into the abbey crypt. When Rogers stood to give a "short address" at noon, his weak voice rose to "thrilling tones" but then fell silent. "That stately figure suddenly drooped," a witness recalled. "He fell to the platform instantly dead."[22] In his last words Rogers was saying that theory had long been separated from application. "Now . . . ," he said, "the practical is based upon the scientific, and the scientific is solidly built upon the practical."[23]

In a simple funeral at the institute on Friday, June 2, Rogers's friend the Reverend George Ellis read the liturgy. He said that science could not speak on the body and soul. Even science's cunning devices were "baffled when they touch that mystery." Yet he likened Rogers—who was called the "Nestor of American Science" as president of the National Academy of Sciences—to a high priest in the temple of science. "He ministered at its altar of nature, unrobed indeed, yet anointed with a full consecration."[24] The next year Virginia's highest mount was given his name.

Rogers today is better known at MIT than in Virginia, where few seem to know who is commemorated by Mount Rogers, which borders North Carolina. In the spirit of science, Rogers and Darwin were united as fellow practitioners. They both accepted transmutation of species, though differently—Rogers by catastrophism and Darwin by natural selection and gradualism. Their greater difference was between the American and European mind-set, one inclined to practical science, the other the theoretical. Hardly a year passes without a speaker at MIT quoting Rogers on "useful knowledge." In contrast, Darwin tried to answer the "mystery of mysteries" with the *Origin of Species* and, in 1871, the *Descent of Man*. In America Rogers pondered how to use Virginia's chemical deposits for fertilizer. He was Massachusetts's first gas meter inspector, and he set scientific standards for the readings. The National Academy of Sciences was founded amid a debate between elite science and its freelance practitioners, and if its first presidents favored a European and purist approach in the academy, Rogers as president emphasized its "obligation to bring to the attention of the government scientific matters relevant to the public welfare."[25]

In modern America this older divide between applied and theoretical science has blurred, though it still adds fuel to the evolution-creation debate. Many creationist elites are in the applied sciences. They look askance at so much evolutionist philosophizing in natural history. Evolutionists argue that pure science is profoundly different from mere technology. Modern science is seamless with nature's past, the elites of evolution say, suggesting that "creationist engineers" lack the scientific imagination to understand.[26]

．　．　．

The elites of Thomas Jefferson's day were the established clergy. Ousting them from government power, in fact, was his way of addressing two more American issues that today drive the evolution-creation debate: the control of knowledge by a special class and the collection of taxes.

Tax battles in the states were first waged to end church hegemony, such as Anglican levying of taxes, or curtailing of dissenters like Baptists, in Virginia. The enactment of Jefferson's Bill for Establishing Religious Liberty in Virginia in 1786 solved the immediate problem. Jefferson's words, however, remain problematic for the teaching of evolution, let alone sectarian religion, in tax-supported schools: "To compel a man to furnish contributions of money for the propagation of opinions which he disbelieves and abhors, is sinful and tyrannical."[27]

By unseating the religious elites, Jefferson also sought to secularize education—and put it under new arbiters of culture. His vision for the "diffusion of knowledge" is applauded today by the National Academy of Sciences, which quotes his assertion that "no other sure foundation can be devised for the preservation of freedom and happiness."[28] Jefferson had begun his secularization attempt when, on the board of governors at William and Mary College, he succeeded in "erasing theology from the curriculum." He founded the University of Virginia in 1825 as an "academical village." When its secular vision was met with social protests, an ethics professor was hired to teach "the proofs of the being of a God, the creator [and] author of all the relations of morality."[29]

Jefferson's secular ideal would not blanket American colleges and universities until the 1960s, when the natural sciences and evolution were already firmly established in higher education. By then, of course, the sage of Monticello's concession to what one historian calls a "nonsectarian religious education and moral formation" on campus was viewed as incredible. Yet soon after Jefferson's death, a spirit akin to the French Revolution swept Virginia, curtailing the role of theology professors in education, clergy in politics, and churches in landholding. "The Jeffersonian tradition in Virginia, while admirably zealous for the separation of church and state, often treats religion as so much a private matter that it should have little to say in the public realm," lamented the white-haired Episcopal bishop Peter Lee of the Diocese of Virginia in 1998. He could have spoken for all of America when he described a cultural tension in his own state—"independent, Bible-centered congregations with inherited suspicion of cities, universities, and contemporary culture."[30]

By the 1970s, America as a whole nurtured the same cultural atmosphere. It spawned an evangelical Christian revival and a new evolution-creation debate. The ferment put Jimmy Carter, the nation's first "born-again" candidate, in the White House and prompted *Newsweek* to dub 1976 the "Year of the Evangelical." On the conservative wing of the revival, meanwhile, was born the new Christian right, the new creationism, and a mass media vehicle for both—religious broadcasting. Virginia was center stage. Baptist pastor and broadcaster Jerry Falwell, born on the banks of the James River, founded the short-lived Moral Majority in 1979. Pat Robertson, the son of a U.S. senator from Virginia, built up his Christian Broadcasting Network audience to the point where he could run in the 1988 presidential primaries. "When people ask me if I believe in teaching creationism in schools, I ask them, 'Do you believe in teaching the Constitution?'" Robertson said during the New Hampshire primary. "Just imagine how it would sound, 'All men are endowed by the primordial slime with certain inalienable rights.'"[31]

Though the Robertson allusion had a generic quality to it, a very specific kind of creationism was in prominence at the outset of this politically colorful period. Its trademark names varied, from "flood geology" and "young earth creationism" to "creation science." But the common point was that science could corroborate an earth only several thousand years in age and a global flood as described in Genesis. First advocated by a Seventh-day Adventist teacher in *The Modern Flood Theory of Geology* (1935), this model of creationism was revived by an engineering professor, Henry Morris. His book *The Genesis Flood* (1961), coauthored with an Old Testament theologian, carried what Morris called a "strict creationism" to an ever wider Protestant audience.

That turning point came when Morris was professor of hydraulic engineering at Virginia Tech. He finally became chairman of the university's civil engineering department. "When I came in 1957, Virginia Tech was pretty small," said Morris. "It did grow quite a bit while I was there. We did get a lot of government funding." Morris had moved from being a lukewarm Southern Baptist evolutionist in Texas to national leader for "creation science."

Ever since his first book, *That You Might Believe*—published in 1946 to help students reconcile the Bible with history and science—Morris had worried over students' souls. "I'd seen the devastating effect that evolutionary teaching had had on so many young people from Christian homes," Morris told me in his office at the Institute for Creation Research in Santee, California, outside San Diego.[32] When at Virginia Tech, "I tried to help by forming a church." He also met the young Falwell, who in Lynchburg, just east of the university, founded a Baptist congregation in 1956 and gave his first radio sermon six years later. When Falwell's Bible college became a university in 1985, it also opened a creationist museum. And while the Baptist pastor was nothing if not political after 1979, it was other activists in the new Christian right who carried the creationist cause into politics and public schools.

Morris did not see politics as the best antidote for the godless times. "My dream, as I used to call it, was a Christian university something like Virginia Tech, with all the external outreaches and research programs, and from a Christian creationist point of view." Never shy in professing his outlook, Morris once packed a Virginia Tech science hall to present the case for a young earth, which ascribed the Appalachians to a global flood and the massive earth movements linked to it. By insisting that God could act directly upon nature, he defied the uniformitarianism of modern geology and revived the catastrophism that had dominated Western science until 1830.

Morris's controversial lecture at Virginia Tech was no catastrophe. "Even that kind of confrontation did not hurt anything," he said. "It might have crystallized the opposition among the faculty. I do think that all led to the reasons I left Virginia Tech." Morris departed in 1970 after thirteen years on the faculty to found his institute, a place where he could write and publish and train a hoped-for army of future flood geologists.

.   .   .

Traveling southwest from Virginia Tech, the rolling Interstate 81 crosses the Appalachians, intersecting Interstate 40, which leads to the university town of Knoxville. Deeper still into the Tennessee River Valley is the city of Dayton, with its courthouse-turned-museum.

Back in 1925, when the city's iron and coal smelting suffered an economic slump, the town fathers pulled off one of the great feats of American boosterism. They staged the John T. Scopes "Monkey Trial" to attract attention—and business. The economic benefit was negligible, but the "sleepy little town among the hills" gave America its "trial of the century."[33] The trial also gave America its historic memory of antievolution laws. Tennessee got its law in 1925, and it lasted until 1968, when the Supreme Court struck down a similar Arkansas statute.

But Tennessee is 43 percent Baptist. It was a dissenting minority when Jefferson had defended it but now is America's largest single Protestant group. Baptists make up the largest cluster of religious identity in twenty states, giving creationism a boost by geography, including in Tennessee.[34] So the lawmakers reinstituted a Tennessee antievolution law in 1973—an action vacated by the courts—and then tried again in 1996. This time they pushed a provision to discipline teachers who taught evolution "as more than a theory." That effort was killed by a committee vote. But it brought to Tennessee an army of lobbyists and film crews, stirred a slumbering national media, generated a month of headlines about "Scopes II"—and drew some famous names in evolution.

Days before the vote was to take place, the British evolutionist Richard Dawkins rolled into Knoxville as part of a three-stop U.S. speaking tour. He ended up in Atlanta to receive the 1996 Humanist of the Year Award. "Science has all the virtues of religion, but none of its vices," he exhorted the assembled members of the American Humanist Association. "The main vice is faith."[35] The short and dapper Dawkins, whose good breeding and Oxford University chair prompted someone to call him "Darwin's greyhound," conveyed to America the Old World esteem for theoretical science. "I shall be making lots of such tours," he told me in the Atlanta hotel lobby after accepting his award, conjuring images of a bygone era, the Gilded Age when Thomas H. Huxley's 1876 tour between Boston and Washington, D.C., was celebrated as a "royal walkabout."

Huxley had been called "Darwin's bulldog," and he had spread the newly minted concept of agnosticism. "The evangelism of science was beginning to produce its own Great Awakening," says historian Desmond.[36] But Dawkins did Huxley one better. "I mean, you have to be agnostic about fairies," he said, "but we all know they don't exist, and that's the way I feel about a deity."[37]

One hundred and twenty years separate Huxley's and Dawkins's tours of the United States, but both men stand as popularizers of evolution for their age. Huxley played his evangelistic role at Chickering Hall in New York City, where he famously rolled out four fossil horses from small to big as "the demonstrative evidence of evolution." British science was viewed then as far superior to the American version, so it was no small prize that "Huxley was applauding the United States" for unearthing in Nebraska the best proof to date of evolution—the horses. The American awe of British science has waned, of course. And so while a prominent citizen gushed that "the whole nation is electrified" about Huxley's visit during America's centennial in 1876, the proper metaphor for Dawkins's tour in 1996 was the computer age.[38]

Dawkins's popular book *The Blind Watchmaker* (1986) included a coupon for a computer program to produce "biomorphs"—creature shapes that evolved on the computer screen as the viewer selected a particular crossbreeding and number of gene mutations. It was evolution by computer selection. I asked Dawkins what he thought of the assertion by the U.S. National Academy of Sciences that religion and science were "mutually exclusive" ways of knowing the world but were not in conflict. "I think it's a cop-out," he said. "And it's a cowardly cop-out." The evolutionist concession that religion is a valid kind of knowledge is simply "an attempt to woo the sophisticated theological lobby and to get them into our camp and put the creationists into another camp." It may be good politics. "But it's intellectually disreputable."[39]

Soon after the Scopes II spectacle had died down, the Tennessee Darwin Coalition organized itself and gave birth to its centerpiece event, a statewide Darwin Day on February 12, 1997, the 188th anniversary of Darwin's birth. On the second Darwin Day in 1998, a promotional flyer deemed evolution "part of our common cultural and educational heritage—not just the domain of an elite group of scientists. We need to be sure that evolution is freely discussed in classrooms and at the dinner table, and not just locked up in an ivory tower."[40] For its second commemoration, financial support flowed from the federally chartered American Institute of Biological Sciences. Evolutionists at other universities inquired about imitating the University of Tennessee model—films at the student union, publicity on twenty-three evolutionary biology courses in its curriculum, and a high school essay contest that asked, "Why should all Tennesseans support teaching and learning about evolution?"

To cap Darwin Day 1998, Cornell University historian of biology William Provine was the keynote speaker but not the only major name in evolution drawn to Knoxville for the celebration. On Darwin Day eve, high school teachers were invited for a training session that included Eugenie Scott, a midwesterner who was director of the National Center for Science Education near Berkeley, California, the leading anticreationist group. Creationism, she explained in her overview, evolves strategically. Once calling itself "creation science" or "abrupt appearance" theory, it may now show up as a demand for textbook disclaimers that evolution "is only a theory" or a request that the "intelligent design" idea be included in classroom biology. She and Provine are peerless as naturalists in science who promote the grand theme of evolution. But at the Knoxville crossroads, they parted ways on how evolutionists should deal with America's religious culture, its populist politics, and the uneasy status of scientific elites.

Scott represents the first approach. By 2000 she had spent a quarter century in this debate and had worked closely with science and educational groups, from the National Academy of Sciences to state teachers' unions. She tells them that in America people cannot be forced to make an "either-or choice" between religious belief and evolution. "That's part of my message to scientists," she said. "You have to allow people to accommodate their religious views to science; otherwise science is going to lose its attraction." She calls this a "statesmanlike" approach. It grants respect to religious faith in hopes that religion need not enter the public science classroom—where it can only slow that learning process. "In my opinion," she writes, "using creation and evolution as topics for critical-thinking exercises in primary and secondary schools is virtually guaranteed to confuse students about evolution and may lead them to reject one of the major themes of science."[41]

At the conclusion of Darwin Day 1998, the university auditorium filled for the address by Provine, a Tennessee native reared as a Presbyterian and son of a philosopher. As one of America's most candid evolutionists, he represents the second evolutionist approach. Seeing Scott in the audience, he points out the contrast. "She works tirelessly for evolution," he says of Scott. "And since she's here with us on Darwin Day, she will tell you there is no conflict between 'good religions' and science." Provine's colorful PowerPoint projection, seasoned with humor and musical ditties, shows the Gallup poll finding that just four in ten Americans say God "guided" evolution. Nearly half of Americans, however, are creationists who could not possibly reconcile evolutionist science and religion, as the Scott approach prescribes. So Provine recommends brutal honesty.

"Evolution is the greatest engine of atheism," he says. Attempts to join evolution with God are futile, as seen in beliefs that God is simply natural law itself or that God created but now is silent. "Those gods, frankly, are worthless," Provine says. "They don't give life after death, they don't answer prayers, they don't give you foundations for ethics. In fact they give you nothing." In case the audience still was unclear about the meaning of evolution, Provine shows an image with cheerful banjo accompaniment: "When you're dead, dead, dead, you are gone, gone, gone." About 10 percent of Americans are at home with this belief: that there is evolution, but there is no God.

Following Provine's view, the public should know that evolution is a slippery slope to disbelief, but in a democracy, such ideas must win by evidence and persuasion, not scientific dogma. So the best classroom pedagogy is to let creationist students speak out and let the youthful debate begin: it only makes dull science class exciting, Provine told me. "You can't shut up a half or three-fourths of the kids in your class," he said. "The creationist kid can go home and say, 'Mom and Dad, you should have seen how I put down that evolutionist in class today!' Why should creationist parents be upset at that? Why should the parents of kids who believe in evolution be upset with that, when they haven't brought up their kid to know enough about evolution so their kid could refute the creationist?"

A scientific optimist, Provine believes evolution will win in the end. Scott worries that stirring such classroom conflict will only baffle students and rob America of its future scientific minds. She, too, has an idealistic goal: that Americans understand the scientific method and its bona fide theories, from gravity to evolution. Provine would not disagree, but he is not the kind of person who is asked to sit on diplomatic federal science panels, a common experience for Scott.

Before his keynote address at Darwin Day, I asked Provine, "What truths about evolution must be taught in school?" He said, "I think you'd be very hard-pressed to tell me the uncontested truths of modern evolutionary biology." What about the fact, then, that nature must come from nature? "Oh, OK," he said, pretending he was impressed. "Does that solve the problem of species?" Evolutionists still do not agree on what a species is, he said, and speciation in the wild has hardly been observed. "A book about that would occupy maybe ten pages. A book about all we know about natural selection in the field, with best examples now, would be a book about yea thick." He showed a gap between his fingers. "Less than a half-inch thick. Big print!"[42]

The evolution-creation debate in the United States began with a book, *On the Origin of Species*. To chart the relationship between the Darwinian legacy in biology and religious belief in twentieth-century America, two greater books of the Western mind take prominence: the book of Scripture and the book of nature.

2. THE TWO BOOKS

The American Museum of Natural History has stood at Central Park West and Seventy-ninth Street in upper Manhattan since 1874, a kind of Evolution Central for America. To prepare for the end of the twentieth century, the museum opened an exhibit on biodiversity. One display declared, "Evolution has produced life's stunning array of species." Another used the most pessimistic numbers to warn that thirty thousand of them—mostly insects, fungi, and bacteria—are wiped out annually.

To help save this biological web of life, the museum also hosted in 1998 the "Religion and Ecology" conference. It represented a congenial bridge between the two sides, or "two books," of Western culture, Scripture and nature, which together have been credited with no mean feat: "The conception of two books, authored by a single all-powerful and all-wise God, became a major presupposition in the development of modern science."[1]

Though penned first by an early church theologian, the two-books metaphor was shepherded into nascent science by the statesman and essayist Francis Bacon. The Enlightenment was happy to separate the two books, though, and today the New York museum draws three million visitors each year to read from the book of nature alone. The "single-book" approach is not exclusive to science, however. In some segments of Christianity, the Bible trumps science in any conflict over the facts of natural history.

Between these polarities—*sola Natura* and *sola Scriptura*—American science and religion have tested an array of two-book combinations. The dichotomy is never so simple, but a survey of how Christian belief and biology have interacted in twentieth-century America illustrates the dynamic. That survey can unfold in two periods, the first from the 1925 Scopes trial to the 1959 Darwin Centennial. The second begins in 1960, when the Broadway play *Inherit the Wind* was adapted for the silver screen, and continues to the present day.

FROM COURTHOUSE TO CENTENNIAL

The antievolution crusades of the 1920s seemed to fulfill a prediction made thirty years earlier of a growing "warfare of science with theology." The protagonists of the Scopes "Monkey Trial" might have come from central casting: Clarence

25

Darrow, the great skeptic, defended evolution as enlightened thinking; William Jennings Bryan, the Presbyterian and Democratic populist, spoke for simple people of faith.

Apart from that courthouse contest, Bryan had another great rivalry going—with the Christian evolutionist Henry Fairfield Osborn, director of the American Museum of Natural History. Their public dispute was more emblematic of changes in how America would view science and religion than even the eight-day legal battle in Dayton, Tennessee. Osborn had denounced Bryan's antievolutionism in the *New York Times* and then in a small book, *The Earth Speaks to Bryan*. In effect, Osborn was saying that science was the province of professionals, not commoners with Bibles. Bryan, however, was loath to give up on science as a democratic endeavor, and so the politician paid his five dollars in 1924 to join the American Association for the Advancement of Science.[2] Bryan's private letters showed that he accepted the evolution of plants and animals, but he drew the line at humans, and on the witness stand he argued from the Bible, the great democratic book open to every American.

In putting God behind all evolution, including that of humans, Osborn argued from scientific knowledge. "So there was the dean of American evolutionists saying all you need to do is go outdoors and you can see the Creator," says Provine. "I don't believe the split between evolutionists and creationists was so big. It was more between creationists and theists"—between a biblical God and a deity behind all natural processes.[3] The clash between Osborn and Bryan mirrored the chasm that was opening in Protestant America, a division of modernist and fundamentalist believers.

Two great pamphlet crusades of that period show how Protestants took sides. Nobel laureate Robert Millikan, a physicist at the California Institute of Technology, was a modernist Protestant. During Bryan's crusade for the Bible, Millikan led a group of scientists to declare the "two books" were not in conflict. One book was for fact, the other for morals. Between 1922 and 1931, he joined with five other scientists and two theologians, both liberal Baptists, to write nine pamphlets on science and religion. Their theme: science made modern faith possible. The project was underwritten by John D. Rockefeller, a Baptist layman whose money founded the University of Chicago. Rockefeller's wealth was also building a modernist cathedral in Manhattan for Baptist divine Henry Emerson Fosdick, who wrote two of the pamphlets, one of which was titled "Religion's Debt to Science."[4]

The orthodox Protestants, meanwhile, had already launched a pamphlet crusade. Between 1910 and 1915, their best thinkers had drafted ninety doctrinal essays, collected as *The Fundamentals*. But only two of the essays were on science, and even they were devoid of "sharp polemics against biological evolution." Darwinism was rejected, but "the overall discussion allow[ed] some limited room for development of species."[5] In 1920 a Baptist newspaper editor drew on the essays to coin the term "fundamentalist." The entire project, a "testimony" in the face of modernism, was paid for by California oil magnate Lyman Stewart, whose Union

Oil Company competed with Rockefeller's Standard Oil. In financing *The Fundamentals*, Stewart also had fueled a new theological movement called "dispensationalism," which put Scripture over science by focusing on the imminent return of Christ; the world became a sinking *Titanic* and science just one of the proverbial deck chairs. Dispensationalism had a monumental impact on conservative Protestant views of science and on the rise of late-century creationism. To parallel the predicted cataclysmic end to the world, the dispensationalist theology required an equally dramatic beginning—the cataclysmic six-day creation.

Looking on in these years was another group of Protestants, the Unitarians. Having discarded the supernatural, a cadre of them invested their idealism in science and reason. The result was "religious humanism," a movement in which "religion adjusted to an intelligent naturalism." The faith extolled science but remained religious "because a concern for human values has always been at the heart of religion." For those who liked credos, the movement was summarized in the Humanist Manifesto of 1933; many evolutionists signed or tacitly agreed. A century of liberal theology had already made the term "religion" nonsupernatural, but the 1930s saw it being captured by science. British zoologist Julian Huxley appealed in 1927 for "religion without revelation." And in his "secular sermon" at the Rockefeller Chapel for the Darwin Centennial in 1959, he proposed that evolution provide "the lineaments of the new religion." Today, agnostics in science quite readily warm to "religious naturalism." A National Academy of Sciences statement of 1999 is typical: "Scientists, like many others, are touched with awe at the order and complexity of nature. Indeed, many scientists are deeply religious."[6]

The ascendance of science in the 1930s, a period that some historians call America's "religious depression," had raised worries that naturalism would co-opt religious faith. In 1939 a group of theologians, led by Rabbi Louis Finkelstein, president of the Jewish Theological Seminary, inaugurated a project to assemble scientists and religious thinkers for a great public discussion. The next year, the Conference on Science, Philosophy and Religion opened against a backdrop of spreading fascism in Europe; it met at Finkelstein's school in upper Manhattan, gathered under tents in the outdoor courtyard. The rabbi's key ally was Harvard astronomer Harlow Shapley, who as the nation's best-known stargazer once bemoaned "the spectacle of one highly trained successful scientist after another becoming soft and religiously traditional." Yet Shapley was now intrigued by this "exploration" between science and religion. His credibility helped draw top scientists to the New York City event each year.[7]

A speech by Albert Einstein, read by proxy at the opening conference in 1940, revealed the underlying tension within the enterprise. "In their struggle for the ethical good," Einstein said, "teachers of religion must have the stature to give up the doctrine of a personal god—give up that fear and hope which in the past placed such vast power in the hands of priests." *Time* called Einstein's message "the only false note" in a forum where 650 people were trying to reconcile the two priesthoods of clergy and scientists.[8] Others, like Mortimer Adler, wanted the annual conference to "repudiate" the materialist ideology of "scientism or positivism which dominates every aspect of our modern culture." Adler, a Jewish intellectual who taught Catholic philosophy at the University of Chicago, already had called Darwinism a "grand myth," "conjectural history," and "full of guesses which are clearly unsupported by the evidence."[9] A preponderance of Catholic neo-Thomist intellectuals showed up at the annual proceedings, which grated on secularists, and the event limped into the next decade, when its topics were diverted by war, ideology, and politics. It reorganized in 1951 as an institute of fellows, yet a formal American dialogue on science and religion had begun.

Nuclear weaponry, not biological evolution, was the science-and-religion issue of the immediate postwar era, and it boosted into public consciousness William G. Pollard, a Manhattan Project physicist who, while at the Oak Ridge National Laboratory in 1950, began his "transformation from a modern pagan to a priest" of the Episcopal Church. The years preceding his 1954 ordination, chronicled in the *New Yorker*, drew other scientists toward religion and certainly helped theologians feel science could be on their side. "He attracted much publicity," said Edward LeRoy Long, a Presbyterian ethicist who was chaplain at a Pollard retreat in 1953. "Appeal to scientists was, at the time, more likely to impress people than is probably the case today."[10]

But the torchbearer of the science-religion dialogue appeared farther to the north, in Boston, where in 1954 the topic arose in conversations among Harvard academics. The force behind this newest initiative was Ralph Burhoe, a Northern Baptist, a former Sunday school teacher, and an eventual convert to Unitarianism. In his role as executive director of the American Academy of Arts and Sciences, Burhoe organized a meeting between the academy's Committee on Science and Human Values and an ecumenical group called the Coming Great Church. That retreat session on Star Island, an archipelago off the coast of New Hampshire, gave birth to the Institute on Religion in an Age of Science (IRAS). Since 1954, the institute has met every July at the island getaway—run jointly now by Unitarians and the United Church of Christ—and produced many of the "science and religion" literati of today.

Burhoe's work has been called a search for a "rational religion" or a "scientific theology." But one of his colleagues, theologian Philip Hefner, said Burhoe valued traditional religion the most. "That's what survived the process of natural selection. Traditional religion contributed altruism and that, he said, made it possible for apes to become humans."[11] Earlier popularizers of evolution such as Princeton biologist Edwin G. Conklin, who wrote the tract "Evolution and the Bible" (1922), had already made this conceptual tie, but Burhoe added the energy of an organizer. He drew Harvard astronomer Shapley and participants from the old New York events back into the movement, even as his motives were questioned by some of his Unitarian humanist colleagues. "I was never convinced," said one, "that Burhoe's effort was more than a sophisticated form of theistic apologetic."[12]

In 1980, Burhoe was the first science-minded American to receive the Templeton Prize for Progress in Religion, worth more monetarily than the Nobel Prize.

. . .

This formative period of the "science and religion" movement—a span from 1939 to 1959—took place within the century of physics. Evolutionary biology had stumbled through the first decades of the century, but by the end of the 1900s, people spoke of the coming "century of biology." By 2000, for example, the federally chartered American Institute of Biological Sciences boasted 125,000 biologist members (through member groups) compared with 100,000 physicists with the American Institute of Physics.

The turning point for evolutionists came between 1937 and 1947, when the so-called New Synthesis gave evolutionary biology its first claim to be studying laws, as in physics. The synthesis defined evolution as natural selection acting on new variations in creatures that sprang from genetic "mutation"—a term that can stand for hidden diversity in the genes or a replication mistake in genes. Evolution happened in part by heredity, and the particulate nature of mixing genes—the "beanbag" principle—rooted biology in mathematics. When a few talented mathematicians began to calculate how genes would mix, mutate, and spread in a population, they founded the field of "population genetics," a keystone of the New Synthesis.

The elusive gene was being found out as well. In the so-called fly room at Columbia University, the spread of gene mutation was being studied in the prodigious genus of fruit fly, *drosophila*. The mechanism of mutation was finally made clear in 1953 with the discovery of the double helix structure of DNA, a molecular chain in which all the different combinations are made up of only four chemicals. When male and female produce offspring, the DNA helix splits to replicate its chemical codes, a process that can produce errors—mutations. So life, after all, was mechanical. For the life sciences the conclusion was clear, says DNA codiscoverer James D. Watson: "No vital forces, only chemical bonds, underlie life."[13]

Even before the helix, however, the work of Russian émigré biologist Theodosius Dobzhansky in the fly room had by 1937 drawn the main conclusion for evolution. A great genetic variation, he said in that year, was hidden in an organism's genotype, or genetic program, and that variety would express itself in future generations—and be tested for survival by natural selection. "For the first time a geneticist talked the language of a naturalist. And I was much impressed," says Ernst Mayr. "He was the first to bring together these widely separated fields of the origin of evolutionary diversity and the origin of adaptation."[14]

Neither Dobzhansky nor Mayr could decipher the purely mathematical models of the population geneticists. So the New Synthesis got its mathematical prowess from the likes of Sewall Wright, an agricultural biologist turned geneticist at the University of Chicago. Wright produced a mathematical measure of fitness in evolution—the "adaptive peak." Picture a three-dimensional map drawn by a computer, the map's malleable surface rising into peaks with curved valleys between them. The peak stands for a statistical point at which an organism's genes and environment allow it optimal survival. An organism located down a slope or in a valley was statistically less likely to survive, given the genes and the environment. This gave evolutionary biology its first scientific diagram. For nearly a century, evolutionists had had only Darwin's branching sketch in his *Origin of Species*, the flowery "tree of life" drawn by artists, or the obtuse mathematical formulas of population geneticists. Mayr founded the Society for the Study of Evolution in 1946, and the next year America's leading biologists ratified the new seamless "synthetic" garment of evolution at a Princeton University conference. Says biologist Francisco Ayala, a student of Dobzhansky, "By 1950 acceptance of Darwin's theory of evolution by natural selection was universal among biologists, and the synthetic theory had become widely adopted."[15]

The New Synthesis had a social impact on the field of biology as well, shifting the status of its careers and restricting its philosophical horizons. For decades, physicists had looked upon evolutionists as amateurs who collected specimens and painted evolution murals on museum walls. With the new mathematical hard edge of genetics, however, evolutionists' "desperate desire to be taken seriously as professionals" was fulfilled, says historian Michael Ruse. Yet it also created new class divisions within biology, upgrading geneticists to researchers and consigning naturalists to be "stamp collectors" in museums. "The geneticists said that paleontology had no further contribution to make to biology," is how paleontologist George Gaylord Simpson put it in 1944. The rift between genetic "experimentalists" and the Darwinian "naturalists" manifested itself in "competition for funds, students, and prestige." The rift was widened by new technology as well, beginning with electrophoresis in the 1970s. This technology, which used electrical currents to separate the parts of DNA, siphoned students off nature preserves and into laboratories to study molecular family trees. Under these circumstances, says one paleontologist, "A biologist would never trouble to go out [in nature] and see what's there."[16]

The New Synthesis also dictated limits on what kind of law a biologist, if philosophically inclined, could impute to evolution. Before the 1930s, many evolutionists had searched for a biological life force. Osborn of the New York museum rejected Bryan's biblical creationism, but he clung to a theory of "vitalism," a progressive life force he called "aristogenesis." The New Synthesis shut the door on these. "All the mechanisms of evolution that were in any way purposive disappeared from evolutionary biology in the thirties and forties," says Provine. "So I call the evolutionary synthesis really the evolutionary constriction. They constricted out all those things that were not mechanistic."[17]

. . .

Long before the New Synthesis expelled spirit and purpose from evolution, modernist Protestants had lost their Darwinian idealism. The progressive evolution championed by Osborn had foundered on the First World War and the Great Depression. Modernists turned to neo-orthodox "crisis theologies," which rejected any attempt to find God in nature. Theologian Karl Barth warned of the futility of natural theology, for the transcendent God was met only in the gospel proclamation of Christ and the existential cry of the soul. Says one historian, "The neo-orthodox theologians, in fact if not by intent, built a wall between science and theology."[18]

Most of America's scientists had been born into Protestant households, but across the mid-1900s their adult church affiliations were either nil or hardly orthodox. A reading of the 1930 edition of *Who's Who in America* found that "the liberal Congregationalists and Unitarians are especially numerous as natural scientists," and a 1959 study of *American Men of Science* noted that most "Protestant scientists were affiliated with churches having a rather liberal doctrine."[19] Such trends raised the question of what kind of religious belief could be held by a science-minded American. Neo-orthodoxy provided a God in the bigger cosmic picture, a "ground of being," as theologian Paul Tillich offered, a deity addressed as "a personal God" only in symbolic language. Theologian Reinhold Niebuhr, counseling believers not to look for divine action in nature, had said: "We see that nature, whatever may be God's ultimate sovereignty over it, moves by its own laws."[20]

Theologians of that era were not enamored of science, said Edward LeRoy Long, who with a science degree went to New York City to be a minister. He completed his doctorate at Union Theological Seminary in 1951. "Thinkers like Niebuhr and [Union theologian and president John] Bennett interacted intellectually with politicians and statespersons more than with scientists," Long recalls. "Tillich interacted more with social activists and with arts types." In fact, the only religious thinkers who seemed interested in scientists were the "process theologians," who built upon the process philosophy of Harvard mathematician Alfred North Whitehead.[21]

Process theology appealed to liberal Protestants who felt that the idea of a transcendent God in a mechanistic biological world did not grapple with the mystery of theodicy—if God is good and almighty, why is there natural and moral evil? The other view that responded to this question was the spiritual evolution of French Jesuit paleontologist Pierre Teilhard de Chardin. Both Whitehead and Teilhard in effect said: God is good but not all-powerful, as evident in the freedom of evolution and the "evil" that freedom makes inevitable. Teilhard said matter evolved from coarser states of chaos and suffering into higher complexity, taking nature and humanity toward a final embodying of God—the Omega Point. Whitehead redefined reality not as "being" but as a string of "events," each one a place where God and entities freely participated in creativity, the ultimate good. This evolutionary theology (or metaphysics), produced by two practicing scientists, is still the benchmark today. But it says nothing about the Bible.

The great Dobzhansky, who criticized Bible fundamentalists but never lost his ties to Russian Orthodoxy, liked Tillich's religion as "ultimate concern," but he embraced Teilhard. He served as president of the Teilhard Society and wrote *The Biology of Ultimate Concern.* By now, his good friend Mayr advised him that "Teilhard is a step in the wrong direction," and materialists called his spiritualism "crazy" or "pious bunk."[22]

But he was in good company, for no less a pioneer evolutionist than Sewall Wright, the geneticist, became a process biologist. Wright let slip his flight from a purely mechanistic view of life in his 1953 president's address "Gene and Organism" to the American Society of Naturalists. Not a few of his materialist students thought he had lost his grip.[23]

Though Mayr was an atheist, the New Synthesis brought him into close friendships with believers like Dobzhansky and also the British ornithologist David Lack. "An absolute true believer," Mayr says of the Englishman. In 1947, at the height of his evolutionist fame, Lack had abandoned agnosticism for an Anglican evangelical faith. His 1938 study of finch survival and change in the Galapagos Islands was key empirical evidence for the New Synthesis. Yet like the neo-orthodox, Lack preferred to compartmentalize his faith and his science. Unperturbed that a good God could reign over nature's deathly struggle, he said that "man is surely unqualified to judge whether this [natural] ordering is in any way evil, or contrary to divine plan."[24]

.   .   .

Around the time Lack was studying finches in the Galapagos, the newly christened American "evangelicals" had split from fundamentalism to also engage modern science. The fundamentalists had formed Bible institutes for their "battle royal" against a Darwinist world, but evangelicals founded liberal arts colleges with science departments. When they organized the National Association of Evangelicals in 1941, they shared with neo-orthodox Christians a middle ground between the modernists and fundamentalists at the ideological extremes.

Some fundamentalists, however, believed science could at the least be an evangelistic tool—"science proselytizing." Though dispensationalist in outlook, the Moody Bible Institute in Chicago hired evangelist Irwin Moon to expedite this soul winning after learning of his crowd-pleasing road show, "Sermons from Science." Later, in 1945, the Moody Institute of Science opened in West Los Angeles, essentially a movie studio for the production of science films that often ended with an invitation to Christ. Even the U.S. military used the quality films. The institute closed, but already in 1941 Moon had organized a lasting fellowship of evangelicals in science called the American Scientific Affiliation (ASA). His "Sermons" had rarely addressed evolution, but that was to be the first big argument in the ASA. For more than a century, orthodox Protestants had abandoned the idea of a young earth for a view that Genesis's six days were ages (the day-age theory) or that millions of years of cataclysm on earth preceded the more recent Garden of Eden (the gap theory). In the 1920s, however, the Seventh-day Adventist itinerant teacher George McCready Price revived the young-earth belief. His writings on "deluge geology" claimed that a global flood had shaped the recently created earth only thousands of years ago, and the idea was being adopted by a circle of believers in the ASA.[25]

The ASA officers viewed "flood geology" as a threat to the group's scientific credibility. "This deluge geology bunch seems to go out with a chip on their shoulders and dare science to knock it off," wrote an officer in 1944. Three years later, ASA leaders commissioned one of their best-credentialed scientists to "present a comprehensive destruction of Flood Geology." Flood geologist leaders such as Henry Morris, however, were resilient. His rejoinder came in a bold conference paper, "Biblical Evidence for a Recent Creation and Universal Deluge" (1953), which the ASA journal declined to publish. That was the year before Baptist theologian Bernard Ramm published *The Christian View of Science and Scripture*, in which he said of young-earth advocates: "The hyperorthodox have made a virtue of disagreeing with science." Evangelist Billy Graham praised the meticulous volume. And Ramm claimed to have achieved "the harmony of Scriptures with geology" by saying the six Creation days were divinely inspired "pictorials" given to Moses (the presumed author of the Book of Genesis). He described the Bible as being in moderate "concordism," or harmony, with geological science and advocated a "progressive creation" across millions of years.[26]

The flood geologists finally bolted the ASA, forming the Creation Research Society (CRS) in 1963. They organized its board "to prevent any future takeover by evolutionists."[27] The manifesto for "strict creationism" was Morris's book, *The Genesis Flood* (1961), which polarized conservative Protestants, generated a groundswell of interest in the churches, and produced a "stunning renaissance of flood geology."[28] Though it was rare for young-earth creationists to have doctorates in science, the ten who formed the CRS were exceptions. And *Evolution, Creation, and Science*, a book by CRS biologist Frank Marsh—the first Seventh-day Adventist in America to earn a doctorate in biology—was rigorous enough to prompt Dobzhansky to call Marsh the "only living scientific antievolutionist."[29]

Some advocates in the young-earth camp remained skeptical about too much reliance on science. They frowned on a "two-book" harmony that might slide into a "two-revelation" belief, as if human science—especially on events of the past—could ever be equated with revealed truth in Scripture. Despite this "one-book" doctrine, creation "science" had been launched. Flood geology believed it could conform the empirical facts to the plainspoken words of Genesis.[30]

The ASA was less literal. It became home to evangelicals who mostly embraced theistic evolution or progressive creation, and it never again saw events so colorful as the young-earth split. The affiliation's boldest stroke—a proposal to hold the International Congress on Science and Christianity in 1969 or 1970—never materialized, and its officers have speculated on why it has not succeeded in shaking American science or religion. "The ASA has never taken an antievolution (or any other) position," said one president. To avoid doctrinal controversy among members, the ASA published almost no books. In another view, the organization hewed to "safer areas of evolution, Adam, and flood geology" and missed the social battles over the atomic bomb, pollution, and genetic engineering. Moreover, "The ASA is graying"—an idea that its younger wing contests. Their dedication to their science fields, they say, makes them a less visible generation, not quite the ASA "movers and shakers" of former times. One ASA member is Francis Collins, an evangelical with no small visibility as head of the U.S. government's Human Genome Project. He says, "I think theistic evolutionism actually has a lot going for it."[31]

FROM BROADWAY TO BIOPHILIA

Two major cultural events, one theatrical and the other intellectual, set the stage for the interaction of science and religion after the Darwin Centennial of 1959. The cold war had produced a dramatic boost in funding for American science. It also had ushered in the anticommunist era, when national security concerns were pitted against freedom of belief and association. That theme might have taken on any number of artistic expressions. The one that transfixed the nation, however, was the Broadway play *Inherit the Wind*, a dramatic rendition of the already overblown legacy of the Scopes trial.

The play suggested that the battle between enlightened science and prejudiced religion was happening at that very moment, or at least had occurred "not too long ago. It might have been yesterday. It could be tomorrow."[32] Hollywood turned *Inherit the Wind* into a movie in 1960, and Spencer Tracy, who played the freethinking, science-backing Clarence Darrow, was nominated for an Oscar. Two years after the movie's release, its tale of tolerant science versus divisive religion seemed to get legal support: the Supreme Court banished organized prayer and Bible reading from public schools. The play is still read widely in high school English classes. Critics of evolution say the drama puts them in a no-win situation: the news media view every jab at Darwin as a resuscitation of the Bible fundamentalists from the black-and-white movie.

The hard sciences suffered their own cultural blow in the 1960s, the decade of Thomas Kuhn's *The Structure of Scientific Revolutions*. The book was the popular apogee of the newfangled sociology of science, which intimated that scientists, too, are driven by economic and social prejudice and are not merely stoics in pursuit of "the truth." Kuhn's work, which has been interpreted variously, claims that a cluster of social, economic, and intellectual forces create a staid "normal science" that is unseated only by a "paradigm shift." The ebullient 1960s embraced this human view of scientists with a Freudian spin: "The contemporary sociology of science stresses especially the irrational source of scientific inquiry. It tends to regard science as a by-product of man's neurotic strivings."[33]

Philosopher of science Karl Popper cast such doubts on evolution when, in his 1976 autobiography, he called it more "metaphysical research program" than real science. Only theories that could be "falsified" by a test were scientific, he said, and so the proposition that dinosaurs evolved into birds was outside science. Before he died, in a much-cited letter to the editor from 1980, Popper conceded that evolution was valid on other lines of scientific logic, but his famous doubts are part of science's cultural history.[34] All of this debate added to the "demarcation problem" that haunts science today. Who has the authority to demarcate what is true science from what is not—especially in the theoretical realms of the evolutionary past? This intellectual fallout of the 1960s gave creationists an important tool.

What the 1960s gave liberal Protestantism, meanwhile, was the God of evolution and the environmental movement. When physicist-theologian Ian Barbour published his influential *Issues in Science and Religion* (1966), it for the first time laid out a coherent spectrum of science-religion issues.[35] But questioning the validity of evolution was not one of them. If Barbour's endorsement of process theology further distanced theological liberals from the evolution-creation debate, it meanwhile introduced them to a new one—the immanence of God in nature and what that says about ecology. Thereafter, Christians increasingly viewed ecology as the moral content of evolutionism. When historian and churchgoer Lynn White argued publicly that Saint Francis taught a brotherhood of species, the theme was clearer still.

White's lecture, given in 1966 to the Washington meeting of the American Association for the Advancement of Science, was titled "The Historical Roots of Our Ecological Crisis" and gained wide reading as a reprint in *Science*. In trying to separate a Saint Francis ethic from the rest of Christian thought, White supplied evolutionists with a long-lived argument (which he softened later) that Christian belief itself corrupted the natural world. At fault was the doctrine that humans are made in the image of a transcendent God, with "dominion" over nature. The ecological signs showed that human science and technology "are out of control," for which "Christianity bears a huge burden of guilt," White said.[36]

Environmentalism became more visible in its activist forms than in its theoretical look at God and nature. Similarly, the kinds of issues Barbour posed lived a quiet life, far from any debate with creationists. This was the world of the science-religion dialogue, manifest in the journal *Zygon*, founded by Burhoe in 1966, experienced at the Star Island annual retreats, and finally given financial momentum by the John Templeton Foundation beginning around 1990. Templeton, a Presbyterian elder influenced by the New Thought beliefs of his mother, sought a fairly traditional God in nature and by avocation was a natural theologian. Templeton funding, however, was given out after peer review and allowed much academic independence, so the resulting programs were thoroughly evolutionist.

A prime example was funding for the Program on the Dialogue Between Science and Religion, a first-of-its-kind division in the American Association for the Advancement of Science. Dedicated to evolution, the program held conferences and published proceedings. Meanwhile, the Center for Theology and the Natural Sciences, founded in Berkeley in 1981, mounted two of the largest "science and religion" projects of the era, the "Science and the Spiritual Quest" forums in 1998 and

2001, and five joint conferences over a decade with the Jesuit-run Vatican Observatory on the theme "divine action" in nature. Though the neo-orthodox thinkers of years earlier would find such juxtaposing of God and nature anathema, neither the "quest" nor "divine action" were anything close to creationism.

The Vatican link was only natural, for the Roman Catholic Church has never condemned evolution, except when it was used ideologically by atheists. The first Catholic criticism of evolution had arisen in England, where, in 1871, biologist St. George Jackson Mivert enraged Darwin by doubting natural selection in his *On the Genesis of Species;* he also emphasized God's infusion of the soul. Since then, theistic evolution has had the church's blessing in the United States, expressed in the 1890s by Notre Dame University chemistry professor John Zahm, a priest who wrote *Evolution and Dogma* at the request of the hierarchy. Those were to be contentious decades between American bishops and Rome over two heresies called "Americanism" and "Modernism," and mostly for that reason Zahm's work was put on the Vatican's Index of banned books.

Catholic leaders in America were primarily concerned about Darwinian eugenics, which during the European immigration of the early 1900s presumed to identify inferior and superior races. Otherwise, the *Catholic Encyclopedia Dictionary* (1931) said, "Catholics are free to believe in moderate evolution [that is, of animals], excluding the evolution of man." And across two papal documents on evolution—the 1950 encyclical *Humani Generis* and its 1996 repetition by John Paul II—"the origin of the human body as coming from preexistent and living matter" was accepted with only one caveat, as stated in 1950: "Catholic faith obliges us to hold that souls are immediately created by God." Though a priest from Notre Dame was a speaker at a side session of the Darwin Centennial in Chicago in 1959, in that year Catholics were "underrepresented" in *American Men of Science*: they made up 25 percent of the citizenry but only 7 percent of the science profession.[37]

Science professors in U.S. Catholic higher education often were priests, but not until 1968 did they create a significant Catholic forum for discussing the contentious issues of God and nature. In that year, two scientists at the Jesuit-run St. Louis University formed the Institute for Theological Encounter with Science and Technology (ITEST). "We got started because Catholics had no other place to go," says Robert Brungs, a Jesuit and physics professor. ITEST wanted to be an "early-warning system" to keep the church abreast of scientific developments. The conferences, beginning with one titled "Toward a Theology of Nature," focused on the impact of technology, not the clash of scientific and religious worldviews. Put off by the vague deity that seemed to dominate the science-religion field, however, the institute insisted on a trinitarian "Christian God" in whom the two books of nature and Scripture were in harmony. "We are interested in dogma and doctrine. We believe in revelation and faith," says Brungs.[38]

The Catholic Church spoke broadly on science and technology at the Second Vatican Council of 1962–65. A new Secretariat for Non-Believers was opened in

Rome, and engagement with scientists was included under its talks with systems and movements of nonbelief. Not until 1977, however, did the American bishops declare, at a synod in Rome, that "the world of science cannot be ignored." So in Washington, the National Conference of Catholic Bishops established the Committee on Science and Human Values. Evolution and creation were not at issue. The nun who led the unit mostly attacked nuclear arms and rampant technology, so the bishops shut it down. Revived in 1984, it emerged as originally intended— as a bridge-building project with science, now in the bishops' Evangelization and Missions Department.

For all these efforts by liberal Protestants and by Catholics, the thunder of the 1970s onward was stolen by the young-earth, creation science, flood geologists. They enjoyed a fifteen-year heyday, beginning with their first two campus debates in 1972 and ending with the Supreme Court's ruling in 1987. "In 1960 there were no creationists. Now there are tens of thousands," says John Morris, son of Henry Morris. Whatever the number of strict creationists among trained scientists, they did stir an entire nation of evolutionists. Geologist Brent Dalrymple said that before the young-earth movement, the geological age question in science was taken as passé. "We now know better," he said. "There was no work available that summarized with any degree of thoroughness the evidence for the [ancient] age of the Earth." So, in 1991, Dalrymple updated the last such work from 1931. Paleontologist Niles Eldredge also looks on the bright side. "In a way, creationism was good for evolutionary biology," he said. "It made us articulate our basic precepts more clearly. And it reminded us that we have, after all is said and done, more in common as evolutionists than we have issues that drive us apart."[39]

. . .

Evolutionists, like any professionals, are loath to display publicly their internal divisions. The split between laboratory geneticists and field naturalists ebbed and flowed, but now it was the mathematicians' turn to raise the tough questions for Darwinism. To them, it still seemed improbable that the mere shuffling of genes could yield such combinations as a DNA molecule of the human brain, or move through populations and produce dramatically new species. In 1966, these skeptics organized a symposium on the math issue at Philadelphia's Wistar Institute of Anatomy and Biology that drew evolutionists of some note. The mathematicians and the biologists agreed to disagree. But in the age of computers, Darwinism had become a puzzle with astronomical numbers and mind-bending random probabilities. It would hardly wow the public, but, says mathematician David Berlinski, "The Wistar Symposium does mark a before and after."[40]

Another milestone along the evolutionist road was reached in 1972 when Eldredge and Harvard paleontologist Stephen J. Gould published a provocative new theory called "punctuated equilibrium." It did not report a discovery, but rather made a candid observation. The fossil remains of organisms almost never showed them changing over their history, a situation they called the "fact of stasis." They added, however, that the placid fossil record is punctuated by rapid appearances of new species. The duo's punctuated equilibrium has been taken as the most prominent new direction in evolutionary theory since the New Synthesis, which emphasized genetic gradualism and the search for a fossil record of gradual transitions. It has merited mention in new textbooks, though critics have said it offers nothing new. Historian Ruse cautiously has given it "paradigm" status. At the least, he says, it "seem[s] to have polarized evolutionists in such a way that punctuated equilibria theory has defining paradigm properties at the social level."[41]

That social level has to do with the professional split the New Synthesis already had revealed between geneticists and paleontologists, for example, and the different attitudes on whether evolution works gradually or in dynamic bursts. There also is a generational aspect to the debate, for young biologists did not want to simply rubber-stamp what their graying mentors had theorized. Gould, in effect, led the charge with his article "Is a New and General Theory of Evolution Emerging?" (1980). He said the New Synthesis was "effectively dead."[42] The punctuationist group defied Dobzhansky's assertion that macroevolution (a major evolutionary change or development of new species) is simply an enlargement of microevolution (the gradual accumulation of mutations that leads to new varieties within a species). Instead, the punctuationists recognized abruptness at some levels of evolution. They also rejected blanket claims that gradualistic natural selection and adaptation explained all, or even most, evolutionary changes.

Eldredge crystallized the new alternative in what he called "hierarchy theory"—that evolution at one level is separate and different from evolution at another level. He would draw a distinction between the "ultra-Darwinists," who connected all levels of evolution to genes, and the naturalists, who took a more pluralistic view. While evolution could come by gene mutation and natural selection, it also might come by environmental change, different animal behaviors, genetic inheritance untouched by natural selection, or natural disasters. By the 1990s Eldredge was speaking of the "sloshing bucket" menagerie of evolution—hardly a stately march of minute genetic accumulations.[43]

In 1980 some American naturalists thought it was time to sort out this disagreement. A group of them at the American Museum of Natural History in New York and the Field Museum of Natural History in Chicago convened the "Conference on Macroevolution." The October summit, a closed meeting, was the most eminent gathering of British and American evolutionists since the Darwin Centennial of 1959. "It was an intellectual knock-down-drag-out," recalls one organizer, ornithologist Joel Cracraft of the New York museum. "There were no empirical issues to be resolved. These were conceptual issues. A megathink on ways of knowing the world."[44]

To creationists on the outside, the debate was evidence that high-end macroevolution—not just the microevolution of finch beak size or of insect immunity to pesticides—was suffering a theoretical crisis. To the scientists at the table, however, the issue was a contest between technical disciplines. "It was centered around the question of whether there is a separate science of macroevolution," says paleontologist David Raup, the senior scientist at the Field Museum. Could biology study evolution at any levels apart from genes in the population? The immediate fallout of the conference was a protest over a *Science* magazine headline, "Evolutionary Theory Under Fire." Letters to *Science* complained that the report was "advocacy" for the nongeneticists, or "fossil zealots." The protests made two points: the New Synthesis never had been a single, rigid theory; and both gradualism and sudden leaps had been part of the original Darwinism, so what was the controversy? Yet the public relations damage was done. Two Illinois biologists noted wryly, "We are sure the creationists will be delighted to have an opportunity to cite *Science* in apparent support of their cause."[45]

A temporary truce in this continuing debate on how to study and analyze evolution was proffered by two supporters of the New Synthesis, biologist Francisco Ayala and botanist G. L. Stebbins. They wrote the next year in *Science* that evolution may be studied at different levels—no problem. Still, the theme of the "unfinished synthesis" resounded more and more in the following years. Said one biologist, "The illusion of a finished synthesis created an intellectual inertia that is unique in the history of science." Certainly by the 1990s, says Raup, "The hegemony—if that's what it should be called—of the New Synthesis has been broken within the field. And it's come from many different quarters. It's come from molecular biology, primarily." The electrophoresis technique, which found the sequence of bases in DNA, could be thanked for that, but there were also new branchings into the study of mass extinction and, for some, the application to biology of chaos theory. This subfield of mathematics tried to understand intricate—but deterministic—laws in unpredictable realms such as clouds, magnetic fields, ricocheting particles, or teeming biological life in an ecological niche.[46]

Evolutionist Lynn Margulis, a pioneer in cell biology, followed her own lights outside the New Synthesis and ended up with symbiosis, not struggle for survival, as an engine of evolution. She conjectured before the centennial gathering of the American Society of Zoologists that, from the viewpoint of the symbiotic approach, the New Synthesis would become "a minor, twentieth-century religious sect within the sprawling religious persuasion of Anglo-Saxon biology." Official science has stood its ground on evolution in the broadest possible terms. "Scientists continue to debate only the particular mechanisms that result in evolution, not the overall accuracy of evolution as the explanation of life's history," the National Academy of Sciences says. In trying to explain life's history, says Raup, "Well, in effect, we don't have other naturalistic alternatives."[47]

Religious thinkers in the United States have taken the mixture of shuddering and shows of confidence by evolutionary science in different ways. The creationists, who read and liberally quoted Thomas Kuhn on revolutions in science, said the establishment was desperately resisting a "paradigm shift" soon to dethrone orthodox Darwinism. Around 1990, the mostly conservative Protestant "intelligent design" movement arose to revive natural theology; they wanted to find evidence, or "inference" at least, of God's intellect in natural processes.

Biology's internecine squabble struck liberal Protestants and Catholics a little differently: to them it was a sign that mechanistic reductionism finally had revealed its limits. That antireductionism was reinforced by the release in 2001 of the human genome sequence, which seemed to have one-third to one-half the number of genes expected. The logic that one gene produced one trait collapsed, said geneticist-turned-Anglican priest Arthur Peacocke: "There isn't enough stuff there to explain the complexity of the big systems and of the human brain and the culture that results from the human brain's interaction with its environment."[48] Just as liberal theology had given up insistence on God's control over nature, science was being asked to give up its belief that chemical mechanics explains life itself.

Two of the century's great American popularizers of the reductionist view of science, however, in its last decade were unwilling to concede to religion what religion had conceded to science. The astronomer Carl Sagan, who died in 1996, extolled the "awesome machinery of nature." He took steps in his last years to bridge the "two books" of nature and religion, but surrendered no ground on reductionism and conceded no turf to the supernatural. What he acknowledged, at minimum, was that religion had moral power. In 1990 he broke the ice between scientific skeptics and church liberals by issuing his "Open Letter to the Religious Community." The topic was saving the planet. The next year in New York City, Sagan stood beside a robed Episcopal bishop in the Cathedral of Saint John the Divine to issue a "joint appeal" on the environment. In his last year of writing, he labored to debunk modern-day "superstitions" such as the existence of UFOs and upheld the veracity of science against what he portrayed as the wishful thinking of religion. Yet he did not go out of his way to denigrate the creationists or Bible belief. "You get older and you mellow," he explained. "I think it's counterproductive to criticize beliefs we don't happen to share and to not try to understand what the beliefs do for the individual believer."[49]

The other great popularizer, Harvard biologist and insect expert E. O. Wilson, has not emphasized a dialogue with religion. He has instead explained religion as a by-product of genetic evolution, and a difficult by-product from which to wean Homo sapiens. "The human mind evolved to believe in the gods," he writes. "It did not evolve to believe in biology." Yet Wilson, too, says the human inclination toward religiousness will in time pay off for the ecological needs of the planet. Genetic and cultural evolution will work together to imbue humans with a devotion to nature itself, a powerful moral sentiment once aimed at gods but finally replaced by "biophilia"—"the inborn affinity human beings have for other forms of life."[50]

The American Museum of Natural History is where Osborn began his debate with Bryan. It is where Simpson and Mayr found their place in the New Synthesis, and it is where Eldredge hosted the "Religion and Ecology" conference—subtitled

"Discovering the Common Ground." The museum displays contain no mention of religion, but the institution has made a slight turn in welcoming religious voices—on matters of ecology at least. At the end of the century, science was saying that religion, if not a book of reality, was still a useful book. "There's an ecological component to all concepts of God," said Eldredge.[51]

Creationists of course presume that their concept of God is about something real. From that common ground, however, their differences begin to multiply.

3. LOOKING FOR BOUNDARIES

Who exactly are the creationists, and what do they believe? What about the evolutionists and their convictions? Simple though these questions may seem, the definitions can be as elusive as liquid mercury, especially if the goal is to satisfy everybody. An exchange in Seattle one rainy February day in 1997 dramatized the dilemma.

It was the first conference held by a new unit in the American Association for the Advancement of Science (AAAS) dedicated to issues of science and religion, and Harvard historian of science Everett Mendelsohn was comparing the young-earth creation science movement of the present with religious tenets prevalent in England of Darwin's day. "Twentieth-century creationism is not an Anglican theology, as it was in 1860," Mendelsohn said. "This new challenge to evolution is not coming from the scientists, primarily, but from schismatic parts of Christianity." The respondent was Protestant theologian Nancey Murphy, a Fuller Theological Seminary professor who took exception to her Harvard colleague's generalizations. "All conservative Christians do not hold the same views of the nature of creation, creationism, etcetera," said Murphy, who researches and writes on theology and scientific method. "Listening to this paper, you would think that perhaps 70 percent of Americans are creation scientists. Actually, there are very few creation scientists. There are a variety of kinds of creationists."[1]

Faced with the challenge of identifying groups in a fluid society, social scientists have developed a tool called "boundary theory." In this view, every group defines itself by a boundary, which is lived out mentally, culturally, and in symbols rather than by holding a membership card. A group's identity may be revised, but the boundary is always an outer limit, separating outsiders and insiders. Even within, however, groups have differences, giving rise to inner boundaries as well. All of this may be applied to the creationists and evolutionists. When the two sides meet professionally, their inner and outer boundaries take on a rare clarity.[2]

EVOLUTIONISTS

Differences among evolution's adherents in America have surfaced in scientific meetings since the dawn of Darwinian theory. Just thirteen years after publication of *Origin of Species*, Harvard botanist and AAAS president Asa Gray, a good friend of Darwin's and an adherent of Christianized Darwinism, expressed his belief in evolution during the association's 1872 meeting. Yet in Gray's generation of American naturalists in the National Academy of Sciences, "the boundary separating creationists from evolutionists long remained blurry." The American Society of Zoologists, formed in 1902, was evolutionist at its inception, but it was hardly Darwinian. Zoologists liked evolution but were not yet sold on natural selection. As historian Neal C. Gillespie said of the first generation or two after Darwin, "He made them evolutionists; but, ironically, he could not make them [natural] selectionists." When the Society for the Study of Evolution was formed in 1946, it largely required the field to work on the principle of the New Synthesis, natural selection working on mutations.[3]

Evolutionists are also bound together by the political camaraderie of battling antievolutionists. The American Society of Zoologists began a campaign in the early 1980s to promote "science as a way of knowing," a terminology that was increasingly picked up to make distinctions against creation science. The advocacy work can have an air of celebrity about it as well. When the National Association of Biology Teachers met in 1998 in Reno, Nevada, it feted two biology teachers who won Supreme Court victories for evolution. Susan Epperson's 1965 lawsuit in Arkansas had extinguished the last state statute that banned evolution in public schools, and Don Aguillard's 1981 suit in Louisiana had stopped the "balanced treatment" effort to include creation science in curricula that covered evolution.[4]

Surprisingly, however, the most notable aspect of natural scientists in assembly is how little they focus on evolution. Its day-to-day irrelevance is a great "paradox" in biology, according to a *BioEssays* special issue on evolution in 2000. "While the great majority of biologists would probably agree with Theodosius Dobzhansky's dictum that 'Nothing in biology makes sense except in light of evolution,' most can conduct their work quite happily without particular reference to evolutionary ideas," the editor wrote. "Evolution would appear to be the indispensable unifying idea and, at the same time, a highly superfluous one." The annual programs of science conventions also tell the story. When the zoologists met in 1995 (and changed their name to the Society for Integrative and Comparative Biology), just a few dozen of the 400 academic papers read were on evolution. The North American Paleontological Convention of 1996 featured 430 papers, but only a few included the word "evolution" in their titles. The 1998 AAS meeting organized 150 scientific sessions, but just 5 focused on evolution—as it relates to biotechnology, the classification of species, language, race, and primate families.[5]

Still, if the ghost of Darwin appeared at any one of these gatherings to check membership cards, these groups and individuals universally would probably announce themselves to be evolutionists. So the question becomes: What boundary renders them all alike?

The starting point usually has been Darwin's central topic, the origin of species. "Darwin considered the mutability of species as fundamental to evolutionary theory," said biologist Paul Thompson. "Belief in it—rather than in natural selection—constitutes the 'paradigm shift' (to use modern language) required in order for one to be an evolutionist."[6] Nevertheless, a single definition of species remains elusive. Creationists, moreover, accept varying degrees of mutability of species, some fairly close to those accepted by evolutionists.

In other words, no single measure such as belief in species change draws a definitive boundary for the purest form of evolutionism. That purity is defined only by the rejection of four proposals—four ideas that any thoroughgoing evolutionist will consign to oblivion. First is the rejection of supernatural intervention in nature. Second, a purist evolutionist rejects any interruption in the regularity of natural law and, third, the idea that nature has any ultimate teleology (that is, a force or principle that guides the end purposes of all things). The fourth repudiation is of preordained "types" in biological life. What evolutionists reject now, in fact, used to be the four darling concepts of natural philosophers. The biblical writers said God not only interrupts nature but formed creatures in "kinds" with a divine purpose. Plato matched every living thing with an eternal Idea. And Aristotle argued that each organism (and even objects) contained a seed of purpose that guided its destiny.

Darwin is credited with liberating biology from all of these concepts, and foremost, says Ernst Mayr, "In the *Origin* Darwin no longer required God as an explanatory factor." But in the real world, not everyone wearing an evolution badge is a purist. While negation of all four ideas draws an ultimate boundary, behind it there are plenty of internal differences. An evolutionist may indeed want to keep one or more—but never all—of the four propositions. For instance, some evolutionists affirm regularity in nature but still look for direction or type in biological matter. Others deny direction and type but admit to jumps in development. A purist like Mayr, however, says all such pick-and-choose approaches commit the error of "essentialism," the wrongheaded search for fixed essences, types, jumps, or direction in biology. His favorite errant is Thomas Henry Morgan, the hard-nosed atheist and biologist who discovered chromosomes. "He was a typologist," Mayr said. "He may not have known that he had philosophical ideas, but this typology was a philosophical concept. He came from the developmental field, and they were all very typological. Big jumps, you see. Mutations. New types developing."[7]

The idea of "big jumps"—leaps that change a major biological feature or turn one kind of organism into another kind—is likewise anathema to evolutionists who reject typology. Darwin and his friend Huxley disagreed on this one, for Darwin had put his faith in the Latin creed *Natura non facit saltum*—"nature does not jump"—and Huxley believed jumps were the only way to explain nature's bounty. Today in evolutionary circles "saltationism," from the Latin term for "jump," is outside the comfort zone, other than occasional exceptions such as Richard Goldschmidt, a geneticist at the University of California at Berkeley. At midcentury Goldschmidt proposed that ebullient mutations must have produced "hopeful monsters"—a first bird born from a reptile egg, for example. For this, he was subjected to what Harvard's Gould called the "two minutes hate," alluding to George Orwell's *1984*.[8]

By dint of his monthly bully pulpit in *Natural History* magazine, Gould has become an important boundary setter among American evolutionists. His *Natural History* essays led to a panoply of books and a public-speaking circuit rivaled by no other evolutionist in America. The public should not be blamed for taking Gould's Darwinism as garden-variety evolution. But his critics within the evolutionist fold grind the issue more finely. Some of them have crowned Gould, a talented man of letters and thus a popularizer, "evolutionist laureate." He has reciprocated the sarcasm by calling them "Darwinian fundamentalists." There is a boundary at issue here, and Gould has helped set it in two ways. First, his fame has provoked others to set boundaries against him. They ally with him, they say, "because he is at least on our side against the creationists," yet they question his deviation from the New Synthesis. Gould calls it his "pluralism."[9]

Generally, though, the second way that Gould sets a boundary is widely embraced by evolutionists. In essay after essay and book after book, Gould argues that there is no teleology in nature, only its opposite, contingency. His best-selling book *Wonderful Life* (1989) drives home the point: it was pure evolutionary luck that put humans on earth. Contingency is so important that Gould has no room for Darwinians who seem to deviate from its blind fate and give natural selection the power of a divine hand. In its pseudodivinity, this hand of natural selection is made all-powerful in arranging for every living thing to have a precise adaptation, a "just so story" like the ones Rudyard Kipling wrote for children. "So it's really a debate over the exclusivism of natural selection in its adaptationist mode as an evolutionary mechanism," Gould explained to me as he sat in a frayed chair in his Harvard office, a labyrinth of tall bookshelves. "It's just this very strong adaptationist bias that goes back to Darwin's generation and before. It goes back to some of the theologians. There has always been that thread of hyper-Darwinism, if you will."[10]

Gould's prime example is the "fatal flaw" of Alfred Russel Wallace, Darwin's codiscoverer of evolution by natural selection. "Wallace's hyper-selectionism led right back to the basic belief of an earlier creationism that it meant to replace—a faith in the rightness of things, a definite place for each object in an integrated whole," writes Gould. His alternative to hyper-adaptationism (a kind of twin to hyper-selectionism) is pluralism, some "adaptation" here and a little "exaptation" there, exaptation being the accidental formation of an organic feature—fortuitous accidents such as the panda's thumb or the big brain of humans.[11]

While Gould, with his pluralism, and the proponents of the New Synthesis all claim common descent from Darwin, other evolutionists are trying to leave Darwin's shadow. To them, evolutionary theory has bogged down in genetics, mathematics, and zoology at the expense of a "holism" or "organicism." Those concepts may better describe the world of evolutionary change in bacteria, fungi, and plants, says biologist Lynn Margulis. She predicts that the New Synthesis, because of its narrowness, will "look ridiculous in retrospect." Evolution is about chemical and cellular mutuality and symbiosis, not genetic and animal struggle, she says, and it takes place on an earth that is "one continuous enormous ecosystem, composed of many component ecosystems."[12]

Margulis occupies one room in the mansion that British biologist Brian Goodwin, author of *How the Leopard Changed Its Spots,* calls the "new biology." In terms of boundaries, the new biology moves dangerously into worlds of types and essences and even directedness. Foremost, it is a biology that abandons natural selection as an external law that can explain novelties in nature. Novelty is a matter of new, functional, biological shapes, says Goodwin. "You need field theories like those used in physics to explain spatial order," he says. "In a developing organism, these are morphogenetic fields, the fields that generate form during embryonic development from the egg to the adult."[13]

Two other evolutionists who depart from the conventions of genetic gradualism and natural selection are Santa Fe Institute biologist Stuart Kauffman and the biochemist Michael Denton, a British national living in Australia. In their separate ways, they both posit an organic unfolding of life based on deeper, more elusive laws in nature. Kauffman uses mathematically based chaos theory, which seeks to find in randomness an actual order, a "strange attractor" that is pulling, pushing, or guiding biological systems forward and in unison. Denton, in turn, has applied to biological life the anthropic principle used in cosmology: that a finely tuned big bang universe seemed destined to produce carbon-based life and thus humans.[14]

In *Nature's Destiny* (1998), Denton holds out a cosmic teleology. He resurrects a popular evolutionist of the past, Robert Chambers, whose *Vestiges of the Natural History of Creation*—the first popular evolutionist book in England and the United States—was published in 1844. Chambers was derided as an atheist by clergy and a dilettante by scientists, but he won popular acclaim by arguing that a God who did not "interfere personally" in nature nevertheless endowed it with "the organic arrangements" governed by the same natural laws as "cosmical arrangements." This kind of organic universe sits fine with Denton. He also is comfortable with Kauffman's arguments for "self-organization" and "emergent properties" in biological systems and with the core idea of Kauffman's *At Home in the Universe* (1995)—that human life was inevitable. "There is certainly more than a whiff of teleology about Kauffman's arguments," says Denton. "[H]is overall conclusion is consistent with my own when he claims, for example: 'We will have to see that we are all natural expressions of a deeper order.'" That deeper order is part of an evolutionary outlook, says Denton, and is "entirely opposed to that of the so-called 'special creation school.'"[15]

THEISTIC EVOLUTIONISTS

For those evolutionists who choose to add God to the New Synthesis or even to the "new biology," another boundary is drawn, and herein lies theistic evolution.

It also draws a boundary with special creation. The "deepest gulf" between the wings of theism is their two "different views of how God acts," said Murphy in her AAAS talk. "In part of this group, divine action is seen as a contrast with natural causation. In the other set of positions, divine action is seen as always happening in and through natural processes."[16]

Theistic evolutionists, who view God as working "in and through" nature, do not like to speak about jumps. Accepting jumps in nature would make them sound like creationists. An important boundary for them is to fall within the scientific model of a gradual, continuous, and even random natural history. Theistic evolution has been a mainstream alternative in America since Darwin published *Origin of Species*. From the nation's modernist pulpits in the early part of the century was heard, "Already the time has come when almost everybody exclaims, 'Evolution? Certainly! Why, I always believed in theistic evolution.'"[17]

There was a Victorian ease to this joining of a rational God who creates order with seemingly random natural processes. The eminent Gray explained it plainly: God ordains that water flows downward, but the random terrain decides its path. Today, when contingency, or happenstance, is the byword of science, Gray's metaphor may seem too poetic and even creationist. At the AAAS event in Seattle, for example, the conference title was "Cosmology and Teleology," but teleology was strictly limited to short-term cause and effect in nature.

The tone was different at the "Science and the Spiritual Quest I" (SSQ) conference at the University of California in Berkeley, which in 1998 gathered monotheist scientists interested in seeing divine intent in an overall teleology of nature. In a Public Broadcasting Service (PBS) documentary aired at SSQ, physicist-theologian Robert Russell, whose Center for Theology and the Natural Sciences sponsored the event, said: "God created by evolution." Such theistic evolution goes beyond issues of mere design, Russell said, to deeper questions of purpose, evil, and human destiny: "I don't think there's much at stake in design theory." This was a Templeton-funded showcase of the science-religion movement, worthy of funding because it bore a "mature theology of Creation." Soon after SSQ made the front page of *Newsweek* as "Science Finds God," Gould was promoting his own *Rocks of Ages* book, which prescribed separation of science and religion. He described the spiritual quest in Berkeley by theistic evolutionists as a foolish "syncretism"; it tried to join bits of science here, metaphors from theology there. At SSQ, said Gould, "absolutely nothing of intellectual novelty has been added"— only "the same bad arguments."[18]

Theistic evolution is borne most forcefully in moderate to liberal Protestantism and in mainstream Catholic thought. "So I believe in a form of theistic evolution," said Cambridge physicist John Polkinghorne, an Anglican priest and the only cleric in the Royal Society, the elite association of scientists. But he is no deist, with a distant, uninvolved God that can go happily with theistic evolutionism. He wants more than "purely a naturalistic story in which God's only role is simply holding the world in being. I think that God has a role in the world somewhat more than that," the tall graying physicist told me. How much more is the $64,000 question. Polkinghorne believes that the new chaos theory provides empirical science with an "openness" that allows for God's "top-down" influence. Reductionist science argues that all life is "bottom-up"—from genes to molecules to organisms. Still, Polkinghorne does not like jumps in God's top-down presence. "What we know about natural history is that it looks pretty continuous," he said. "I think God has chosen to act in that sort of way, interacting with the world without overruling it. A God who is lawlike is going to act that sort of way. It's not going to be by a magician, and certainly not by a forced manner, that you could see something totally new [appear in nature]."[19]

By putting the creationists' God in the "magician" category, Polkinghorne sets a clear boundary. Yet there are still other fences to erect among his fellow theistic evolutionists. He has erected them in a little book, *Scientists as Theologians* (1996). He lines up three of them on a spectrum from conservative to liberal: first himself, then biologist Arthur Peacocke, also an Anglican cleric, and finally theologian-physicist Ian Barbour. "I see a sort of scale which I call from 'assimilation' to 'concord,' depending on how greatly you allow theological content to be affected by, modified by, scientific concepts," he said. "I'm more traditional, more orthodox." Barbour is most willing to modify Christian theology to adapt to science. "So Ian wants the special status of Jesus Christ, but he sees it as an evolutionary emergence, a spearhead." In between Polkinghorne and Barbour stands Peacocke. As a biologist, Peacocke must meld his theology with the messy organic world of Gould's contingency. The result? "God has to be conceived of as creating through chance events operating in a lawlike framework," Peacocke told the SSQ gathering. "This is a long way from the artificer Creator, but perhaps nearer to a composer God weaving the fugue of evolving forms." Science, he added, has made "increasingly implausible any talk of God intervening in the world to change the course of events."[20]

On the conservative wing, however, Polkinghorne still struggles with that intervention. He said as a scientist he wants to find a "causal joint"—the point of contact between God's supernatural mind and the matter of the universe. He also calls this the problem of "agency," or how the free human mind directs its physical body and how an absolute God shepherds the material universe. Polkinghorne is no creationist. But he allows that in one "new thing" God defied all the laws of natural history. About two thousand years ago, Polkinghorne said, God resurrected the biological corpse of Christ and made it a new kind of substance. In all, Polkinghorne stands for and sees a "great revival of natural theology," what Murphy describes as a "third way." The revival, says Polkinghorne, "is taking place not on the whole among theologians, who have lost their nerve in that area, but among scientists."[21]

The SSQ conference was held in cooperation with a decade-long project with the Vatican Observatory, which has its American outpost in Arizona. Catholics comfortably wear the hat of theistic evolution. Neo-Thomism, a modern version of Thomas Aquinas's understanding of Aristotle, has made this possible. Neo-Thomists argue that by a principle of "double agency" in the world, God is the primary cause, and nature's laws are secondary. Science looks at the secondary cause and finds Darwinian evolution. The primary cause is a question for theology alone. Two years before SSQ, Pope John Paul II told the Pontifical Academy of Sciences that the church acknowledged evolution as "more than a hypothesis" (the event that prompted Gould's *Rocks of Ages*). Amid the spiritual and materialist versions of evolution, the pope said, the faith insists on divine creation of the human soul—a jump in nature. "Thus," says one Catholic theologian, "the point at which evolution would have moved from a hominid to a human species would have required at least that level of divine intervention."[22]

Cell biologist Kenneth Miller, a textbook author and evolutionist, says his atheist colleagues find his Catholicism untenable because it does at points allow the supernatural to intervene in the natural world. In their textbook, Miller and his co-author Joseph Levin endorse the classic stance that "evolution works without plan or purpose." Yet Miller would still allow God to intervene on, or behind, matter, he told me one summer in his Brown University office: "At some point, during the evolution of the human species—maybe it was *Australopithecus*, or *Homo erectus* or *Homo habilis*—all of a sudden somebody had a soul. So therefore, there being no genes for the soul, at some point you have a new individual who has a soul, whose parents didn't have one. How can you reconcile that? Isn't that an absurdity? My answer is essentially no. It is not an absurdity. I look to evolution as an explanation of the physical ancestry of the human species. I look at God for an explanation of the spiritual ancestry." The sacrament of the Eucharist he takes at weekly mass proves his point, he said. The bread and wine are matter, but they also are a vehicle of spirit.[23]

PROGRESSIVE CREATIONISTS

Charles Darwin, a day after *Origin of Species* appeared, was already speaking of "some immovable creationist!" The term was a new coinage that did not gain real currency until the latter 1900s. Historian Ronald Numbers says that Darwin's immovable critic could have been any of three different types of creationist, despite Darwin's desire to simplify his enemy in *Origin of Species* as "the ordinary view of creation." Among top American naturalists in 1859, the three kinds of creationism "spanned a conceptual spectrum ranging from a virtual infinitude of miraculous interventions . . . to perhaps only three."[24]

No one has formally proposed how many interventions by God in nature are required to move from theistic evolution to "progressive creation," but efforts to create a semantic unity between the two are common in groups such as the American Scientific Affiliation. "Evolution can be considered without denying creation; creation can be accepted without excluding evolution," said Richard Bube, a Stanford University scientist who long edited the ASA journal. With a "prophetic voice," biologist Keith Miller said the melding of creation with evolution comes by rejecting the "'wisdom' of our time, cloaked in scientific authority, which states that natural causation excludes the divine."[25]

Yet for progressive creation, which accepts an ancient earth, the semantics are not enough. It demands two things: some historic authenticity to the sequential Genesis days, and that divine intervention may override "natural causation" in a singular act of creation. These were common goals in orthodox Protestantism after Darwin, eclipsed only in the 1960s by young-earth flood geology. By the late 1980s, however, the ancient-earth position was enjoying an intellectual revival, and not because of geology. Its impetus was the search for design in nature via physics and biology, both well represented at the "Mere Creation" conference in 1996, a benchmark in the progressive creation renewal. Design has been the bane of all evolutionists, and particularly atheists such as Nobel Prize–winner Francis Crick, who warned: "Biologists must constantly keep in mind that what they see was not designed, but rather evolved." Yet in the 1990s, the most common reason educated Americans believed in God was because they saw "good design/natural beauty/perfection/complexity of the world or universe."[26]

"Mere Creation," which gathered at Biola University, drew 194 mostly conservative Christian thinkers, mainly in the sciences, from fifty-two institutions of higher learning. Participants called it an "unprecedented intellectual event." And unlike SSQ—whose "Sunday school politeness," the New York Times said, allowed "none of the impassioned confrontations expected from such an emotionally charged subject"—"Mere Creation" was ready to skirmish under the banner of "intelligent design." "What is uniting people is the idea that divine action is real and, from our point of view, it is also detectable," said one "Mere Creation" presenter, philosopher of science Stephen Meyer.[27]

Old-earth progressive creationism has traditionally laid out its task as "finding correlations between features in the biblical text and phenomena in nature, including the question of the antiquity and unity of the human race." Yet one salient feature of the "Mere Creation" event was its utter lack of papers alluding to the Bible. It focused instead on what Murphy had called God's action "in contrast to natural causation." The "Mere Creation" presenters divined that "contrast" in a fine-tuned universe that produced carbon-based life, the rise of the first cell, and the Cambrian "explosion," an event beginning 530 million years ago when all basic body plans, called phyla, seemed to have arisen. God's action over millions of years was also inferred in distinct phyla, such as insect, bird, and mammal. The contentious debate on the human fossil record, its dating, and various theories of branching and geography was also fair game. At a bare minimum, progressive creationists find divine action in the appearance of human consciousness—a point at which their boundary with theistic evolutionists blurs.[28]

Having established a scientific agenda to find this divine contrast, "Mere Creation" had set up quite a challenge for itself, said biochemist Michael Denton, a noncreationist and Darwin critic. "All the talks are becoming criticisms of evolutionary theories," he offered at one point from a side aisle. "When you carry the message out of this hall, and if this conference is going to be successful, it has to defend special creation." That defense is mainly being built philosophically. In the "Mere Creation" setting, ideas such as "theistic science" and "theistic realism" arose as rational grounds to presume God's intervention in the laws of matter. Mathematician William Dembski, who studied at the University of Chicago and MIT, elaborated on the design-oriented philosophical realism as editor of the conference papers. "Science, we are told, studies natural causes whereas to introduce God is to invoke supernatural causes," he writes. "This is the wrong contrast. The proper contrast is between undirected natural causes on the one hand and intelligent causes on the other. Intelligent causes do things that undirected causes cannot."[29]

Theistic evolutionists have accepted the anthropic principle of a universe mathematically apprehended by the human mind, and an order of atomic forces seemingly aimed to produce humans. But they will not go down the creationist road of intelligent design, with its ideas such as "irreducible complexity" and the "explanatory filter." In the former, certain kinds of tightly coordinated biochemical complexity is taken to be far beyond the reach of what evolution can do: intelligence might be inferred as being behind the complexity. That level of inference is reached by means of the filter, which proposes that once natural law and random probability have been filtered out as explanations for a phenomenon, the third level of the filter could infer intelligent design. Dembski said with optimism, "Scientists are beginning to realize that design can be rigorously formulated in a scientific theory." All creationists, including old-earth progressivists, insist on God's freedom to act in nature, a freedom that a "theistic realism" stance would seek. But Dembski also draws a moral boundary for the believer. He argues that the belief that God is forever hidden in nature's laws is not compelling to ordinary people. Moreover, scientists view it as weak-minded religion. That is why "full-blooded Darwinists" find theistic evolution so "contemptible," he said: they find "the theism in theistic evolution is superfluous" and its adherents lacking "the stomach to face the ultimate meaninglessness of life."[30]

Though the Bible was not a topic at "Mere Creation," it is an inescapable issue for progressive creationists, as seen in the "battle for the Bible" among American evangelicals in the 1970s. That clash over the "inerrancy," or historical truth, of Genesis generated three councils over a decade. At the second council, convened in a Chicago hotel in the fall of 1982, a plenary of a hundred Bible scholars considered stances on the Bible and natural history. Science professor Walter L. Bradley of Texas A&M University argued that progressive creation was the best fusion of "the biblical and scientific particulars"; it encompassed both of God's "mechanisms in creation," process and miracle. Evolution is process only. Sudden creation is all miracle. The two-mechanism argument won the day, and participants signed the Chicago Statement on Biblical Hermeneutics, which said that inerrancy did not require a recent six-day Creation. Since then, young-earth proponents such as Duane Gish note "a great deal" of agreement with old-earth believers, especially compared with theistic evolution, but call the new enthusiasms over design old hat. "We did that years ago," Gish said. "Maybe these people were just not aware of it." Still, the age question tends to overshadow all others. It certainly did in 1982, when flood geology leader Henry Morris drew a boundary; he refused to sign the Chicago Statement, calling progressive creation an "old, time-worn, compromising hermeneutical system."[31]

YOUNG-EARTH CREATIONISTS

Since its founding with a call for academic papers in 1984 and its first meeting in 1986 at Duquesne University, the quadrennial "International Conference on Creationism" (ICC) has become the preeminent meeting of its kind in the world. Its goal of providing "a peer-review forum whereby the young-earth, global flood, creation model could be rigorously developed" has made it the spearhead of a more intellectually exacting young-earth creationism.

By the time the 1998 conference was held at Geneva College, north of Pittsburgh, "the quality of the submissions had improved exponentially," according to Robert E. Walsh, a mathematician active in the Pittsburgh host group, the Creation Science Fellowship, and editor of the symposium papers. About fifty papers had been selected—a third of the total number submitted—and most presented a hypothesis for a global flood or young planet, not just an attack on evolution. An internal discussion in 1998 concerned populist creationism, a slipshod variety purveyed by neophytes that obscures the rise of a sounder neocreationism. "We need a board of evaluators," said paleontologist Kurt Wise, a Harvard graduate and professor at Bryan College in Dayton, Tennessee. In Sunday schools, Christian schools, and home schools, he said, "The poor kids are being taught garbage." While that comment was meant for slipshod young-earth instruction, a similar disdain is shown for the way antiquity and evolution were smuggled into the first attempt at a national creationism conference, held by a local creationist group in Baltimore in 1982. "We were appalled," said Walsh. "It was warmed-over theistic evolution. It was all compromise."[32]

From the strict creationist side of the boundary, Genesis is taken at face value: a sudden Creation over days and, according to the genealogy in Genesis from Adam to Moses, from six thousand to ten thousand years ago. The fourth ICC, in 1998, gave a prime evening slot to intelligent design, but in this creationist circle earth science is always foremost. "From Genesis we have had three main events that have affected the world tremendously," said Danny Falkner, professor of astronomy at the University of South Carolina. They are Creation, the human Fall, and the global flood the Bible says Noah survived. These physical events—physical in the spiritual Fall because that introduced physical death to nature—set the ICC agenda: to research a young earth and recent man, the deterioration of a pre-Fall earthly paradise, and the geology of a flood. The papers often analyze the effect of water on land, irregularities in dating rocks, and, more recently, the tectonic geology of the planet. The localized studies could meet the expectations of conventional science guilds. But the larger presupposition—that God "*used processes which are not now operating anywhere in the natural universe*. We cannot discover by scientific investigation anything about the creative processes used by the Creator"—is what astounds.[33]

The ICC assemblies have always met in the East. Yet the best-known presentation of young-earth creationism is to be found in the desert West, at the Museum on Creation and Natural History. Visitors to this institution between San Diego and the desert pass through a series of exhibit rooms that explain death and disorder in nature this way: "The universal Second Law is the scientific reflection of God's curse on His created world because of sin." The most problematic part of the creationist platform, however, is God's creation of "the heavens and the Earth," depicted at the museum with images of distant stars aglow against black space. "The biggest problem we have, in my estimation, is the light-travel-time problem," astronomer Falkner said. He knows that the speed of light dictates a universe older than biblical chronology allows, for it takes at least *tens* of thousands of years for light from distant stars to travel to the human eye. "I'm still looking for a good answer on this one," he said.[34]

Thorny as the age question may be, paleontologist Wise made the stakes crystal clear in his plenary talk at the ICC of 1998, "Why a Young Age Model Is Important." A few years earlier, I visited Wise at his Bryan College office, where he already was peer reviewing papers for 1998, and he explained how fickle the boundaries can be. "There is a slippery slope from the young-age creationist position down to atheism," he said. "And to stop somewhere along that slope and accept the theistic evolution position is a tenuous position at best. What are the trees on that landscape you can hold onto?"[35]

. . .

If the evolution-creation debate sets up boundaries in the worlds of professional science and professional theology, the rest of America has its boundaries, too, if a bit more shifting and amorphous. The four significant social boundaries are between scientists and the public, between theological traditions, between degrees of attentiveness in the general public, and between science and today's "postmodern" or New Age public attitudes.

Scientists and the public share a substantial boundary only within the stance of theistic evolution: God and evolution in harmony. When scientists were queried in 1996 with the same three-choice Gallup poll frequently given the citizenry, the following contrast (see Appendix, table 3) appeared:

1. Humans developed over millions of years from less advanced forms of life. God had no part in this process.
    Scientists: 55%
    Public: 10%

2. Humans developed over millions of years from less advanced forms of life, but God guided this process, including humankind's creation.
   Scientists: 40%
   Public: 39%
3. God created humans pretty much in their present form at one time within the last 10,000 years.
   Scientists: 5%
   Public: 44%

(7% no opinion)

After the roughly 40 percent overlap of scientist-citizen comfort with theistic evolution, the next most striking finding is how few Americans hold the naturalistic view favored by most scientists.

Not surprisingly, almost no scientists accept that humans arose in their "present form at one time within the last 10,000 years." But it is telling that 44 percent of Americans do. Most of these respondents doubtless hold the young-earth view of biblical literalism, but certainly not all. Many in this recent-human-origins group might have ducked the issue of the earth's age. Or they may believe in an ancient earth with relatively young life, a view found among progressive creationists. What is more, believers in recent human origins could not all be Bible fundamentalists and conservative evangelicals, for those sectarians make up only a quarter to a third of the U.S. population. Of the 44 percent who like the ten-thousand-year option, many must be mainline Protestants, Catholics, and Jews.

Such beliefs are not deeply pondered. Creationist Kurt Wise finds that most believers are not science buffs but "concerned consumers" who follow the advice of a pastor or friend. "They just choose what feels good to them," he said. Historian Ronald Numbers portrays a public inattentive to such detail: "The great unwashed don't know one creationism from another." Young-earth creationism, whose simpler formulation had wide appeal and succeeded in uniting conservative Bible believers after decades of fractious disputes on Bible interpretation, stole the headlines for a quarter century, he said. "The so-called flood geologists have been so successful that they have co-opted the term."[36]

A major survey from 2000, commissioned by the advocacy group People for the American Way (PFAW), confirmed the public's inattention. Half of the respondents had never heard of creationism. Those who had mostly equated it to creation "exactly as the Bible says." The rest (about 40 percent of those who knew the term) conceived of it generally; God created, though "not exactly" as the Bible said. Nearly all respondents knew the term "evolution," but only half were "very familiar" with the concept.[37]

The major religious traditions also set out theological boundaries, and though individuals may pass across, traditions stay fairly rigid. "Those theologies make a huge difference in the analysis of evolutionary proposals," says historian Mark

Noll. Catholicism has no formal boundary against evolution, as long as God is there, so America's most prominent theological lines are drawn among two Reformation traditions: Lutheran and Calvinist. Every tradition with a strict view of the Bible has resisted evolution, and so it was that Seventh-day Adventists and Missouri Synod Lutherans were key founders of the young-earth Creation Research Society. Still, the public debate is stirred most by Calvinist traditions, especially Presbyterians and Baptists, but also Anglicans, Congregationalists, and Methodists (the dominant tradition in Dayton in the days of the Scopes trial). The Calvinist-Lutheran boundary may spring from the different ways that John Calvin and Martin Luther, both towering theologians, had attacked the God-and-nature question. "Calvin was interested in the natural world and the knowledge of God," said theologian Edward LeRoy Long. "Luther spoke more of the experience of God, justification by faith."[38]

By way of the English Reformation, Calvinism came to America. It brought a face-value approach to knowledge, or a notion that a Christian mind could plumb nature better than an unsaved mind. What remains today is a Reformed discontent with those who would entirely extricate knowledge of God from nature. Reformed thinkers attempt a "straightforward apologetic" to harmonize natural science with Scripture, said Lutheran theologian Philip Hefner. "Sometimes they actually will try to challenge evolutionary science." Conservative Lutherans have also done this, but mainly within the church and rarely in the public square. Modernist Lutherans avoid conflict with evolution by taking Genesis symbolically. Both wings, however, work under Luther's doctrine of "two kingdoms"—one of faith, one of the world—and let secular authority decide questions of evolution in public schools or in science.[39]

The newest boundary in America defies just the right terminology, though it now is being called "postmodernism" or by its older name, relativism of belief and of knowledge. There are soft and hard kinds of the new relativism, but the one that worries science most is the antiscience variety. "For the first time in centuries, there are thoughtful persons who are not morally certain that even our greatest [scientific] achievements do, indeed, constitute progress," opined the president of the National Academy of Sciences in 1980. Two decades later, the president of the Scientific Research Society suggested that not only the "nature and practice of science" was under siege, but now there were "attacks on the possibility of any reasonably certain human knowledge."[40]

This soft relativism has not been amenable to the scientific claims made by evolutionary biology. The PFAW poll in 2000, for example, found that just three in ten respondents considered evolution "completely" or "mostly" accurate; only four in ten said evolution is "close to being scientifically proven." A third who accepted evolution nevertheless said, "You can never know for sure" on its factual reality. What has been called "a scientist's worst nightmare" showed up in the musings of a Kansas student in 1999. "No one was there that's still alive today that actually witnessed creation or evolution," the boy said on National Public Radio's

Weekend Edition. "It's just what a person believes. I mean, we have no right to say what exactly is true."[41]

Evolutionist and educator Paul Gross warns that this relativism of knowledge has reached high into colleges of education, even into the sciences. Here may be a case in which evolutionary thinking has backfired on the empirical claims of evolution. Gross traces the new academic assault on objective knowledge to the Swiss educator Jean Piaget, who drew on Darwinian ideas for a theory of evolutionary, or developmental, learning in humans—an idea that swept teachers' colleges. The theory says that knowledge is not a transfer of true information but a personal construction of reality. The "constructivist" view of learning has spread rapidly, said Gross: "The literature of science teaching has been, during the two decades past, a flood of constructivism." He said the "astounded layman" might want to ask, How can natural science be taught when teachers doubt there is objective scientific truth?[42]

Well within the relativism boundary also fall New Age beliefs, which embrace evolution with enthusiasm. But it is the evolution of mystical forces and intuitive feelings. Nearly all Americans tell Gallup polls they believe in God, but roughly a third of them are really acknowledging an impersonal "universal spirit." Interest in Asian beliefs, *Star Wars* movies, and the new environmentalism have forged this amalgam of cosmic force and biological evolution. HarperSanFrancisco, the publisher that dominates the market of spiritual and New Age books, declared evolutionist Pierre Teilhard de Chardin's *Phenomenon of Man* to be the top spiritual title of the twentieth century. And if this boundary needed a watchdog of evolution, it might be Deepak Chopra, an M.D. and best-selling author. "We are in a phase of evolution which we can only call metabiological evolution," he told the National Press Club in Washington in 1997. "Biological evolution was about survival of the fittest, and metabiological evolution will be about survival of the wisest." Post-Darwinian evolution "is going to be evolution of consciousness and ultimately evolution of consciousness of consciousness."[43]

Only thirteen years after *Origin of Species* was published, one American scientist exulted that "the modern theory of evolution has been spread everywhere with unexplained rapidity, thanks to our means of printing and transportation."[44] Theories now seem to move as quickly as electrons. To put a face on that information flow, evolutionists and creationists have produced savvy personalities who, with logic and affability, try to persuade the public to their side. Two of the best at this have made their persuasive cases from the same location, the environs of the University of California at Berkeley.

4. HEARTS AND MINDS

From their separate bases at Berkeley, physical anthropologist Eugenie Scott and law professor Phillip E. Johnson have been standard-bearers in the war over evolution and creation. Theirs is an intense, if friendly, rivalry, typically played out on radio talk shows and in print. Six days before Christmas 1997, however, it erupted into the national consciousness as they faced each other on public television's *Firing Line*. That night's audience, which was more than half a million households, was the largest number of Americans ever to share ringside seats at a modern showdown over the question of origins.

The debate was about both science and belief—but mostly about winning hearts and minds. "Hearing Phil define evolution is a little bit like having [celebrity atheist] Madalyn Murray O'Hair define Christianity," said Scott. Lawyer that he is, Johnson held up an exhibit, the "Darwin fish," an evolved version of the "ichthus" fish often used to symbolize Jesus Christ. Both have found their way onto car bumpers. "This is what we call evidence of intelligent design," he said, asking rhetorically whether the Darwin version was meant to "mock the Christian fish symbol." His Exhibit 2 was a fund-raising letter from Scott's National Center for Science Education (NCSE) that offered a refrigerator magnet in the evolved-fish shape as a token of thanks for a fifty-dollar donation.

"I have seen the Darwin fish and the Christian fish facing each other on bumpers," Scott parried. "It's a rather ecumenical car, I think." Her implication— that many Christians are not offended by Darwinism—was clear. Not at all chagrined by mention of the magnet offer, she acknowledged that her nonprofit center, an educational enterprise that sometimes even leads tours to the Galapagos Islands, was dependent on donations—and she said she regretted not having an 800 number to hold up for the TV audience.

Two hours later the cases for and against evolution had been laid out, but moderator Michael Kinsley despaired of a verdict, saying, "I guess we will know in a few million years who's right and who's wrong about this."[1]

. . .

Most days in Northern California, a morning fog rolls into San Francisco Bay. The blanket of gray creeps over the hills on the eastern shore, covering the picturesque university town of Berkeley. Eugenie Scott knows the town well. She runs the

National Center for Science Education, which until 2001 had made its home in El Cerrito, a sleepy city that borders north Berkeley. When I visited in 1996, the center was a one-story storefront divided into offices and a warehouse for literature. In a small room in the back, Scott, the executive director since 1986, gestured about. "This is the nerve center of anticreationism," Scott said. "Both rooms."

She is a tall, studious-looking woman with short brown hair and large glasses, a "nontheist" with a natural wit she uses to good advantage in her role as advocate. Paired with a theologian who said nature shows that God works by both law and chance, Scott quoted a famous British scientist on what nature suggests about that God: "An inordinate fondness of beetles." More seriously, though, she calls her work an uphill battle. "Yes," she said, "I've got all of mainline science behind me, but they're not paying any attention."[2]

Rank-and-file scientists historically have stayed out of social debates, including this one. The public, meanwhile, is not equipped to take sides—just one in nine Americans feels "well informed" on science issues, one survey found. The masses may watch nature documentaries, but that could be their closest brush with details about biology or evolution. They are riveted instead by reports of medical discoveries, environmental news, and space flights. The public's fullest response to the evolution-creation debate arises from its democratic spirit: roughly two-thirds say that out of fairness, public schools should teach both views.[3]

This is only the start of Scott's worries. Creationists also have the airwaves: evangelicals operate about a tenth of American radio stations and have wide viewerships on cable television. "Who gives a rip about mainstream science when you are reaching how many hundreds of thousands of people on the radio every day?" she complains. A "Who's Who" of evolutionists is listed on the national center's stationery, yet none is a foot soldier for the cause.

"The only reason a scientist would have to fight or argue with religion is when either one steps on the other's turf," Scott said, her humor resurfacing. "Most of the time there is no religious position on how fluids get through a cell membrane." What is more, she said, biologists and earth scientists labor under a double standard. "Why is it that evolution is singled out for this kind of treatment? No one is pressuring physicists, nobody is pressuring astronomers, to come into the class and say, 'Now we're going to study the solar system, and historically people thought the sun went around the earth, but now heliocentrism is our best theory—but you don't have to believe it.'"

Despite such social challenges for evolutionists, the national center is light-years ahead of where Scott and her allies found themselves in the early 1970s. In those days few biology teachers had detected the tremors of a renascent creationist movement. William Mayer, director of the Biological Sciences Curriculum Study—the federally funded textbook publisher—was the evolutionists' early Cassandra. Counseling urgently in 1973 that public awareness was the best defense, he advised: "Maximal publicity is required."[4]

In the year of that sage advice, Scott was only several months away from completing her doctorate at the University of Missouri. Her mentor, James Gavin, had given her examples of creation science literature, which she collected like curios. Her nonchalance ended two years later, however, as she and about fifteen hundred other spectators at the university watched Gavin be undone by two young-earth creationists, the first she had ever seen in action. "It was terrible," Scott recalls. She had been teaching anthropology for a year at the University of Kentucky when she returned to Missouri and witnessed Gavin's humiliation. His opponents were a famous duo, Henry Morris and the avuncular Duane Gish, the most polished presenter on the creationists' fledgling debate circuit.

Textbook controversies over evolution had been percolating through the 1960s, the first state creationism bill was introduced in 1971, and as early as 1975 a Henry Morris *Impact* pamphlet touted a "resolution for equitable treatment of both creation and evolution."[5] Gish himself, however, as a Berkeley-trained biochemist, would embody for official science the brightest of the new creationist vanguard. By persistence he had secured a two-hour program slot at the annual meeting of the National Association of Biology Teachers in 1972 and there allied with a Michigan State University professor to present the genetic and fossil evidence for creation. Even more of a heads-up for the evolutionists, however, was the publication of Gish's talk in the house organ *American Biology Teacher*. It appeared in the same issue as Mayer's "maximal publicity" warning. For evolutionists higher up, this was too much. The next editor swore in print that creationists would never get a voice again. The entrenchments had begun.[6]

The 1975 debate between Gavin and Gish in Missouri stayed in Scott's memory, and then in 1981 she had the opportunity to switch from spectator to participant. In Lexington, Kentucky, the five-member Fayette County Board of Education faced a vote on whether to put creation science into the high school curriculum. Given her familiarity with the literature, Scott led the opposition—and had another awakening. This time, the eye-opening was not so much academic as political. "While it did matter what the scientists said, that was not the real issue," she said in her El Cerrito offices. "What really matters is not what's in people's heads, but what's in people's hearts." Another principle stood out as well. "It wasn't the scientists who turned the tide in Lexington at all," she said, "it was quite honestly the mainline religious community." Its clergy could be for God and for evolution. For budget reasons as well, it had turned out, the Fayette board rejected a creationism curriculum.

University scientists had learned to organize against local threats, but their efforts did not go much further—until the creation science movement touched national politics. That happened in 1980, when the conservative Ronald Reagan ran for president. The previous year, Gish's workplace, the Institute for Creation Research, had again issued a model resolution school boards could adopt for "balanced treatment" of "creation science" and "evolution science." The new Christian right was beginning to emerge in America, and in concert with that development, creationists turned the resolution into a model bill and, in time, introduced it into more than twenty state legislatures. Two months before Election Day, moreover, Reagan spoke to evangelical ministers in Dallas and answered a question about evolution with, "Well, it is a theory, a scientific theory only, and it has in recent years been challenged in the world of science." Walter Hearn, an evangelical evolutionary biologist with the American Scientific Affiliation, remembers the mood: "And then the scientists said, 'Hey, this guy is going to get elected and he's going to start giving grants to the creationists.'"[7]

The "maximal publicity" counseled by educator Mayer seven years earlier became a national concern. Wayne Moyer, a veteran biology educator who had taken the reins of the National Association of Biology Teachers (NABT) in 1979, proposed in *BioScience*, the journal of the American Institute of Biological Science, establishing a fast-action network to monitor creationists. He warned of their "highly organized, well-financed effort to legislate" the creed in schools, which threatened "an American equivalent of the Lysenko affair," when Marxist ideology was forced upon Soviet biology. The essay stirred several letters, whose authors became a nucleus for retired Iowa biology teacher Stan Weinberg to create a network known as the Committees on Correspondence, a name coined by Moyer that recalled colonial resistance to British rule. It aimed to have committees in each state take the heat off schoolteachers by sending in biology professors and researchers to testify at school hearings or in statehouses. As a case in point, Eugenie Scott had hooked up with Weinberg while fighting the Lexington battle. "The network was intended to keep people from having to invent the wheel every time," she said. "It was interesting, the small network of people who existed back then." Moyer, who met Scott in those halcyon days, agrees: "Though it was never really a large group, they were all members in NABT or had been involved in teacher education through the years."[8]

The creationist forces seemed to converge in 1981. In that year, the Reagan administration began to disclose projected budget cuts for science education, and Arkansas passed Act 590, which mandated "balanced treatment" in public schools. The American Civil Liberties Union sued, and as the case moved toward its December trial, the national media became fixated on a new voice in politics. Jerry Falwell's Moral Majority, which had only a part-time network of independent Baptist pastors, was "in a real sense 'made'" into something bigger by media excitement and was hugely magnified as a social force once the elite media took an opposition stance.[9] On that momentum, in October 1981 Falwell's Liberty University staged the best-attended evolution-creation debate on record, drawing three thousand to see a duel between Gish and biology professor Russell Doolittle. It was taped and set for national distribution. Evolutionists knew by now that such public displays never turned in their favor, but it was hard to reject the creationist dare. Doolittle had misspent his allotted eighteen minutes, never getting to his slides and conclusion, and emerged from the debate saying, "How am I going to face my wife after making such a fool of myself?" Gish came out mimicking evolutionist Thomas H. Huxley's victorious debate words from 1860: "When something like this happens," Gish said, "all I can say is that, 'The Lord delivered him into our hands.'"[10]

Creationists clearly were masters of the public rhetoric, though evolutionists would prove to have the upper hand in the courts. "We allowed one lone academician to go into this debate on their turf, in their auditorium, with their audience," Moyer said at the time. "We should have taken the thing a lot more seriously."[11] Later in October, Moyer and his NABT staff called a Washington summit of anticreationists. They met in the friendly sanctum of the National Education Association headquarters, a few blocks from the White House. In the same week, behind closed doors, the National Academy of Sciences held its first strategy meeting on how to deal with the creationists. Years later, after Moyer had retired to a thirty-two-acre "farmlet" in Tidewater Virginia, he easily recalled the dire mood of the early 1980s: "It was a growing realization that these people were very serious about their mission to rid the world of evolutionary dogma."

Scott was a bearer of some good news in that time of embattlement. She took her message to the two Washington conclaves. "Monday was the National Academy meeting, and also Monday night was the Lexington board meeting. Tuesday morning I was supposed to give the NABT presentation, and I didn't have a punch line," she recalls. Monday night, her husband gave her a blow-by-blow telephone account of the final school board meeting—and its defeat of the creationist curriculum. "So then I could report this at the NABT meeting. And it was very dramatic. It was really quite exciting," Scott said. The NABT ended its meeting with three goals: to counter the Falwell broadcast with a media blitz, develop anticreationist literature, and improve the state committee network. By the time of the National Academy of Sciences meeting, its council had already issued a resolution saying science and religion are "mutually exclusive," but still to come was this idea: turn an academy legal brief that lawyers and scientists had prepared for a possible appeal of the Arkansas trial into a mass-distribution booklet. When *Science and Creationism: A View from the National Academy of Sciences* came out in 1984, it flooded the nation's schools.

No meeting of scientists, however, matches for fanfare a gathering of the American Association for the Advancement of Science in the capital city. During the Vietnam War era, protesters lambasted science's partnership with the Defense Department, and in 1978 activists in the left-wing Science for the People dumped a pitcher of ice water on E. O. Wilson, the founder of sociobiology—the theory that genes determine culture—as he sat onstage. Yet in 1982, when the sprawling AAAS conference again was held in Washington, courts and activists both seemed to be on science's side. With uncanny timing, Arkansas circuit judge William R. Overton at midweek handed down a ruling favoring the American Civil Liberties Union in its challenge to Act 590. Overton's decision that it was unconstitutional

to mandate the teaching of creation science set off great rounds of self-congratu-
lation at the AAAS. The convention that year was dominated by sessions on evolution, with special forums on the creationist threat.

"Exponents of evolution seemed to be preaching to the converted, since there
were no dyed-in-the-wool creationists in evidence," reported the association's
Science magazine. The magazine had missed one: biologist and evangelical Hearn.
He gave Christian evolutionists a presence amid the January media frenzy. "We
quickly drafted a press release with the heading 'Creationists can be evolutionists,
too.' They disappeared like hot cakes. What we said even showed up in some stories," he recalls.[12]

Beyond the media spotlight, Weinberg held a strategy meeting, during which
he counseled that the intoxication resulting from Overton's decision should not
cloud reality: the war was not over. "To win their battle, the creationists do not
need legislative success or favorable court decisions," he said. Court victories
"make good propaganda, but they are almost irrelevant to the decisive confrontation, for which the arena is public opinion in local communities."[13] Weinberg
proposed that the Committees on Correspondence, by then in forty-nine states,
be transformed into a permanent national institution with a grassroots voice.

That winter sojourn in Washington was an omen for Scott, who returned to
the choppy waters of her academic life. Later in the year, she lost a tenure tussle at
the University of Kentucky and then moved to Berkeley with her husband, he a
lawyer and she pursuing postdoctoral work at San Francisco State University and
rearing their infant daughter. From 1984 to 1986, she commuted to branches of
the University of Colorado to teach, and in Colorado engaged in two more school
district flaps over creationism. About that time she answered an advertisement
for the NCSE, which had a board and budget but needed a day-to-day director.
Weinberg's vision of a national group had not left the drawing board until 1986,
when the Carnegie Endowment had come through with $150,000. Scott flew to a
Newark Airport hotel for an interview with board members, who hired her from
a small field of candidates. Now the center had a location, too—the basement of
Scott's home in Berkeley.

In theory the national center had a natural constituency of 1.7 million Americans, trained evolutionists who worked in natural science, taught it at universities, or were earning advanced degrees in the subject. Naturalists are not joiners,
however. The AAAS, for example, averaged an impressive 140,000 dues-paying
members in the 1990s, but that was in a sea of nearly 5 million U.S. scientists, engineers, and technicians. Having girded itself for a culture war, the center had a
limited pool for recruitment, even if its founding vision was expansive.[14] "We
were supposed to have publications," begins Scott. "Audiovisuals. Teacher training. Public information. Obviously, putting out the brush fires and helping people cope with creationism. We had two major thrusts. One is the evolution-creation education thing. The other is science as a way of knowing. How can we get
across to the public what science is, what its strengths and limitations are? Be-

cause that's part of the creation-evolution thing." The founding board was "a little naive" about creationism, she adds, kindly enough. "They thought creationism would be a blip, and they could move on. Ha! It really hasn't diminished at all. If anything, it's gotten to be more."

The opening of the NCSE in Berkeley moved the polarity of the evolution-creation debate to California. A century earlier the debate had unfolded in intellectual birthplaces such as Boston and Philadelphia. But by the 1980s science was driven by economics, and only Texas came close to having the textbook- and curriculum-purchasing power of California. The state played host to the "Berkeley crowd" of evolutionists in the north and the "old-fashioned antievolutionists" in the south.

Henry Morris had built the southern outpost in 1970 when he opened the Institute for Creation Research (ICR) as the humble appendage of a Bible college. Fifteen years later it moved into a new two-story facility with offices, classrooms, a laboratory, a bookstore, and the Museum of Creation and Earth History. The institute had become the creationist crown jewel not only for San Diego but also for a nationwide young-earth movement. San Diego became home to key creationist activists, not the least among them two families that waged legal battles with the California Board of Education, catalyzing a state "antidogmatism" policy. For seventeen years the policy softened teaching of evolution in California schools.

Soon after its arrival in Berkeley, the national center plotted its strategy in the trendsetting Golden State. The board of education was key. Before long the center's chairman, Kevin Padian, a paleontologist at the University of California at Berkeley, was sitting on the California Science Framework Committee. With Padian's help, the antidogmatism policy soon was eclipsed by new science standards that pervasively used the term "evolution."

None of these evolutionist feats came like bolts from the blue, for a small band of anticreationists already had been laboring in Southern California for at least a decade. The San Diego chapter of the American Humanist Association (AHA) had been a chief watchdog over the creationists. It also had promoted initiatives by the national group, such as an "affirmation" of evolution as scientific fact that had been sent to the nation's major school districts in 1977. The manifesto bore the signatures of 104 scientists, some of them Nobel laureates.[15]

No city has hosted more public debates between young-earth creationists and evolutionists than San Diego. In the fall of 1972, the first year ICR staff members went on the debate circuit, Duane Gish faced off against the same biochemistry professor, Russell Doolittle, who would fumble his eighteen minutes in the national spotlight at Falwell's school a decade later. Doolittle had also joined in campus debates in 1973 and 1975. His baton was then taken by the president of the San Diego humanists, Fred Edwords, who began his campus disputations in 1976. Edwords had founded *Creation-Evolution*, a journal of vigilance focused on creationists.

The San Diego humanist chapter was like a sentry, but the national center relieved it of duty after 1986. The national center took over *Creation-Evolution*—and inherited the expectations of secular humanists. They became a sizable minority among the center's dues payers, who totaled thirty-five hundred in 1997. The secularists wanted to go after religion itself, but the diversity of the membership made that impossible, said Scott. "We have everything, literally, from atheists to evangelicals. I kind of get hammered a lot from the atheists because they get confused about what we're trying to do at the center. Their feeling is, 'We should be tougher on religion. We should get out there and kneecap the theists as much as we can.' And that's not what we do."

The national center was just finishing its first year when the Supreme Court, on January 19, 1987, ruled in *Edwards v. Aguillard* that a Louisiana "balanced treatment" law was unconstitutional. In a letter to *Nature*, Scott cautioned that the ruling "does not say that no one can teach scientific creationism," but it was nevertheless strong wind in the center's sails. A court ruling that creation science was religion also put creationist institutions, such as the ICR Graduate School, under new scrutiny. For years, California had approved the graduate school degrees as bona fide, though not backed by any accreditation. A group of evolutionists, the national center among them, felt it was time for the ICR to be held to the court ruling. As center chairman Padian put it, "What they teach is not science. It's a bunch of crap." A state review team visited the school in 1989 and, with apparent behind-the-scenes prompting, voted to rescind the approval status. Rather than shut down, ICR and its lawyer, Wendell Bird, sued the state. The creationists claimed violations of civil rights and due process, and the state backed down. "So far as known, this marked the first real court victory for creationism," Morris remarked in 1993, a full thirty-two years after he inaugurated the movement.[16]

The 1980s closed for the national center as a first chapter in the modern evolution-creation debate. For the next decade the center still struggled to make ends meet and to come up with a broad definition of its mission. In 1994, Scott was elected to the California Academy of Sciences, and her comments filled news articles more than ever. In its thirteenth year alone, the center reportedly responded to anticreationist appeals for help in nearly every state. "Whenever there is the slightest odor of creation, she is right there," said Gish, who at age eighty was on the road to yet another debate. "She goes all around the country. If there's any action to put creation in a local school district, she's there."

In her El Cerrito office while we talked, Scott answered the telephone. An Oregon parent says creationism might be stalking the school district. The center will mail her an advice packet. These are the "brush fires," said Scott, from Tennessee's attempt to pass another antievolution law to a case in Georgia, where school officials glued together two pages in an earth science text so students could not read about the big bang theory. In 1995 the Alabama Board of Education passed a measure requiring that a disclaimer be affixed on the inside cover of biology textbooks, and the argument that "no one was present" to see life begin and that "evolution was a theory, not a fact" led to the Kansas State Board of Education controversy in 1999. Yet the work expands. When the seven-part PBS documentary *Evolution* aired in 2001, Scott was an adviser and national spokesperson, and her work was showcased in the last segment—"What About God?"—which looked at the creationism wars. That year the center also opened new offices in Oakland, California, adding more space—and a first-ever sign out front.

Already in 1991, however, Scott's radar had detected a new creationist blip very close to home—in liberal Berkeley. Phillip Johnson, twenty-three years a professor at the university law school, came out with his *Darwin on Trial*, a criticism of Darwinian theory from the bowels of a prestigious secular university. Scott was fifteen years Johnson's senior in the evolution-creation wars. She had witnessed an entire era of creation science, young-earth activism. Johnson, however, wanted to change the subject. "Now we shift over to the more evolved creationists," Scott said with hindsight.

. . .

The university credentials made a remarkable difference, if Johnson's sudden fame is to be understood. His tenure-protected nouveau creationism was the kind that, at least by backhanded compliment, the nation's university-tenured evolutionists could not ignore.

So there was Johnson, a prickly issue, even as the Smithsonian Institution, in April 1998, celebrated E. O. Wilson's great summa, *Consilience: The Unity of All Knowledge.* Wilson's argument—that all knowledge can be tied to biology— would have its critics, biologist Paul Gross said in one of two scholarly responses after Wilson presented his case. Foremost would be the design people, Gross said. They use an "ancient argument," he warned, that goes "back before William Paley, before Descartes, before Thomas Aquinas, but as well straight forward to today in promotions by a Phillip Johnson."[17]

Johnson could be thought of as creationism's bulldog. The short, amiable professor told me his story in 1996 at a dining table outside his small kitchen, munching on popcorn as he explained his work and motivation. He had an office up the hill at Boalt Hall, home of the law school, but his stucco and clapboard house in a midcentury neighborhood is his base for moonlighting as an antievolutionist intellectual and publicist. In a big-windowed study behind the kitchen, he pounds on his computer to produce books—five on religion's clash with naturalism—speeches, and book reviews. He surfs the Internet to join the cyberdebate on origins.

"If you're interested in intellectual questions at all, the question of where do we come from and what is reality really like, I mean, that's the big one," Johnson said. "Why wouldn't anybody be thrilled to get to the bottom of that?"[18] For too long, he said, the question was, What does science think about Genesis? He reversed the burden of proof. His question is, Why is science given carte blanche to dictate the rules of modern knowledge? It is not a spellbinder for a church audience, but it is a topic germane to his main target: universities.

Intellectual theism has not fared well against Darwinian naturalism, he surmises, because it is mired in two well-worn tracks. Young-earth creationists must focus on the Bible. In the other track, "The theistic evolutionists were sort of apologetic, you know, 'Gosh, can't you leave a little room for us?'" Johnson said. "They accepted a very subordinate role. Their theology followed along in the wake of evolution." To keep on message, Johnson is fastidious about avoiding discussion of the Bible or his own beliefs. He is a Presbyterian and a latter-day political conservative—but that, he said, is beside the point. Back on message: evolutionists use "a loaded definition of science that allows only one possibility. The whole thing will just collapse if people look at it a different way."

Before he turned to teaching, Johnson was a Ventura County criminal prosecutor. In courtrooms every day, attorneys tighten verbal screws on opponents in hopes of making them crack. What the creationists needed was that prosecutor's touch. "You had to meet intimidation with counterintimidation in order to move the discussion forward," he said. "Now, that perhaps was the lawyer's contribution." Johnson concedes that he may have the zeal of a convert. Reared a lukewarm Protestant in the Midwest, he studied English at Harvard and law at the University of Chicago, imbibing deeply of what he calls "philosophical naturalism." He saw it in academia and also in the courts, where he clerked for Chief Justice Earl Warren. By then an agnostic, he abided by naturalism's cardinal rule: nothing intellectually credible allows for the supernatural.

However, by 1991, the year of *Darwin on Trial*, he was fifty-one years old and pondering his unexpected arrival at a Christian point of view. He dated his conversion, which followed a failed marriage and was supported by a second one in the church, to about 1980. He attended First Presbyterian Church near the campus and had considered himself somewhat mired in his career track as recently as the mid-1980s.

What a difference a sabbatical can make. Johnson spent the 1987–88 academic year in England, attached to University College, London, but without teaching burdens. Passing a London bookshop, he came upon Richard Dawkins's book *The Blind Watchmaker* (1986), a defense of gradualistic Darwinian evolution, which he read. Then he returned for *Evolution: A Theory in Crisis* (1985), by Australian biochemist Michael Denton. His mind in full gear, Johnson next paid homage to Charles Darwin by taking the twenty-mile day trip to his museum-estate in the city of Down. In his last months abroad, he visited the British Museum of Natural History in London. There he chatted for several hours with vertebrate curator Colin Patterson, a Darwinian, but an openly skeptical one.

While abroad, Johnson met Mark Labberton, an American theology student at Cambridge—and the future pastor of Johnson's own First Presbyterian Church in Berkeley. Labberton introduced Johnson to another American, Stephen Meyer, who was earning a doctorate in philosophy of science. Meyer, a geologist, had abandoned his work in Texas oil after hearing a scientific debate on the origin of life, "the big one," the issue that also had begun to hound science neophyte Johnson. "He got obsessed with evolution," says Meyer. "We walked around Cambridge kicking the pea gravel and talking over all the issues."[19] Before leaving London, Johnson typed a 120-page, single-spaced monograph—the first trace of *Darwin on Trial*. It went through revisions and new gestations for three years, and David Raup used its August 1988 version to play devil's advocate in a University of Chicago graduate seminar.

By the end of 1989, Johnson still did not have a book, but he had developed a reputation. In December a group of church-state lawyers who had spent a decade in the culture war on school prayer and "teaching about" religion in public schools felt ready to take on their next contentious issue. They expressed it in the title of a three-day conference, "Science and Creationism in Public Schools," at the Jesuit Campion Center outside Boston. Among the meeting's fourteen participants were, of all people, Johnson and Gould. Johnson had been invited by the director of the Center for Law and Religious Freedom. Harvard-Smithsonian astronomer Owen Gingerich had urged Gould to attend, and Raup had come on his own to meet the Berkeley lawyer. The group persuaded Johnson and Gould to stage an unrehearsed debate. "It was fairly plain that someone was particularly interested in introducing Phil Johnson's views to the group," says Gingerich. "So that was something of what was behind it."[20] Nearly a decade later, Gould says he hardly remembers the meeting. For Johnson, however, it was "one of the pivotal events" for his becoming a voice in the national discussion—a chance to test his arguments among evolutionists, some of them atheists and others not.

Johnson was at home with American law. He was still an outsider, though, when it came to scientists and even the Christian subculture in science where he might find natural allies. Eventually he cast his fate with the "intelligent design" circle of intellectuals, a circle in formation when Johnson came onto the scene. The group's impetus was a 1984 book, *The Mystery of Life's Origin*. Written by chemist-historian Charles B. Thaxton and engineering professor Walter L. Bradley, it challenged the optimism about discovering how chemicals turned into living cells and opened the way for a mingling of biochemistry, information theory, mathematics, embryology, genetics, and philosophy of science to infer that intelligent design was a valid scientific theory for nature. Meyer was in the circle, so Johnson had left England intending to look up Thaxton and Bradley. Two years later, in 1990, the intelligent design fraternity held a meeting to scrutinize the California lawyer. "It's a question of looking someone over," Johnson said. "I very much approve of that."

Through his widening circle of erudite acquaintances and "science consultants," Johnson honed his manuscript to completion. The release of *Darwin on Trial*—which questioned seven areas of "evidence," the "rules of science," and "Darwinist religion"—made him a national figure. "By the time I got home from grad school, Phil was already famous," said Meyer, who teaches philosophy at

Whitworth College in Spokane, Washington. "His book was out. And his credentials immediately gained him attention. He also got trashed in all the major science journals."

The new public role also drew from Johnson's paternal experience as an academic dean. Now, he tried to open doors for younger scientists whose dissension from Darwinism raised career obstacles. Some of this group was called the "wedge," and as one biologist said, "Phil is the 'sharp edge' and the rest of us are the ever-widening shank." After 2000, when Johnson had retired from university teaching to carry out the wedge strategy full time, his speeches often highlighted its ten-year success. "Whenever I see anybody who's interested in pursuing this question, then I want them to be in our movement," he said years earlier. "We are not the kind of movement that has a doctrine."[21]

Those in the early orbit around Johnson gained what Meyer calls "cultural confidence." They first tested that wherewithal in early 1992 by debating a panel of naturalists in a session convened at Southern Methodist University. "The confidence level went way up," Johnson said. So did the fun. Meyer says the unassuming Johnson tended to surprise his intellectual rivals. "You know, he has a little duckish appearance. A true academic with tweed coat." Debate rivals liked Johnson's backroom Supreme Court tales, erudition, and get-along demeanor, and Meyer spoke of his rare durability. "One thing that made Phil so important was his willingness to take the hits," Meyer said. "There are a lot of people who just do not have a thick enough skin to have people say the kinds of things they said about him."

The science journals and magazines, all of which lined up against *Darwin on Trial*, were especially hard on him. Gould panned the book in *Scientific American*, saying Johnson "has taken a low road in writing a very bad book." Lawyers have every right to weigh in, Gould wrote, but Johnson confused evidence in science and evidence in law and was wrong to claim that Darwinism by necessity leads to atheism.[22] In the second edition of *Darwin on Trial*, Johnson took glee in adding what he believed were blow-by-blow rebuttals to Gould's contentions.

Some Christians in science were no less caustic than the Harvard paleontologist. Johnson owed his fame, they said, to tenure at a big-city school rather than at a church college. "Oh! He's a Berkeley professor!" they exclaimed sarcastically. For a host of evangelicals who were theistic evolutionists, Johnson was repeating the old "God of the gaps" mistake, saying evolution could not create a part of nature, so God must have. Faint praise also came from young-earth creationists, such as Harvard-trained paleontologist Kurt Wise. He had given Johnson's manuscript a thumbs-down. "All you do is attack evolution," Wise recalls saying to Johnson. "You do not propose an alternative. This is a wimp's way out. Now, he and I—we understand my position on that."[23]

In a matter of a few years, Johnson—or rather, one's attitude toward him— had become the litmus test in the American evolution-creation debate. The tell-all question was, Where do you stand on Phil Johnson? The evangelical rejoinders surprised Johnson at first, but then he began to understand. The evangelicals had "grown used to thinking that they own the Christian part of the academic world because the fundamentalists are isolated," he said, "and I've certainly challenged that." More precisely, Johnson challenged theistic evolution, which he called theologically meaningless. In a series of lectures at conservative Trinity Evangelical Divinity School in 1993, Johnson said his critics practiced "theistic naturalism"—a kind of pious atheism when it came to nature. "That term has been a source of great unhappiness," said Johnson, unrepentant.

Where Johnson's Berkeley credentials counted most, it seemed, was among the evolutionists who saw him as a bright and affable iconoclast. Raup says Johnson's book was on the money. "Phil's done a superb job," he says. Raup taught science at the University of Chicago, where Johnson studied, but he was not heaping alumni flattery on Johnson. "He's really done his homework. It's phenomenal the way he absorbed the field. Now, many people would say, 'He doesn't know anything. That's obvious from the first page.'" Raup shakes his head. "That's often said without *reading* the first page."[24]

In American Christianity, Johnson found the greatest receptivity in evangelical campus ministries such as InterVarsity and Campus Crusade for Christ. The latter, in fact, had been a main sponsor of young-earth creationist debate in the 1970s and 1980s, the only alternative to theistic evolution. "They couldn't go either way," said Johnson. He offered a new approach. Campus Crusade reciprocated by organizing speaking events on campuses. Duane Gish says he would not call Johnson an "out-and-out creationist." The design emphasis has always been part of the young-earth repertoire, he says, happy to have Johnson join in, even as "Phillip Johnson says he keeps an arm's length from what he calls the biblical literalists."[25]

As he produced a series of five books between 1991 and 2000, Johnson's radio, television, and public talks increased. He was clearly the "godfather and guru" of a more extroverted intelligent design movement by 1996, when he was emcee and concluding speaker at the "Mere Creation" conference. That same year he helped to establish the Center for the Renewal of Science and Culture at the Discovery Institute, a conservative think tank in Seattle. The center, funded mostly by evangelical philanthropists, offered research fellowships for scientists interested in intelligent design—or in challenging Darwinian orthodoxy. In March 1998, Johnson gave the venerable Norton Lectures on science, philosophy, and religion at Southeastern Baptist Theological Seminary in Louisville—a step toward reaching all six of the Southern Baptist Convention schools, the largest theological system in the world.

By decade's end Johnson's books had sold a quarter million copies, and friends quipped that InterVarsity Press, his publisher, "would be happy to publish his laundry list." Yet the work planned for 2000, *The Wedge of Truth*, had something different in mind. It was foreshadowed at the "Life After Materialism" conference at Biola in 1999. As before, Johnson played host, and his closing talk reviewed his in-process book, like a bellwether of the intelligent design agenda. Ready to escalate the debate, Johnson said he would declare that Jesus Christ was the alternative to naturalism; the assertion came under a subheading near the end of the book, "Who Do They Say That I Am?" "Obviously, it is going to push the envelope," he told the group. The disclosure pleased some allies but surprised others who preferred Johnson's "not the kind of movement that has a doctrine" approach. One conference speaker had argued that "the concept of design does not necessarily lead to a Christian God." But by then, Johnson no more embodied the intelligent design movement than did any other individual: a pluralism maintained.[26]

Views of salvation aside, it was perhaps paleontologist Wise's earlier challenge to "build a theory" that led Johnson to propose his first alternative to naturalistic evolutionism. He endorsed "theistic science" in his book *Reason in the Balance* (1995). It is a view, one sympathetic scholar said, that "merely postulates the possibility of divine action in the natural world. . . . The theistic realist may believe that certain facts of the world are such that only a being with the attributes of God could possibly be responsible for such a system." If such epistemology was too heady, Johnson had also aimed a book at high school and college students. Titled *Defeating Darwin by Opening Minds*, it warned them of three mistakes easily committed in the evolution debate: making the age of the earth the main issue, assuming that God has "retired" because nature has laws, and conceding to evolutionists the claim that they alone use rationality. He was equipping a next generation of antievolutionists, and he believed that cultural trends would be on their side.[27]

"It takes a generation for these things to filter through the culture," said Johnson, continuing our discussion at his home in the summer of 1996, still eating his popcorn. "It doesn't happen quite overnight, but it's only a question of time before a theistic paradigm becomes established." The American population is predominantly religious, he points out, the elite culture notwithstanding. "We definitely have the numbers. And they own the newspapers, the television stations, the public schools, and the government," he said, referring to the anticreationist elite. He is reading a book on China during the Second World War, and a military analogy comes to mind: "The communists have the countryside, and the Kuomintang has the cities, the guns, and the American dollars."

. . .

Neither Scott nor Johnson has been keen on debating too much scientific or biological detail before the public, but both have attacked the theme of the status of science and religion in intellectual culture. They pushed this topic on the *Firing Line* special. "Evolutionary science takes its starting point from a philosophical position known as naturalism, or materialism," Johnson said in opening the TV debate. "It follows, logically, that science must and can explain the origin of complex organisms solely by natural causes, meaning unintelligent causes. God may not create directly, nor may God direct evolution, because God is an intelligent being, and evolution is by definition a mindless process."

That comment is what prompted Scott to mention O'Hair, the celebrity atheist. She calmly explained that evolution has two territories, the evolution of the universe and of biological life. "Now, notice in this definition," she continued, "I didn't talk about who done it and I didn't talk about how, because those are separate issues." The "who" is a religious issue; the "how" is a valid topic for scientific debate, she suggested. "Scientists are very much united on what happened. 'Evolution happened,' to modify a bumper sticker. But how it happened is something that we argue about a lot in science." Science cannot answer the God question: "We can put on our philosophers' hats and comment as individuals, but as scientists we can't deal with ultimate cause."

Ever since Darwin, the debate has raged on whether Darwinism rules out God, particularly a God who dabbles with the nature and the humans he created. Johnson says Darwinism is atheism. That is not necessarily so, says Scott, but she knows some scientists make that leap. "Scientists have to be careful with their words. If you use words in a way that does not allow theists to retain their faith, you are making the same mistake that Morris and Gish, all these old-fashioned antievolutionists, are making. They're saying, 'You've either got to be an evolutionist or a religious person, but you can't be both.'"

This is an issue Scott acted upon in one fairly dramatic episode. During the early fall of 1997, Christian philosopher Alvin Plantinga of Notre Dame University and historian of religion Houston Smith advised the National Association of Biology Teachers that the preamble to its 1995 statement on evolution used two words—"unsupervised" and "impersonal"—that suggested atheism. Their letter was blunt: unless the association dropped the words, the Christian right would use them to say NABT teachers promoted religious doubt in biology class. The preamble read: "The diversity of life on earth is the outcome of evolution: an unsupervised, impersonal, unpredictable and natural process of temporal descent with genetic modification that is effected by natural selection, chance, historical contingencies and changing environments."[28]

In the first round, the Plantinga-Smith proposal was rejected by a vote of the NABT's executive council, which had gathered in Minneapolis in October. Having learned of the vote late in the game, Scott flew to Minneapolis to lobby. Indeed, she persuaded the NABT president to rally a majority for a new vote, and a day later the deletion was adopted.

The wire service Religion News Service broke the story, but evolutionist protests did not arise until December, when the vote was reported in the *New York Times* as a "startling about-face." Dropping the two words "represents the first wedge of a movement intended to surreptitiously introduce religious teachings into our public schools," said biologist Massimo Pigliucci, who led an Internet protest from the University of Tennessee. His petition garnered its first hundred signatures in a matter of weeks. Wayne Carley, executive director of the NABT, welcomed the petition as part of the professional discussion but stood by the change. "To say that evolution is unsupervised is to make a theological statement," he said. In what Scott called "l'affaire NABT," she also circulated a response. "NABT was not knuckling under to creationist pressure, but responding in a responsible manner to a perception on the part of religious Americans that it was making an antireligious statement," she contended. Biology teachers are not antireligious, she added, warning: "Such a perception is inaccurate, but it is also injurious to members of the NABT, the teachers who must teach evolution."[29]

Johnson enjoyed the evolutionist implosion. His Internet salvo said Pigliucci was right: "I salute the Open Letter's candid statement that the American public correctly perceives a direct conflict between neo-Darwinian evolution and the Judeo-Christian concept of a personal God," he wrote. He said that the evolutionists' dilemma—what to tell the public about atheism—was created by their own machinations. "Evolutionary science is a lot like American football with its two-platoon system," Johnson wrote later. The first platoon, he said, assures America that evolution is no threat to faith, as in the National Academy's booklet, *Teaching About Evolution.* The other platoon, however, is at the Smithsonian celebrating Wilson's book *Consilience,* which argues that biology leads to atheism. The first platoon is a diversion, Johnson said, while the second speaks evolutionist candor.[30]

In the years since "l'affaire NABT," Scott has also weighed in on this question of the proper turfs of religion and science. In a benchmark paper, she identified two movements that were not quarrelsome with science: the "science and religion" dialogue, which seeks no conflict between faith and evolution, and the "Christian scholarship" movement, with its straightforward claim that a scholar's Christian belief can influence his or her scholarship. The rules break down, warns Scott, in a third approach: "theistic science." She eschews its blurring of the two disciplines.[31]

This, of course, is the approach backed by Johnson, who was becoming a more wide-ranging writer, now quoting C. S. Lewis, now the Westminster Confession. Niles Eldredge had debated Johnson at Calvin College in 1996 and then watched his profile grow. "I think he's looked at as one of the reigning intellectuals, or at least among the intellectuals, in Christian thinking in the United States now," Eldredge said. On the Calvin College visit, Eldredge had joined a faculty forum in which the school's science professors, evangelicals all, excoriated Johnson's brand of creationism. "So Phil insisted his view of God is *the* view of God," an "interventionist God," said Eldredge, an agnostic. "And it was fascinating to me. The fight is not between me and Phil or me and them. It seems to be among them."[32] Evolutionists enjoy watching creationist implosions, too.

When Johnson came on the scene, his backers touted him as a master of evidence and logic—a worthy mind to show where evolution came up short. Gould, of course, met that thrust by saying courtroom evidence meets a different kind of standard—beyond reasonable doubt. Insisting on an earth full of evidence that fleshes out the modern theory of evolution is far different from looking for a smoking gun. Still, Johnson charges, the theory is so big that it just explains everything, and it does so as biologists disagree on myriad details. "And they aren't details," said Johnson. "As far as I can tell, all evolution means is that God didn't have anything to do with it. After that, everything else is up for grabs. There's a relationship among living things. There's microevolution, and then things just happen. If you're a Gould type, evolution is just 'one damn thing after another.' And if you're a Dawkins type, it's this logical progression of constant micromutations."

Scott and Johnson are two of the most active voices in the modern evolution-creation debate. Behind them, however, are others, experts in the sciences, who may shun the spotlight so they can produce the kind of data on which Scott and Johnson invariably draw. David Raup tells of his father's era, when a self-respecting scientist would not even go to an open science conference because of the detachment expected of scientists. Those days are over—especially in the fields touching on the evolution-creation debate.

5. NATURE ALONE: EVOLUTIONISTS

Science is not a democratic domain, and its principles are not subject to vote, evolutionist Joseph D. McInerney said in his office hard by the Rocky Mountains. Making American schools safe for evolution has consumed half his career, and he knows it can be a battle. Take, for example, the public's desire to pick and choose its own scientific beliefs. His 1993 essay in *American Biology Teacher* lampooned such "voting in science" under the headline "Raise Your Hand If You Want Humans to Have 48 Chromosomes."[1] Scientists, of course, have found only forty-six chromosomes.

At the time, McInerney was director of the Biological Sciences Curriculum Study (BSCS), a leading biology textbook producer. A Wyoming school superintendent had called to complain about the evolutionist stance in the textbooks. He advised McInerney that his district taught both creation and evolution and let students decide by a vote. This conversation spawned the essay, but for the moment McInerney tried to correct the superintendent's error. "So I said to him, 'When they study germ theory, you should also bring in somebody who believes in demonic possession to explain disease. And then let the kids vote on which one is right. If they vote for demons, the school district can save money on immunizations.' And he said to me, 'You know, you're being a real smartass.' And I said, 'No, I'm taking your approach to science and moving it to another level. When you do anatomy and physiology, let them vote on whether the heart has three chambers or four.'"[2]

McInerney, an energetic, dark-haired Irish American, told me the story when I visited Colorado in 1996. He has relished many such debates. He is a former president of the National Association of Biology Teachers and until 1999 was director of the BSCS, established in 1958 to design the first biology textbooks fully treating biological evolution. That was a year before the Darwin Centennial, which used a federal grant to bring one biology teacher from each state to Chicago for evolution workshops (and later produced a small book, *Using Modern Knowledge to Teach Evolution in High Schools*).[3] But it was the BSCS textbooks that most revived public school biology. The textbooks also, far more than the centennial, evoked a backlash of creationist political advocacy.

Though this pioneering aura has faded, BSCS is still in business. Its offices are backed by the dramatic spires of the Rocky Mountains, which tell part of the story of plate tectonics and of "deep time" itself. Having broken away from the land-mass now called Africa, the North American plate drifted into place two hundred million years ago. It had pushed westward at a rate sufficient for friction and the Pacific Ocean floor to rumple its surface, forming the mountains that stretch in-land as far as modern-day Colorado.

"Evolution requires those immense periods of time," said McInerney, who has been trained in genetics, biology education, and public health. "The time scale also puts in perspective all the short-term changes we see around us." The rise and fall of a species or environment, in fact, is a mere blip on the radar screen of earth's history. The genetic code for life, according to current theory, also stretches back 3.5 billion years. But only in recent years has science begun to pro-vide the public with the simplest knowledge of this linchpin of modern biology. The BSCS textbooks, for example, first took up genetics in the 1980s. By 2001, it had received a Department of Energy grant to produce a learning package for teachers on bioinformatics, the new science of reading and analyzing the human genome, or DNA, sequence that was in the public domain on the Internet.

Such rapid expansion of the field of genetics created new needs and opportu-nities. And so it was in 1999 that McInerney finally left his textbook mission to help guide the nation's health workers through the genetic revolution—and to be part of a new field called "Darwinian medicine." Two months after the comple-tion of the human genome was announced in June 2000, and five months before it was published for the world, McInerney was tapped to head the National Coali-tion for Health Professional Education in Genetics, a project of the federal gov-ernment and the medical industry.

Francis Collins, director of the U.S. National Human Genome Research Insti-tute, called McInerney's job of bringing doctors and nurses into the genetics revo-lution a "critical position" with a "very ambitious agenda." The new field was being called "genomics," or the application of genetic discoveries to medicine. But always the evolution educator, McInerney was among the first to note how evolu-tion had been played down in the public fanfare. During the White House an-nouncement in June 2000, only Craig Venter, whose Celera Genomics paralleled the government race to sequence the genome, bothered to explain that, whether bacteria, plants, or humans, "we're all connected to the commonality of the genetic code in evolution." President Bill Clinton spoke of how science was "learning the language in which God created life," and Collins cited the DNA code as "our own instruction book, previously known only to God." Whether these were figures of speech or theological claims, the White House verbiage gave both evolutionists and creationists plenty to seize upon—only showing, as McInerney knew well, that society's debate on evolution would carry over to the medical revolution of the human genome.[4]

Before that genomic revolution, however, it had been the Soviet launch into orbit of the unmanned *Sputnik* spacecraft in October 1957 and the ensuing space race that had mobilized American science and paved the way for science educators like McInerney. In 1950, Congress had formed the National Science Foundation (NSF) to channel millions of dollars into research and science education. One of these funding streams gave birth to BSCS to produce textbooks. Founded at the University of Colorado, BSCS spent five years on consultations and field tests before publishing its first materials in 1963. Half of the nation's schoolchildren soon would use the textbooks.

Virtually every other publisher plagiarized the company's approach; BSCS had "encouraged the copying of its materials." Its early leadership, however, did not make BSCS immune to budget storms and political controversy. McInerney recalls some BSCS milestones: in 1972 it became an independent nonprofit organization separate from the university; in 1982 the staff, which had peaked at fifty-five in 1977, plummeted to two as federal allocations for science education were cut.[5] On the eve of the company's fortieth year, the competitive textbook market forced it to break new ground and initiate its first capital campaign. "We've never gone out and asked people to contribute to the organization simply because it's a worthwhile place to have around," said McInerney.

BSCS textbooks have made a name for themselves, thanks also to vociferous creationist attacks. "It sounds like a platitude, but science teachers look to us to hold the line, to protect scientific integrity," McInerney said about BSCS. "They will say, 'Don't give in. We need you out there doing this.'" To prevent any watering down of the textbook explanations of evolution, BSCS copyrighted the material. It is a stance that often undercuts sales. "We have walked away from some fairly hefty chunks of money," McInerney says, then explains the reason: "Publishers cannot change a word of our material without our permission."

From the start, BSCS published three textbooks to emphasize biochemical, cellular, and ecological biology. One reason was to be flexible on the market; another was to avoid the impression that BSCS was going to use its federal funding to impose a single-book "national curriculum" on the states.[6] "Our focus when we began was the high school biology curriculum, but we have since broadened to having programs from kindergarten up to college," said McInerney. With a focus on basic science for lower grades and biology on up to college, all the texts included learning activities. The biochemical "blue version," for example, had an exercise called "Patterns and Purpose" that asked students to look at patterns in the natural world. Beginning with nineteenth-century theologian William Paley's argument that a watch found on the heath must have a watchmaker, the exercise asked if there must be a pattern maker to give such apparent design to nature. The lesson concluded that evolution can provide such natural complexity. To show how science can change its mind, students learn that it once was believed that DNA was transferred only by cell fertilization, but then came the discovery of transfers of mitochondrial DNA or trinucleotide repeat expansions. "They are nice vehicles for demonstrating to kids how science, how knowledge changes," said McInerney. "Scientists expect knowledge to change. But it changes within very robust theory bases."

This packaging by BSCS of current science occurs in a modern, boxlike building of brown cement in Pikes Peak Research Park, where McInerney's second-floor director's office takes in a panoramic view of the Rockies. On his desk the day I visited was *The Demon Haunted World* (1995), Carl Sagan's paean to science for its virtues of awe and skepticism.[7] McInerney sees in this next-to-last book by Sagan a worthy theme for science education: to teach American students to have skeptical minds—which, he quickly adds, are different from cynical minds. "One of the problems parents have with the kind of science education we do is that we try to create skeptics," he said. "We try to make kids understand the value of skepticism." He paraphrases George Santayana, a father of philosophical naturalism in America: "Skepticism is the chastity of the intellect not to be surrendered lightly to the first comer."

In *Demon Haunted World*, Sagan recalls asking the Dalai Lama whether he would drop even his most central mystical beliefs, such as reincarnation, if they were scientifically proved wrong. Sagan credited the spiritual leader for agreeing that, yes, "Tibetan Buddhism would have to change." Sagan's point was that science will always be willing to change; it will always welcome disproof, a challenge that religions—except perhaps the Dalai Lama's—shun.[8] As McInerney puts it, "If paleontologists uncovered properly dated strata, and we found trilobite fossils above hominid fossils"—meaning humans appeared on earth before simple marine animals—"scientists would be going, 'Whew! Well, we probably should rethink this, shouldn't we?'" So it was with plate tectonics—the explanation for the force that created the Rockies and the Appalachians, thought to be lunacy until the evidence mounted up. "Well, the data prevailed," said McInerney, "and scientists will change their minds." He would argue that creationists are different. "I've asked this of creationists: 'What will change your mind?' The answer is 'Nothing.' Nothing short of the Second Coming. If Christ materialized at the Institute for Creation Research and said, 'You're wrong,' that would work—maybe."

Young people hold misconceptions for other reasons, said McInerney. They tend to think creatures have guidance programs built into them, much as Aristotle believed. Why, McInerney asks students, do ducks have webbed feet? "They needed it so they developed it," he said, mimicking their answer. "They went to the webbing store and got webbing for their feet to help them swim better." Students also stumble on concepts such as genetic variation and the way surviving organisms become predominant. Adaptation—the process by which a creature develops skills or features needed to thrive in its environment—is also an elusive concept. On this, said McInerney, students are unwitting disciples of the early French evolutionist Jean-Baptiste Lamarck; they think giraffes lengthened their necks by stretching for leaves and then passed that length on to baby giraffes.

Since the 1980s, Colorado Springs had become a national hub of Christian organizations, and so McInerney became conversant with evangelicals. An agnostic and lapsed Catholic, he evangelized the estimated 250 home-schooling families in town with free science lessons at BSCS. Genetic engineering in particular captured the interest of the Christian parent-teachers who came. They learned how insulin, a hormone once mined from animals, was now mass-produced by the insertion of a human insulin gene into fast-reproducing bacteria. "They thought this was the coolest thing in the world," recalls McInerney. Why could a human gene work in bacteria? he asked. "Because we are related to those organisms by descent with modification," he told the home-schoolers, adding that 500 million yers of organismal life on earth is written into human genes. "Well, when I said that, the shade went down. They weren't hearing that." Many creationists accept genetic relatedness but argue that God created using the same genetic parts. "My response is, 'Fine. Just don't pass that off as a scientific explanation,'" said McInerney. If creationists have any stocks in pharmaceutical companies, he says, their pocketbooks at least are admitting to genetic evolution.

This area of medicine, which includes immunity to chemicals in organisms, emerged in the 1980s as one of the top arguments for evolution. Insects that survived insecticides did so by helpful genetic mutations, and thus they bred a new generation that was not brought down by the farmers' poison. Skeptics say it is only molecular adaptation, however, not Darwinian evolution of a new species. Still, evolutionists can draw on the principle to discuss human disease. Under "Darwinian medicine," the history of genetic mutation is used to analyze people's receptivity to one ailment or another—or a pharmaceutical or genetic therapy as a remedy. Indeed, one theory about the AIDS epidemic is that the virus was isolated and harmless in its first animal host but spread rapidly once it jumped to a human genetic system that lacked the immunity.

"Genetic medicine is simply another example of the validity of [geneticist Theodosius] Dobzhansky's well-known assertion that 'nothing in biology makes sense except in the light of evolution,'" McInerney said, now in his new field. With the new human genome library available to all, the ethical quandaries have become more urgent. How much gene tampering should doctors do? With what degree of certainty should they predict a patient's genetic predisposition to one ailment or another? How will insurance companies look at people suspected of having, or showing, this "inclination toward" a disease? How far can private business go to capitalize on the new genetic market? Tough human questions all, McInerney agrees. But from a scientific viewpoint, he argues, evolution is "the conceptual glue that can help to bind the disparate aspects of medicine into a coherent whole."

Back during his days in Colorado, McInerney had tried to make BSCS a good neighbor, which can be difficult when the issue is complex science. When he volunteered to review science standards drafted by the Colorado Board of Education, he had to be blunt about their statement that evolution is the change in organisms through time. "No, it's not," he said. "Evolution is the change in populations of organisms through time." In Colorado Springs, Liberty High School was among a few schools that rejected the BSCS blue version, unhappy with a statement in the opening chapter that says "creation science is not a science." "We used creationism as the primary example of pseudoscience," says McInerney. He tried to negotiate the value of the book with an offended parent on the textbook committee but finally had to set her straight. "I didn't create the term 'creation science,'" he told her. "If you had called it creation magic, or creation mysticism or creation mythology, we wouldn't even be having this discussion. But you people called it creation science. Now I have to respond. So I'm sorry you're offended, but I'm deeply offended by your calling what is essentially magic, science."

POPULATION THINKING

Biologist Ernst Mayr has debated a creationist only once that he can recall. Yet he is no puppy dog of evolutionary biology. He has been called variously "Darwin's bulldog," a senior scientist with the argumentative zeal of a graduate student, and, in *The Scientific 100*—a big, glossy book that sits on his coffee table—"an aggressive polemicist." He waves it all off, this widowed father of two daughters and kindly patriarch to five grandchildren and that many more great-grandchildren.[9]

Mayr has avoided the politics of his fellow naturalists, some of whom have dabbled in Marxism and others in the social implications of genetic determinism. Nor has he fought with creationists or disparaged religion. His good friend was British ornithologist David Lack, who died an evangelical, and once when Mayr received a theological question by mail from a student at Bethel College, Mayr curtailed his natural bluntness. "Since I have no intention whatsoever to disturb his belief in God, you see, it is a difficult thing to answer," he said in his German-accented English. "So I more or less weaseled out."

When someone asks, "Who is the greatest living biologist?" the legendary Mayr is often named. In 1998, at age ninety-two, he published yet another book, *This Is Biology* (and still another in 2001: *What Evolution Is*). In *The Scientific 100* hit parade for Western science, Mayr is among only twenty-two life scientists to be named from all of history. As scientist number 65, he is the Darwinian closest behind Darwin himself. The father of evolution comes in as number 4—preceded by physicists Isaac Newton, Albert Einstein, and Niels Bohr.

Mayr is one of the last surviving eyewitnesses and contributors to the New Synthesis of genetics and natural selection. Its founder, the geneticist Dobzhansky, was his "best friend." He ate lunch daily with the "truly brilliant" paleontologist George Gaylord Simpson, who was known during his career as the dean of evolutionists. Simpson was "a man of few words"; Mayr remembered not one serious scientific conversation during their tenures at the American Museum of Natural History, Simpson as head of paleontology and Mayr as curator of the bird collection.

The personalities were mercurial, Mayr says, but the New Synthesis has proved unshakable. "I do not know of a single thing that the synthesis brought out that has been shaken for all the new developments," he said in his Cambridge, Massachusetts, home. Much of the social and scientific status accorded to the New Synthesis was achieved through Mayr, a world-class ornithologist, or bird expert. An early activist for establishing a guild of professional evolutionists to gain federal grants for studies in evolution, he was also a frontline proponent of the New Synthesis.

Sad to say, he laments, his old friend Dobzhansky is hardly remembered anymore. Dobzhansky's contribution was catalytic, but it stopped there. It was up to his colleagues and students to elaborate, and they did: Simpson on the fossil record, G. L. Stebbins on plant evolution—and Mayr on how species form. In his book on the New Synthesis, Mayr detailed how geographically isolated populations can change gradually and give rise to a new species. He called the process "allopatric speciation," and the group that turned into a new, interbreeding species was a "founder population." Speciation occurred not only by isolation, he argued, but also by small wandering groups or even a single female who crosses into a new population. The result: "peripatric speciation." All of this, said Mayr, is part of the pluralism in biology, the weaving together of several theories to explain evolution.

The diversity, however, rests on the even more rudimentary concept of a "biological species," which Mayr had singularly established in science parlance. "It was a widespread concept all along, except people didn't put it into a formula as I had," said Mayr. Still, by one count there are fourteen competing definitions of species, and Mayr readily admits the biological ones do not apply to many plants, bacteria, or one-celled organisms. When a group of younger biologists met in Philadelphia, they could not get past that definition. "All they did was fight about what a species was," says historian William Provine, who adds that the issue is hardly settled. "The problem of the origin of species can't be solved unless you know what a species is," he says.[10]

There was only one major omission in the New Synthesis, intellectually formed between 1937 and 1947, Mayr said, and that was the work of the developmentalists, those who studied embryonic life. "Now, the developmental biologists in the 1970s began to complain and say, 'Oh, these people of the synthesis ignored development,'" Mayr says, but he counters that their exclusion was their own fault. "The developmental people at that time . . . were all completely opposed to Darwinism, to natural selection," Mayr says. "They didn't want to join the synthesis." He throws up his hands. "So naturally, the developmental aspects were not part of the synthesis."

Clearly, biological development has turned out to be a major sticking point for evolutionary theory. Common descent may be seen in the common bone structures of all creatures that walk the earth and in a genetic code that appears to be shared by every organism. Those similarities, however, break down on the segment of life from fertilization to existence of a mature embryo. This has led some evolutionists to suggest that more than one common ancestor may account for the embryo differences; but Mayr insists, as ever, "Nobody says that." Others argue that the embryo itself evolves, creating all the differences. Even Darwin said fitting embryonic development into the evolutionary tree of life "is second to none" in proving his theory. The other great conundrum is the genome, or genetic program of an organism. If the programs have such similarity across all of life, why are the results—a human and a worm or a whale and a spider—so different? The genotype is a tough nut, says Mayr. "The genotype, the brain, the ecosystem—those are the three great frontiers of modern biology."

Mayr was prepared for his own frontiers by a well-educated German family. His father, a jurist, owned three thousand books, and the extended family included many doctors. His parents probably were agnostic, he said, though his mother attended a Lutheran church for the intelligent sermons she heard there. She urged Mayr at about age fourteen to be confirmed by the minister. "And he more or less succeeded in getting me back into a Christian frame of mind, but it didn't last very long."

He studied medicine, but after earning his first degree he became fascinated with stories of expeditions to collect flora and fauna and with the great explorers such as Von Humboldt in South America and Alfred Russel Wallace in south Asia. He turned to zoology to gain the skills needed for his own adventure. The Rothschild Museum in England, which had a large natural history collection, hired him to collect birds in New Guinea and the Solomon Islands, where he spent two and a half years. There he observed how groups of birds became isolated on the many mountains, developed biological barriers to interbreeding, and hence became different species. "The actual job of it was very descriptive, but you can't help but make comparisons, and that leads to generalizations," Mayr said. Generalizations suggest "concepts," and that gives rise to a new scientific theory.

Back in Germany, Mayr was considered for curator of the Rothschild bird collection—the world's largest—but instead he came to the United States in 1931 to classify birds at the American Museum of Natural History. The museum later bought the Rothschild collection, all 280,000 bird skins, a twist of fate that made Mayr its curator after all. In New York his work dovetailed with that of Dobzhansky, and he assisted in shaping the New Synthesis.

Before the 1930s, when scientists began to understand genetics, Mayr had been an adherent of French naturalist Jean-Baptiste Lamarck's theory that acquired traits are passed on to offspring, which explains how variations evolve. At the museum, the sheer bench work of classifying birds restrained Mayr from plunging too deeply into all the theoretical debates among Darwinists. "I had a huge mountain of material to work out. That's what I was hired for," he said. In 1953 he was appointed Alexander Agassiz Professor of Zoology at Harvard; once there, he had time for prolific writing and lecturing and eventually became director of Harvard's Museum of Comparative Zoology for a decade. Since his retirement in

1975, a bright oil painting of the white-haired ornithologist has hung in the Ernst Mayr Library.

Mayr is the classic systematist, or interpreter of natural structures and their evolutionary relationships. He coined many of the field's most popular terms, such as "cladistics," a clade being a branching of the phylogenetic tree—a tree, that is, of genetic-guided shapes of organisms. Looking back, however, he believes his real legacy does not come from his field studies, museum classifications, or organizational leadership. "I think, possibly, and I may be all wrong, my contribution establishing the philosophy of biology will fifty years from now be considered my greatest contribution," he said.

The philosophy of science developed at a time when physics was dominant in Western academe. During a 1953 lecture tour in Europe, Mayr spoke in Copenhagen on the fact that only biologists trace "emergent properties" produced by nature. "At that time, the concept of emergence was highly unpopular everywhere because it was considered something vitalistic, metaphysical, supernatural, so on and so forth. No one would touch emergence with a ten-foot pole." In the Copenhagen talk, Mayr bucked that trend, unaware that Niels Bohr, age sixty-eight and a Nobel laureate for describing the atom, was sitting in the audience. During questions Bohr politely disagreed with the contention that biology had an exclusive claim on emergence. Oxygen and hydrogen join, and water is the emergent property, he pointed out. Mayr had to agree. What he was getting at, however, was this: physical laws produce new systems, but only in biology do new forms of life emerge out of long, complex histories.

Mayr traces his new trajectory to events in 1958 and 1959. Physics was still lulling biologists into seeing organisms as unchanging types, an idea Mayr wanted to refute. In Washington, he delivered an academic paper calling for a framework of "population thinking." It was the antithesis of typology because it said every individual organism is unique. The next year, at the 1959 Darwin Centennial, he made clear his differences with the geneticists. To say that genes were foremost in driving evolution was an impossible reductionism, he insisted. The whole organism—the phenotype—is the struggling entity, not the tiny gene-bearing chromosomes. "Selection operates on the phenotype, the final product of the interaction of all the different genes," he told the Chicago assembly.[11]

Late in his career, after forty years of agreement, he also dismissed the mantra that evolution is "a change in gene frequencies." Again, that puts the cart before the horse; organisms were the front line of evolutionary change, and genetic heredity followed along after. "The change in gene frequencies is the result of evolution," said Mayr. "You get a change of gene frequencies, but that change does not push these other phenomena." He also prefers Darwin's theory of how groups gave rise to human morality over the geneticist theories of kin selection. "If the individual in the group behaves in a benevolent, altruistic way," he said, "then the group as a whole benefits, and any tendency toward such behavior that has a genetic basis will be favorable to selection."

The distinct thing about biology, Mayr said in retirement, is a simple set of propositions, but they bear constant repeating. Physics looks at laws that govern orbiting planets, chemical bonds, and even the four-letter DNA code, and so its leaders "developed a philosophy of science that stressed laws and predictions," he said. "It had no room for historical factors or elements, because they don't play any role in regular physics." In biology, however, a conceptual framework is required: "Theories in biological sciences are based on concepts," and evolution is the greatest biological concept of all.

Mayr's next step is to define two realms of biological science. The first is mostly mechanical: How does the body of a bird work? How does a cell divide? How does the eye see? This is biology of the present tense. It asks the "what" and "how" questions. Evolutionary biology, however, demands a look into the past, where the "why" questions might be answered. Why can one creature be seen in the zoo today, while another creature is reduced to fossils? The past is the story of shifting heredities and changing environments, which are measured in probabilities, not laws, he explains. The biologist, in short, must use "historical narrative."[12]

How such Mayrian clarity will be passed down remains to be seen, for the time when legacies could be guaranteed somewhat by teacher-disciple relationships has passed. A scientist who attracts apostles needs both character and the right circumstances. Mayr holds up Dobzhansky as the "charismatic" type, and indeed he had many students. Simpson, famous in his day, stood aloof from paleontology students. Mayr says he balked at sharing his knowledge. "Simpson could have developed a real school and had lots of students who would have affected the field very much if he had been willing to take students." For his part, Mayr may have had charisma, but he did not have the circumstance to draw student-heirs.

"I was chairman of this department when Ernst retired, and I organized a party," recalls geneticist Richard Lewontin, sitting in his Harvard lab. "Among others things I tried to get hold of students to come to the party, but they don't exist."[13] Because Lewontin has taught molecular biology, he has produced about 120 students, but because Mayr looked at the larger organism, he missed the curve after 1953, when the "year of the helix" inspired life science students to stampede into genetics. "All the bright young fellows immediately got into molecular biology," Mayr said. "Why didn't I attract these brilliant ones? Well, my way of doing things with speciation was already considered by these students probably as old hat, and they wanted to be at the new frontier."

Rather than debate creationists, Mayr has preferred to take on his colleagues in the sciences, especially the physicists. In his only public debate with creationists, which took place in a three-part television production in London, the physicists did poorly, he said. In the first segment a young-earth creationist imported from the United States faced off with Sir John Maddox, the famed editor of *Nature*. "He's a physicist," Mayr said. "Much to my amusement, the creationist, as far as the general public was concerned, had the better of Maddox." Then Mayr and British biologist John Maynard Smith got a turn. As biologists, he said, "We massacred this creationist."

DECONSTRUCTING SCIENCE SOFTLY

Professor Michael Ruse, a friendly Englishman raised as a Quaker and then lost to a self-described "knee-jerk Darwinism," appreciates the thick skin of, say, reptiles. You need thick skin to be a freewheeling philosopher in the partisan world of evolutionary biology. Some would be surprised to hear that biology, so materialist a science, even needs philosophers. Since midcentury, however, all of science has acquired both philosophers and historians of science.

As a college student in England, Ruse studied math, physics, and philosophy, but only around 1967, when he was past his midtwenties, did he read *Origin of Species*, which was a "love at first sight" experience. He moved to Ontario, Canada, for a teaching appointment at the University of Guelph, and there was taken by a new discipline that had also been coming of age in the 1960s. "Amongst other things at that time, other than just lots of sex and drugs, there was very much the influence of Thomas Kuhn's book *The Structure of Scientific Revolutions*," says Ruse.[14] He sits in his campus office thinking back, pulling at his cropped graying beard. "There's no question that, as the sixties drew to an end, I and one or two others in my cohort, people like [Northwestern University science historian] David Hull and others, realized that the history of science was going to be an important aspect of what we were doing."

Kuhn's book says science is defined by the people who govern it, and that it often requires new people with new ideas to shift "the paradigm," as when Copernicus put the sun at the center of the solar system. Ruse spent a year training in the history of science at Cambridge, and he came out a moderate in this new approach. Call him a "soft" constructionist. He believes the cultural setting of scientists shapes their "truths," but the scientific method is nevertheless a firm taskmaster of objectivity.

Having grappled with the nature of biology in his dissertation, Ruse wrote his first book, *The Philosophy of Biology*, in 1973. That made him a leading voice in the field, but nothing like what he would become eight years later, in 1981, when the New York office of the American Civil Liberties Union (ACLU) recruited him to testify in its suit to overturn Arkansas's law mandating the teaching of creationism alongside evolution. The muscular Manhattan law firm representing the ACLU needed a philosopher of science to take the stand, define science, and make clear to a federal judge hearing the case that creation "science" was not what it claimed to be. Ruse had just completed a critical paper on creationism, and he felt ready for the challenge.

Once he had agreed, "the people from ACLU kept asking me for position papers, and they wanted them literally the next day. Again, this was in the early days before Federal Express. I had no idea you could find a courier to take something from Guelph to New York overnight, but they did."

The day Ruse testified in the Little Rock courthouse, ACLU lawyer Jack Novik asked a question that is etched in judicial history: "What is science?" Ruse said it had five characteristics: (1) it is guided by natural law, (2) it explains by reference to natural laws, (3) its theories can be tested in the empirical world, (4) its conclusions are tentative, and (5) it is falsifiable. "I was the one who pretty much plundered the textbooks for this," Ruse says years later. His definition was central to Judge William R. Overton's ruling. "Creation science as described in Section 4(a)" of the Arkansas balanced treatment statute "fails to meet these essential characteristics," said the judge. He ruled that creation science was not science but religion.[15]

The evolutionists sang lusty songs of victory at a parting victory dinner in Little Rock, and Gould, who also had testified, wrote essays lionizing the judge for his deftness and courage and mourning his death from cancer soon after the trial. Ruse, however, got mixed reviews. One of his peers reported that his five-point definition "generated polite but fierce controversy among philosophers of science."[16] What is more, the statement, which some considered a faux pas, was part of a precedent-setting ruling. Another peer said, "It is probably fair to say that there is no demarcation line between science and non-science, or between science and pseudoscience, which would win assent from a majority of philosophers."[17] This commentator adds that creation science can indeed be tested by scientific methods—and simply proved to be "very bad" science.

Ruse concedes the diversity of opinion but is unapologetic. He tells his peers that trials are strategic: lawyers know exactly what they want. And was not the desired result achieved? "I felt I was doing the job of the profession," he says, and he notes that the criticism had a benefit for him. "Half of me was rather pleased, because it kept my name . . . you know, nothing like being in a controversy." Many books later, Ruse settled on a new definition of science: it has (1) predictive accuracy, (2) internal coherence and external consistency, (3) unifying power, (4) fertility to expand research, and (5) simplicity or elegance.[18]

Ruse's first public encounter with creationists came in 1977, when he was invited to join a campus debate at Northwestern University in Chicago. One such experience was enough to show him what a crowd this topic could draw to a basketball gymnasium. "I swear to God there were three thousand people in the audience, ten of whom were evolutionists," he recalls. The creationist argument, he says, was the same as in Darwin's day, so as a historian of the subject he felt better equipped than even his biologist teammate. "I took to it like a duck takes to water," he says. "It was, 'Um, gee! Where have I been all my life?' Waiting for this."

The joy that coursed through him may be attributed to his Quaker genes, for Quakers are social activists. "I always felt rather dissatisfied with my life as a philosopher," he says. "It wasn't doing much social good." Like other witnesses for the prosecution at the 1981 Arkansas trial, Ruse was sought as a speaker and writer. Ten years and nine books after the Little Rock showdown, his name was synonymous with public espousal of Darwinian naturalism. In 1998 a measure of that eminence was his debut as speaker at the annual Institute on Religion in an Age of Science conference on Star Island. Yet as a public figure, a slightly different Michael Ruse was about to emerge. Some call it a "dramatic conversion," which bemuses him, and trace it to early 1992, when he had debated Phillip Johnson and his team of intelligent design creationists at a Texas university. Ruse calls his change in view gradual, beginning after 1979, when writing *The Darwinian Revolution* had focused him on the issue of "values in science." In particular, he had "embarked on a historical study of the whole question of [belief in] progress in evolutionary biology." By the 1990s he had simply "become much more vocal in my opinion."[19]

Vocal was the proper description ten months after the Texas debate, when Ruse summarized for an American Association for the Advancement of Science session, "The New Antievolutionists," his experience at the Texas debate. "As I always find when I meet creationists or nonevolutionists or critics, I find it a lot easier to hate them in print than I do in person," he told his audience. On science, he continued, "I think that we should recognize, both historically and perhaps philosophically, certainly that the science side has certain metaphysical assumptions built into doing science, which—it may not be a good thing to admit in a court of law—but I think that in honesty that we should recognize." His use of the word "metaphysical" caused scandal among some evolutionists. More sympathetically, biologist Arthur M. Shapiro asked, "Did Michael Ruse give away the store?" He did not think so. Niles Eldredge was less sanguine on the metaphysics comment. "See, he's giving away the whole thing," Eldredge protests. "And he's not a scientist either."[20]

Ruse thinks the bench science critics simply missed the "sophistication" of his point, which tries to save "professional evolutionary biology" from its slide into unprofessional storytelling. As the founding editor of the journal *Biology and Philosophy*, he used its pages to set some of these matters straight. In a postmortem on his AAAS talk, he wrote: "I am not sure that I carried many of my audience with me—evolutionists tend to be as fervent believers as creationists—but we did have a good discussion, and nobody accused me (to my face) of being a traitor to the cause. . . . For the record, I think evolution is true and natural selection is the main mechanism."[21]

Before his appointment to the University of Florida at Tallahassee in 2000, Ruse, his second wife, and their three children had lived in a stone house not far from the Guelph University campus. Its front door was a cheerful yellow, and it had the appearance almost of a Victorian cottage that a poorer Charles Darwin might have owned, overgrown boughs and all. Before our formal interview, Ruse sat me in his bright remodeled kitchen, fed me lunch, and said that the problem with most evolutionists is that they never studied mathematics. He did, and that moved him to his reductionist approach to biology, an approach that says the smallest law and units in matter dictate the large laws of the material universe. He protests the label of materialist, though. "Materialism seems to me to be a bit silly, a bit like atheism in the sense of it was OK in the nineteenth century, but how can one be a materialist with electrons? I would describe myself as a naturalist."

He has been a frank defender of sociobiology, the position that genes greatly determine human behavior. In keeping with all these views, Ruse's book *The Philosophy of Biology* (1973) brought the discipline under the management of physics, mathematics, and mechanistic laws that work from the bottom up, from genes to organisms. The book expressed exactly what Ernst Mayr had been fighting against. "Michael Ruse started in that field with all the wrong ideas," Mayr says with blunt collegiality. "It was all based on physicalist thinking. So he didn't make a contribution at all to the philosophy of biology. In fact, theoretically, he held up the field."[22] Ruse hears this with a hearty laugh, showing the thick skin he so values.

Whatever his merits as an original thinker, Ruse has been credited with two other qualities. Theists like his way of setting out the reductionist view clearly and merrily. "He does it with humor. He acknowledges his own humanity."[23] Others appreciate his willingness to analyze the cultural factors shaping science, even when it comes to disagreeing with friends and contemporaries. For example:

1. Gould's opposition to sociobiology, with its implicit racial determinism, arose from his Jewishness and abhorrence of eugenic notions harbored by the Nazis. "I mean, most of the critics of sociobiology are Jewish," says Ruse. "Don't misquote me as saying, for instance, that I am suggesting that Gould is now pushing Jewish paleontology."

2. Mayr is bent on setting aside biology as its own special science, a move partly calculated to secure grants and professional distinctions usually heaped on physicists. "Ernst has just spent the last twenty or thirty years of his life arguing for the autonomy of biology. Why? Because the molecular biologists are coming along, ultrareductionistic, and saying give us all the grants, give us all the students, because this is the way to do biology, and previous biology was all stamp collecting. So what Ernst is trying to do is argue that biology is different. It's got its own models. It's autonomous. It's organicist [rather than lawlike physics] and has teleonomy [emerging properties]. And that therefore it cannot be reduced."

3. Evolutionists have not been able to escape the cultural baggage of believing in progress in nature and life, and that affects their research conclusions. "Some of the most significant of today's evolutionists are Progressionists," he has written. "Evolutionary thinking of the most professional or mature level is still pervaded by metaphors sympathetic to progress, such as 'tree of life,' 'adaptive landscape,' and 'arms race.'"[24] In other words, evolution as science travels with a lot of cultural baggage.

This thesis, Ruse says, is the result of his pondering since 1979 on science and values, and it comes across in the "really big book," *Monad to Man,* that E. O. Wilson urged his friend Ruse to write. Ruse has a cultural interpretation of Wilson's science, too. "Ed is a secular humanist cum evolutionary progressionist cum [a scientist with a] world picture." The drive toward a world picture is rooted in Wilson's southern evangelical upbringing, now eclipsed by science. Wilson "has this evangelical style, this preacher's style, a born-again Christian who then found evolution. He's looking for a myth of religious dimension." To what effect? "I think it affects his style of science." Wilson has envisioned the unity of all human knowledge as the "epigenetic" product of genetic evolution.

In *Monad to Man*, Ruse traces belief in progress to Victorian England and shows how every major evolutionist since Darwin has attached that belief to the empirical science of evolution. He says evolutionary science has its best chance of escaping progressionism at the micro-study level of mathematics and genetics. Yet all in all, if it were not for the cultural baggage of Darwinism, Ruse might have lost interest. "Evolution is a metaphysics subject," he says. "It's not just metaphysical, but I think it's a subject with lots of metaphysical implications. And I'm a philosopher, so that is naturally what interests me." When Gould declared that science and religion were separate and distinct realms, Ruse said that is fine, but only intellectually honest and truly egalitarian if the scientists let theologians try to make sense of nature, just as evolutionists constantly try to make sense of religion.

One might think Ruse's next big horizon was the Human Genome Project, with its goal to unlock the DNA code and its implications for human life and medicine. But no, he says. "There's lots of money there. But it's basically technology." The next horizon took Ruse back to the cultural debate of Darwin's time. He read heavily in Christian theology to answer the question, Can a Darwinian be a Christian? Before the book came out, he answered, "I'm absolutely convinced of that. I think to be a Darwinian and a Christian at the same time means that you've got to think through some problems. But let's face up to it, that's what life is all about."

MAKING SENSE WITH EVOLUTION

The first scientist to take the stand in the 1981 Arkansas creationism trial was described by an observer as a "courtly European intellectual." Francisco Ayala, an evolutionary geneticist and native of Spain, surely had rehearsed one particular line with his friendly interrogator, a lawyer for the ACLU, to guarantee dramatic effect.

"Mr. Ayala, do you know of any other case when the state has sought to prevent, limit, or regulate the teaching of evolution?" the lawyer asked.

Ayala said, "Of course."

"Where was that?"

"Joseph Stalin's Soviet Russia."

With those words "a new solemnity dominated the proceedings," according to one friendly account. Arkansas might be on the brink of totalitarianism or a Stalinist "Lysenko affair"—or so one side in the court drama hoped to suggest.[25]

Good lawyers can coax even the most understated scientist into the theatrical. About sweeping claims in his own field, however, Ayala can be a conservative and even skeptical voice. When it was proposed in 1987 that the ancestor of modern humans was "mitochondrial Eve," one woman who lived two hundred thousand years ago in Africa, he balked. The theory did not meet the test of science. He would rather say there were some hundred thousand Eves much further back in time.

This is the specialized world of those who track the genetic timetable backward from the present. It presumes a steady rate of genetic mixing and then calculates mathematically from the modern diversity of human DNA back to its simpler forms in the past. The theory of the mitochondrial Eve was proposed again in 1991 by biochemist Allan Wilson, who "turned back the clock" by using a specialized computer program to study multiracial DNA samples of mitochondria (specialized cell structures that have their own DNA), nearly always inherited from mothers.[26]

Based as they are on inexact science, however, clockwork theories easily clash. When Ayala runs his computer on a part of human DNA that deals with the immune system, he finds molecules much older than the hypothetical Eve. "That [immune] molecule existed sixty million years ago," he says, "about when the dinosaurs became extinct, before the origination of modern monkeys, let alone humans." By 2001, in fact, Ayala was publishing studies that questioned "whether there is a molecular clock at all," since he found proteins that "evolve erratically," some fast and some slow, despite the neutrality theory that molecules evolve at an even pace—and thus can pinpoint past branching of organisms.[27]

The earliest humans are one of Ayala's many evolutionary interests, and he has proposed a whole theory of human ethics born out of walking on two legs and developing a large brain. Ayala's science career started off on a much less philosophical note, for in the 1960s he did the tedious lab work of tracing fruit fly gene mutations at Columbia University. He was a prize student of Dobzhansky, whom he is wont to quote—"Nothing makes sense in biology except in light of evolution"—and Ayala would one day follow Dobzhansky's footsteps by getting elected to the elite National Academy of Sciences.

Like most biologists, Ayala entered the field preferring the laboratory to the public spotlight. Having been a child in Spain during its civil war (1936–39), he was not seeking a culture war when he came to the United States in 1961. Yet when one beckoned him in 1981, he responded with the duty befitting his office that year, president of the Society for the Study of Evolution.

A decade earlier, Ayala had gained U.S. citizenship and taken a post at the University of California at Davis, where he worked alongside G. L. Stebbins, who had applied Dobzhansky's New Synthesis to botany. In the spring of 1972, Stebbins, a member of the university's genetics department, turned a lecture by the creationist leader Duane Gish into a debate. The two-hour exchange before one thousand students was the first-ever campus "debate" for the creationists. At another university the next year, Stebbins debated a creationist once again. Ayala had seen some of the debates and heard other war stories from Stebbins.

"I became persuaded that one could not ignore these people," says Ayala, a tall, thin man with jet black hair that he combs back.[28] He had helped the ACLU lawyers draw up a witness list for prosecuting the Arkansas creation science case. "This was about education of children in the schools. This could be damaging," he says, because "nothing in biology makes sense without evolution." Ayala is recalling the events at his spacious, well-lit office high in the green biology building at the University of California at Irvine, where he arrived in 1987 and where I first met him.

On the creationist front, Ayala chaired a committee for the National Academy of Sciences to draft its first salvo against the movement, *Science and Creationism*, issued in 1984. The document began as a legal brief in case the 1981 Arkansas creationism trial was appealed, but as it was turned into a academy tool, it is said that Ayala even wrote the opening essay, signed by academy president Frank Press.

Ever since those years, Ayala has been known on the evolutionist grapevine as a company man. "Ayala is extraordinarily political," says Cornell's William Provine.[29] To do committee work, one almost has to be. For seven years Ayala chaired the biology panel of the National Research Council and advised other federal agencies. He had been a president and chairman of the board of the American Association for the Advancement of Science, and when the AAAS accepted Templeton Foundation money to open a unit on science and religion, a somewhat controversial move, the credible Ayala was elected its chairman. His hand also guided an updating of *Science and Creationism*, issued in 1999.

Ayala and Lewontin are said to be Dobzhansky's brightest students. Who will come next is anyone's guess, Ayala says. "I wish I could know who are the next generation of great evolutionists." Whoever they may be, he will tell them they have a generation or two of work to fill in the entire Darwinian picture. "Unfortunately, there is a lot, lot, lot to be discovered still. To reconstruct evolutionary history, we have to know how the mechanisms operate in detail, and we have only the vaguest idea of how they operate at the genetic level, how genetic change relates to development and to function. . . . I am implying that what would be discovered would be not only details, but some major principles."

Though scientific theories may be provisional, he says he does not believe new discoveries will substantially change what already is known. For every riddle, there are many Darwin-confirming discoveries that already have been made. Take, for example, the mathematical probability that new creatures can arise from mutation and natural selection. A well-known experiment by Ayala in 1974 showed that the offspring of fruit flies with a low tolerance for alcohol would evolve toward high tolerance—a test of how genetic variation and natural selection work together.

In his public talks, Ayala also uses the laboratory example of bacterial evolution. Bacteria in a cubic centimeter of culture are given histidine to stimulate growth. In a few hours they multiply from only a few dozen to two to three billion. Then the antibacterial agent streptomycin is added, and only twenty or thirty bacteria survive. Those that do not die have a genetic "mutation," or variation from the others, that gives them resistance. The survivors multiply again into the billions. At that point, the growth-inducing histidine is removed, and only about one hundred bacteria survive, again because of a genetic feature. The survivors multiply, Ayala says, illustrating how natural selection has created a colony of highly resistant bacteria. The odds that a single bacterium's gene mutations would produce the resistance are only four in ten quatrillion (four times ten followed by fifteen zeros). But in a petri dish, under the power of natural selection, the "evolution" of resistant bacteria has high probability.

In his university office, Ayala reaches across his desk to tap on a technical paper, a dry scientific report. Yet he also writes for the public. Issues such as apparent design in nature or so-called progress or teleology in evolution are not out of bounds. After all, Ayala once viewed nature as God's handiwork, back when he was a young Dominican priest and his superiors sent him to Columbia University in New York City to study genetics. Evolutionists still remember him doing lab work in his Roman collar.

Though laicized for years, Ayala still moves easily in theological circles. When the nation's Catholic bishops, who meet in Washington, D.C., every November, hosted a briefing on issues in science in 1997, Ayala gave a talk on human evolution and ethics. A venture a year earlier took him to Castel Gandolfo, the papal residence in the hillsides east of Rome, where biologists and theologians exchanged views on the topic of "divine action" in nature.

One theologian, says Ayala, proposed that God has "supervenience," the power to intervene at a submaterial level (the term traditionally means a higher material order influencing a lower order). "Now, some of my friends there still want to put the finger of God somewhere," says Ayala. He disagrees, but patiently.

In Catholic circles, at least, there is no problem with the New Synthesis. Indeed, around the time Ayala was visiting Italy for the forums on "divine action," Pope John Paul II said that independent efforts to verify evolution had made "a significant argument in favor of this theory." The pope also took a mild swipe at biblical literalism.[30]

Back in Ayala's office, our conversation has verged almost to theology. Does Ayala think God exists? "Now, what I believe or not is for me to keep," he says. Yet he has offered this thought to the theologians: it is a "small God" indeed that must be found in biological details.

When Dobzhansky wrote his *Biology of Ultimate Concern*, a quasi-theological work, Ayala was a reader and adviser, but his only comment in retrospect is on their differing views on evil. The Russian émigré had tasted it in famine, revolution, and tyranny and was driven to search for an answer, which he may have found in the hope for freedom in genetic diversity. When he was a priest, Ayala did not mind his vows of poverty, chastity, and obedience, but he had intellectual problems with "certain issues like, for example, understanding evil." Did Dobzhansky's biological formulation of freedom help? "That does not settle my problems, my demons," Ayala says. "We were friends. We spent many evenings talking about these matters."

In an age of science, when mysterious forces are no longer used to explain nature, evolutionists are giving a new meaning to some theological ideas of the past. Ayala is one of those evolutionists, and he has talked or written about the two ancient ideas of creativity and teleology, the belief that all of life is guided toward a purpose and end. "Obviously, when I say evolution is creative, I don't mean it is creative in the sense that Christianity predicates from the creative act of God, creating everything from nothing and not the laws of nature. That's not the way evolution works," he says.

"Natural selection is not random. It favors precisely those things which are useful to the organisms. Some people, as you know very well, the fundamentalist creationists—I call them the fundamentalist antievolutionists—make a statement like, 'Evolution cannot be explained by chance.' And I say, 'Well, of course it cannot.' Surely there's a nonchance component."

Chance, he says, is in the genetic mutations. The nonrandomness exists in the way nature, according to its laws, selects changes that ensure an organism's survival. That is how evolution creates. "How could a human hand be made so clearly for grasping, and an eye made for seeing, to come about by chance?" Ayala asks rhetorically. "It's not possible."

The belief in a cosmic teleology, or divinely guided path for life, dies a hard death in a Western culture nurtured on it for centuries. It was one of the great debates in biology during the Darwinian revolution, and it continues to be a topic for scientists like Ayala today. On one occasion, while speaking at a conference on "Cosmology and Teleology," he tells his audience that "teleology" is still a perfectly good word, if understood properly. There are many teleologies, or directive forces, Ayala says, beginning with the "external" kind as when an architect designs a building. There is also a natural "internal" teleology, and that is provided by the law of natural selection as it helps organisms to adapt. Natural selection formed the human eye and allowed it to see, for example, a creation once believed to be a teleological miracle but now explained by physical laws alone. This natural teleology is of two kinds, Ayala continues, one being bounded—as when a chick hatches from an egg—and the unbounded kind, and that is the ever-rambling direction of evolution itself.

Such hairsplitting is mainly for the philosophers, but there is a message for the public as well in the teleology debate: the things in nature that appear to be guided or designed are actually the product of blind physical laws. Darwin played no small part in this death of teleology, Ayala says. "Darwin completed the Copernican revolution by drawing out for biology the ultimate conclusion of the notion of nature as a system of matter in motion."[31]

But that does not mean that matter in motion is everything, for it certainly cannot explain to Ayala himself why he is moved by his favorite painting: Pablo

Picasso's *Guernica*, a tableau of war after German airplanes bombed northern Spain in 1937. The painting has a power greater than its canvas, wooden frame, or the paint—the material components that science is limited to explaining, Ayala told his audience. "Science is a wonderful and successful way of knowing," he said, noting that scientific knowledge after the Second World War produced 50 percent of all U.S. economic growth. "All of these remarkable achievements are a witness to the validity of scientific knowledge."

Nevertheless, Ayala continued, "Successful as it is, and universally encompassing as its subject is, the scientific view of the world is hopelessly incomplete." In his courtly European manner, he listed other kinds of "knowing": common sense, philosophical reflection, and religious and artistic experience. "There are matters of value and meaning that are outside science's scope," he says.

THE FACT OF STASIS

Niles Eldredge is a man of many incarnations. At the "Epic of Evolution" conference, held in Chicago in 1997, he stands next to a one-story-tall slide projection of his "intellectual hero," George Gaylord Simpson, a smiling man with a signature goatee. In the new biodiversity exhibit at the American Museum of Natural History, a tiny computerized image of Eldredge introduces thousands of visitors to twenty-eight groups of life displayed in a dramatic exhibit hall. "All organisms on earth are related through a central, common, evolutionary ancestor," the bearded, bushy-haired, and kind-looking man on the computer screen says.

On the fifth floor of the museum, wide-open hallways run for three blocks, connecting Eldredge's office with all the others. By 1997, when I visited, it had been a behind-the-scenes labyrinth he had walked for nearly thirty years, primarily as curator of invertebrates. "If I were king, I'd have a hall of evolution," says the real-life Eldredge, dressed in a black shirt and black slacks. "Absolutely! There's a whole lot of knowledge that we have about how the evolutionary process works, and I'd love to have a hall that exposes that."[32]

It is surprising to hear that one of the nation's great natural history museums is not geared enough toward evolution. Eldredge made sure the biodiversity exhibit trumpets the connection, but the stately museum still is focused mainly on things, not processes. The most forthright statement on evolution comes in the exhibit on man and in the famous horse skeletons in the fourth-floor vertebrates exhibit. "In that floor you trace out the genealogy, and the word 'evolution' is used all over the place," says Eldredge. He notes, however, that the strong evidence for evolution in DNA similarities, or a full-blown look at natural selection, does not make it to the museum walls. Erecting a great hall of evolution is not a new idea, he says, and he predicts the museum eventually will build one, "but we have budget constraints."

A native of New York who was reared a Northern Baptist, Eldredge earned his doctorate at Columbia University, which has a joint program with the museum.

The time was the early 1970s, when the big names in evolution walked the land, but the promising students of his day chafed at living in the past. "Rebellion is part of the game," he says. "You're never going to find anything new if you just do what you did as a graduate student forever, and even worse, just parrot what your professors always tell you."

Keeping the excitement going for graduate students can be a challenge. Every field is highly specialized, and the temptation is to do a doctoral dissertation on yet another "just so story" about how a particular organism evolved the way it did. "Natural selection is a mathematical, basically, law that's been well substantiated empirically," Eldredge says. "So I don't think any fundamental change in our understanding of natural selection is going to come about." More research can be done on the environmental role in species selection. New horizons need to be pushed back continually.

At the New York museum, the switch to cladistics, a system based on theoretical branchings of organisms, was the closest thing to a coup d'état. Simpson and Mayr both disdained it, as did others of their generation. Indeed, when cladistics was adopted at one of London's great museums, one group of Darwinians called it Marxist science, as if the classification method reflected the jumps of dialectical materialism.

Mayr says the system produces weird categories. "To say that birds are dinosaurs is a lot of nonsense. The birds are something very new, special." The older art of systematics tried to connect evolutionary ancestors not only by outstanding features—tails, wings, legs, or skulls—but also by any other similarities; it was a process that required excruciating judgment calls and interpretation. Still, Mayr says, "The best kind of classification is that which is based on the greatest number of characters."[33]

Efforts to produce conformity in a university biology department or museum staff can vary, but Eldredge says there are hardly any such efforts at the New York museum. "Now, is there a thought police in this institution?" he asks rhetorically. "To the extent that there's any kind of urge to conform, or pressure to conform around here, it's about cladistics." Yet a new generation will soon feel cladistics is not their baby. "I see some signs now of people sort of rebelling against that," he says.

It was part of this healthy rebellion in science that in 1971 led Eldredge and his Columbia classmate Gould to cowrite a paper positing the theory of "punctuated equilibrium." Though almost ancient at thirty years old, it remains the only popular addition to the gradualism of the New Synthesis of 1947.

Eldredge says he and Gould only took their elders' insights a step further. Simpson at one point had proposed "quantum evolution," or evolution that included jumps that led to entirely new and major fossil groups, but he had backed away in the face of the massive consensus for gradualism. As a student, Gould noted that his mentor, Norman Newell, had mined a fossil bed containing an immense expanse of time and had found no evolutionary change. He and Eldredge came to call this the "fact of stasis." If nothing changed over vast periods, then significant evolutionary changes must have happened faster than gradualism had presumed—faster meaning in tens of thousands, not millions, of years. These were the "punctuations," and they found them where Dobzhansky and Mayr had pointed, in small groups of organisms isolated off to the side, and where genetic and reproductive changes happened more rapidly.

"So in a way, punctuated equilibrium was putting together the Simpson perspective of the significance of paleontological patterns, particularly this abrupt-appearance business, but taking that down and interpreting it in Dobzhansky and Mayr terms," Eldredge says. "I think that's a fair statement of what we were about. They could have done that work themselves. Why they didn't, I'm not sure."

Critics have called the theoretical detour "evolution by jerks." It upset them first because it questioned gradualism and second because creationists began to arrogate the Eldredge-Gould paper as evidence for the biblical God's jumplike special creation. Gradualist Daniel Dennett berated Gould for the confusion, and "Gould scholars," including Michael Ruse, say he has backpedaled. Ruse says the theory has been revised downward over three phases.

Gould says Eldredge "developed most of the ideas."[34] Eldredge acknowledges: "Part of the problem has been our rhetoric and our desire to say that we have something new here. I think with Steve, he goes too far and he says it's radically new."

Forcing a new approach on a staid science does more good than harm, Eldredge concludes. "There is some confusion that won't go away, and part of that is from the rhetoric." In 1985, elaborating more on his own thinking, he called his forebears' work the "unfinished synthesis." He argued that evolution should be seen not as a smooth, uphill climb but as a hierarchy of lower and higher levels bumping into each other in complex ways. Each level, whether it is a case of genetic mutation or a food shift in the Brazilian rain forest, is worthy of its own rigorous study. [35] "It was my original manifesto, at least in book form, of hierarchy theory," he says.

From declaring the synthesis unfinished to substituting hierarchy for gradualism, the enterprise took gumption. "Basically I was attacking my elders," Eldredge says. His adventure ultimately paid them homage, but in the short run, it was a theoretical snub. "It was taken by them, it was taken by me, and certainly Steve felt the same way, that we were basically warring with them."

British gradualists, led by such men as John Maynard Smith and Richard Dawkins, also begrudged the American revisionists. Eldredge answered that prejudice with a book and an image of the university "high table," the place in England where only the top dons may sit. There sat the "ultra-Darwinists" with genetic gradualism, Eldredge says, but now the "naturalists"—who saw jumps, hierarchies, and extinctions—also had a seat. "I resent the usurpation of the word 'naturalist,'" Dawkins says, noting that many British geneticists also hunted butterflies and insects in the field. "In any case, the alleged difference is exaggerated.

And there isn't that much difference actually between the school Eldredge represents and the school that I or Maynard Smith represents. And certainly when you compare us with creationists, we're about that far apart," he says, showing a tiny gap with his fingers.[36]

Eldredge first met the creationists in the early 1980s in the person of Luther D. Sunderland, a scientist with the National Aeronautics and Space Administration. Sunderland, a creation activist, in November 1981 heard what sounded like an antievolutionist statement at a monthly forum sponsored by the natural history museum. He was taping the forum, and he provided his bootlegged samizdat to fellow creationists, who circulated it. The tape caught Colin Patterson of London's British Museum asking fellow scientists during a provocative speech, "Can you tell me anything about evolution, any one thing, that is true?"

Before that time, Sunderland had come to Eldredge wearing the hat of a researcher for the New York State Bureau of Science Education.[37] He used parts of his interview with Eldredge, however, in testimony before the Iowa School Board in a battle over passing a "balanced treatment" act. Iowa education official Jack Gerlack called Eldredge to ask why he was helping the creationists, which Eldredge was not. Even now, some creationists call Eldredge a friend, primarily for his friendliness. Eldredge considered the Sunderland shenanigan an act of war. "I took up the cudgels because of Luther. What he was doing was lobbying," says Eldredge. "He had me saying that I thought that teaching evolution and creationism side by side in schools was a good idea. I was very angry about that, so I got involved." Eldredge wrote *Monkey Business*, one of five classic anticreationist books, and joined the debate during its most politicized years, from 1980 to 1987.

"There are two big sets of issues that involve my profession, in a very general sense, and the public domain," Eldredge says. "Creationism is one. The biodiversity crisis is the other. I think, intellectually speaking, there's nothing interesting whatsoever about the creationism thing." Biodiversity ranks high among Eldredge's interests, but creationism is on a bottom tier. Nevertheless, he feels challenged to clarify three creationist matters that he says remain garbled in the public mind. First is the insistence that evolution "is only a theory." Second are the persistent attempts to say natural history is a kind of metaphysics. Third is the effort to split natural development into microevolution and macroevolution.

The "only a theory" comment can appear in the strangest places. After the Supreme Court's 1987 ruling against the mandated teaching of creationism in public schools, *New York Times* science editor Nicholas Wade applauded the court but added: "It's true that evolution is only a theory and that some scientists, contrary to scientific method, occasionally treat it as a dogma." Eldredge could not believe his eyes. "Nobody says, 'only a theory,' but creationists do, as if theory were a bad term. And there was Nick Wade repeating that stuff," he says. Eldredge wrote a letter to the editor to set the record straight. "Quantum mechanics. Special relativity. Plate tectonics. All are theories, yet no one, in my experience, ever says they are 'only' theories," he wrote.[38]

The next slippery slope Eldredge sees is the characterization of the study of evolution as conceptual, not empirical, science. Creationists love the distinction. Because no one was around hundreds of thousands of years ago, they argue, evolution and creation are, in effect, rival but equal philosophies.

Eldredge says some scientists, including Ruse, are guilty of admitting to metaphysical assumptions in science. His elders even made fatal distinctions between studying the present and the past, Eldredge adds. "I'm trying to get rid of that distinction," he says. "Mayr and Simpson are guilty of fostering that distinction." He prefers a formulation by Gould that talks about things in nature and law in nature—studying the fossil of a trilobite versus looking at the laws of change over time.

Eldredge has also had to address the commonplace perception that microevolution and macroevolution are worlds apart. Creationists often say the evolution of a finch's beak (microevolution) is a far cry from the evolution of a fish brain to human consciousness (macroevolution), and to jump from one to the other is to make an impossible leap. "If you agree that microevolution takes place, you've given up the argument on macroevolution. Absolutely." One leads to another, he is saying, and in a very gradualistic logic that his own hierarchy theory had tried to correct. In the hierarchy approach, genes do not automatically affect animal behaviors, because different forces are at work at the level of organisms and environments. Still, he added, microevolution proves macroevolution. "You would think that I wouldn't say that. But in point of fact, there is one set of homologies, a nested set of resemblances, that makes up all life's systems."

Over the years, Eldredge has become a popularizer of natural science, and evolution in particular. When this overlaps with the topic of ecology, he also ventures into the question of human value systems—and that means cultural evolution and religion. He is tired of debating physical origins, but he offers the hand of camaraderie to religionists who care about the environment. He says the Genesis story, for example, was created by "very smart" people. These people were among the first of their species to shift from hunting and gathering to agriculture, which marked human control of the environment.

The control of nature has now become overwhelming under a mass civilization of industry and technology. So Eldredge has proposed an updated "creation myth," or story of meaning, for humankind: "The People came from the earth and were linked through the recesses of time with all other creatures," according to his 314-word naturalist scripture. Problems arose, but ingenuity prevailed. "And they embraced sustainable development, matching economic growth to the carrying capacities of their surroundings. The people lived. And it was very good."[39]

SURVIVAL OF THE LUCKIEST

When David Raup wrote a book about the "ways of science," its dedication struck some as mismatched with his establishment credentials. Raup was dean of science at Chicago's Field Museum of Natural History, a member of the National Academy of Sciences, and an author of standard textbooks on paleontology. Yet his dedication reads, "To all the mavericks who keep challenging the conventional wisdom."[40] Raup is known for his ameliorative ways among the sharp-edged doyens of science and for his quest of the offbeat theory. He is associated with mass killings—of dinosaurs, that is—and is known for his affair with a star.

Raup tried to work with the minutiae of evolutionary biology but found that only the big picture spoke to him. "I've done some sort of bench science," he says, "but most of it has been pretty broad-brush, large-scale, which has solicited a fair amount of contempt among my more traditional colleagues. I've never described or named a new species. So this is really not good form. I've gotten away with it."[41] Naming a new species is often the hoop that doctoral students must jump through to earn their evolutionist credentials.

Raup is a short, brown-haired man with a deep voice and bags under his eyes. He has a fatherly demeanor, but hardly a merry one. He knows he smokes too much, even in the calm of retirement in the lakelands of Wisconsin. A cradle naturalist, he is the son of botanist Hugh Raup, who until 1967 headed Harvard's forestry program. The son is still an everything-is-matter man and a "devout evolutionist." Yet he admits it can be "fun" talking with academically trained creationists.

A love of speculation guided his scientific career. He graduated from the University of Chicago and Harvard, worked a short stint with oil companies, and settled at Chicago with appointments in geophysics, evolutionary biology, and the conceptual foundations of science. "I think it's very likely that we're missing enormous natural phenomena," he says of evolutionary studies. He finds the probability of God's existence to be extremely low but also says the chances are pretty good that although naturalistic science and evolution will endure, Darwin's theory may not.

"I think that's probably true," he says. "The history of science is loaded with cases like this, where despite our inevitable feeling that at any given time we know most of what there is to be known, the next ten or twenty years show that we were dreadfully wrong in some area. . . . The Darwinian model works as far as it goes, but this may only represent some substantial fraction of the whole."

Raup pioneered the use of computer models in biology. In the late 1950s he used FORTRAN, an early programming software, for his "highly theoretical, mathematical studies of the evolution of life." He drew pictures of ideal snails and other mollusks on computers, then simulated an evolution process to see how the shapes changed. He asked questions like: "What is the range of all possible snail shells, including those that never appeared on Earth?" He even second-guessed the master himself, asking, "What would evolutionary patterns have looked like if Darwin had been wrong?" His textbooks made him safe for academia, but he says his experiments were "a little jarring" to fossil-collecting colleagues. Fossils are nice, he says, but it is the big picture that exhilarates.[42]

He was in the big-picture mode in 1981 when he and a few colleagues instigated an unprecedented conclave at the Field Museum, calling together the top British and American evolutionists for the "Conference on Macroevolution." The event gathered representatives of two clashing schools, those who started with genes and extrapolated outward, and those who studied all levels as independent forces in evolution. The second group, made up mainly of paleobiologists like Raup, says, as he explains it, "that there are qualitative differences between evolutionary processes at low and high levels. And so, you can't really understand the evolution of life without looking at phenomena that act between and among species, and even higher groups."

Despite being on the paleobiology side, Raup regrets that the New Synthesis has shut down speculation on other paths. Previous eras, by comparison, yielded an exciting diversity of thought, he says. "In the post-Darwin period, particularly the early part of this century, with the discovery of genetics, there were all kinds of varied views and ideas and theories and models about how evolution works."

With his interest in the big picture, Raup was intrigued by a theory, new in 1980, that an asteroid's collision with Earth killed the dinosaurs. It was explained in an article in *Science*, titled "Extraterrestrial Cause for the Cretaceous-Tertiary Extinction," by Luis Alvarez and three colleagues at the University of California at Berkeley.[43] Alvarez's theory, like that of plate tectonics, was not taken seriously at first, but the evidence kept coming in. Meteorites are the source of the element iridium, and traces of it were showing up around the world in the same stratum, as Alvarez had predicted.

Current theory puts the crash-down point off Mexico's Yucatán Peninsula. The impact is thought to have been sufficient to cloud over the biosphere and cut off sunlight from the food chain—a wrenching blow in biological evolution, not the gradualistic rise of life seen on classroom posters. Mass extinction also contradicted classic Darwinism, in which gradual death was twinned with gradual evolution. A theory that made death so sudden and universal created yet another puzzle for those who posited evolution by slight accumulation. Raup was among the most candid on how such old-fashioned "catastrophism" was now a thorn in the side of conventional evolutionists, for "Lyellian uniformitarianism might thus have to yield to Cuverian catastrophism." The great George Cuvier, a nineteenth-century French creationist, attributed biological change to major planetary or volcanic disruption.[44]

But Raup and a younger Field Museum colleague, J. John Sepkoski, were looking for more. Interest in catastrophe had been fueled in part by public worry about extinction of the Brazilian rain forest. Among specialists, theories also arose about giant craters and mass extinctions appearing on a time line, showing a pattern or "periodicity" to events. Raup and Sepkoski had also come to such a speculative conclusion in 1983 by massive computer analysis of nine or ten likely extinctions of the last 250 million years. They proposed a 26-million-year cycle and by 1984 had, in effect, challenged astrophysicists to try to explain it. As Raup would put the case, "A great big clock in the sky is controlling biological destinies on Earth."[45] He knew not what it was—"something in the solar system or galaxy"—but, again, he put himself in opposition to the establishment: Darwinian paleontologists favor the gradual, undirected view of natural history, so "most paleontologists have reacted negatively to periodicity."[46]

The media loved it. So did a few physicists, astronomers, and atmospheric experts willing to propose what caused the pattern. Most engaging was the mid-1983 theory that a companion star to the sun had regularly orbited into the solar system, deflecting to Earth rocks from the Oort cloud of comets whose orbit was beyond Mars. The theory put the "little star" about two light-years away from Earth, moving farther away, but in a few million years due to turn earthward again, a journey back of a dozen million years. By 1984 it was given a name and popular fame. "A Death-Star Theory Is Born: Nemesis," *Newsweek* declared.[47]

Whatever the cause of such "large body impacts," they might have cut off the dinosaurs' food supply and in earlier times extinguished the vast population of trilobites, or marine arthropods. The theory of mass killing had a political byproduct. Astronomer Sagan borrowed the computer modeling of how atmospheric contamination caused extinction, and in the last gasps of the cold war, he warned that a thermonuclear battle's "nuclear winter" could do the same.

Mass killing defied evolutionary gradualism, letting one meteorite blow do what eons of natural selection were supposed to have done. Survival was not by the fittest, Raup mused, but by the luckiest. Even on this, however, Raup's natural skepticism comes into play: "It still could have been that mammals . . . went around eating dinosaur eggs."[48] Either way, according to conventional theory, the demise of dinosaurs allowed the rise of tiny mammals—and Homo sapiens.

For our interview, Raup and I met in a well-lit atrium in the basement of the Smithsonian Institution castle, scene of a paleontological society meeting. Two of his longtime friends walked up, Gould and Steven M. Stanley of Johns Hopkins University. Their common cause is exploring the punctuationist view of evolution, in which jumps in nature—such as those caused by meteorites crashing into Earth—are just as likely as gradual transitions. Stanley proposes that a terrestrial catastrophe opened the door for human evolution. A volcanic or seismic shift, he argues, closed what is now the Isthmus of Panama, and cold ocean currents swerved north, bringing the Ice Age that forced man's ancestors into the dry savannas. Here is a trio of evolutionists who like the big picture, as well as a modicum of iconoclasm.[49]

"It gets a lot of people into trouble, because if you carry iconoclasm too far, you don't accept anything," Raup says. Scientific self-testing should supply adequate doses of iconoclasm. "But unfortunately it's not all that common. The challenge of presumptions is not as common as we'd like to see it. The scientific method is supposed to have that built in, but I don't think it really does."

What is there to be skeptical about these days? The New Synthesis, for one thing. "It was crafted by a relatively small number of people," he says. "It was mainly zoological groups, so that the botanists may have gotten short shrift. There are a lot of differences between plants and animals. It was a finite group of people with genetics as their common interest. So you could even call it a historical accident, or it's also possible that they found the truth."

Raup has said he was greatly influenced by Simpson and Mayr, two men who applied population genetics to biological life. Thus inspired, he tried to apply the same analysis to the fossil record once computers and other new tools were available. "I've since learned that it simply can't be done," he says.[50] The mixing of genes in populations helped to explain smaller-scale adaptations, he says, but not the big picture.

After forty years, he continues to say that the "hegemony" of the New Synthesis has been broken, and new ideas and approaches are gaining acceptance. "All of a sudden, in the last few years things have just opened up. Everything was once Dobzhansky-style, interbreeding populations and allele frequencies [how often alternate forms of a gene show up on a chromosome]. It's now a wide-open game. And I think one reason for this is that a lot of people in biochemistry and molecular biology got into the evolutionary field without the brainwashing of the New Synthesis." He names names, but his main point is that a good number of students missed the dogmatic professors. "They were naive to the dogma."

The creationists bring their own dogma to science, Raup believes. Yet he also thinks some of his colleagues protest too much. Creationists long have cited his 1979 article in the *Bulletin of the Field Museum of Natural History*, in which he said there are "fewer examples of evolutionary transitions than we had in Darwin's time" because "more detailed information" has debunked some fossil claims of the past.[51] In Raup's defense, anthropologist Laurie R. Godfrey denounced creationist exploitation of Raup's statement, saying it was torn out of context.[52] Raup is less worried. "So that quote was picked up, and I read that passage many times, and it is not quoted out of context," he says. Next he volunteers a study by one of his students that concludes that ten times more transitional fossils should be found than paleontology actually turns up. The student proposed that better techniques are needed to determine what is a transitional fossil.[53]

Raup's early encounter with serious creationists came when Kurt Wise was doing undergraduate work at the University of Chicago. Then he experienced the musings of Phillip Johnson. In the early 1990s, he attended an "intelligent design" meeting in California. "On this evolution-creation question, I think, though not religious, I have to grant the possibility of some design in the normal creationist context," Raup says. A true skeptic, he has to allow probabilities of almost anything. Thus, he gives a low probability rating to a supernatural realm but wonders whether what people call supernatural is not really some netherworld of nature.

When Raup speaks of design, he does not mean something conceived by a supreme being. The options in science are space aliens, or some form of higher order, or self-organizing laws at work in matter. Meanwhile, he says, Darwinism remains the best material explanation—even if it explains only three-tenths of natural reality.

"The big seven? I don't know," he says, referring to the seven-tenths of knowledge yet to be discovered. "That's the fun part. Now, many people are very uncomfortable with that. They're very uncomfortable with that seven-tenths that's missing, if my numbers are anywhere near correct. This, for some reason, doesn't bother me a bit. In fact, I find it challenging."

What some people call the supernatural or an intelligent Creator, Raup would entertain as possibly something at the front end of matter, something beyond our senses now but still with a "naturalistic base." That something could be called God, he says. "It's unfortunate that we have this dichotomy between the naturalistic and God, as usually defined." Having brushed shoulders with a few believers and creationists, with their belief about beginnings, he has found himself looking at his own beliefs about that same topic. It can be a kind of bias, he suggests, just as white folks had never tried to understand the plight of blacks until the civil rights movement forced them to. The assumption that all is material "just runs our lives," he says of fellow scientists.

Living in rural Wisconsin rather than urban, intellectual Boston or Chicago has also allowed him to see the contrast of the scientific intellectual and the rest of society. "I'm learning quite a bit about the rest of the country," says Raup. "There's a big difference. Of course, I grew up in Boston with all of the usual East Coast prejudices." Teaching in the Midwest, however, Raup had already found that, like many naturalists, many creationists also grew up with a youthful fascination toward nature. They, too, struggled to fit science to a personal outlook on life. One of his students, Kurt Wise, an Illinois native whose father was a newspaperman, was a perfect example.

6. GOD AND NATURE: CREATIONISTS

Paleontologist Kurt Wise looks out over the craggy vastness of the Grand Canyon and sees the hand of God—and evidence of the biblical flood described in the story of Noah's Ark. That makes this Harvard-trained scientist a heretic among his peers in academia but, paradoxically, not a hero to other creationists. To many of them, he is an elitist, "an evolutionist in sheep's clothing," as some have intimated.

Wise not only teaches science at Bryan College in Dayton, Tennessee, but also wages battles with scientific theories. "I've been trained to be a theoretical paleontologist. A theoretical scientist is in the job of creating and destroying theories at will," says the tall, plainspoken, and mustachioed professor. "I can take any evolution theories and wipe them out, rather easily. But that is ridiculous. It's cutting one head off the hydra. There will be fifteen more that will come up in its place, because there are a lot of smart people out there who devise such theories."[1]

From the window of his campus office, Wise looks out at ridges of limestone, sandstone, and shale that make up the southern end of the Tennessee Valley, made famous by the Scopes trial in 1925. He says the geology of the valley is thousands of years old, not millions, which puts him at odds with conventional geologists. He is a neocreationist and holds that the earth is very young. For him the universe was created when God spoke his Word. Creatures were formed as "kinds" and then diversified or degenerated according to natural laws. The fossil record and the hills outside his window were shaped by a global flood and other terrestrial upheavals, all suggested in Genesis.

"If you accept that the earth is old, and if you accept the chronologic column, then Genesis one through eleven—20 percent of the book—must be rejected," says Wise. You must also reject the 15 percent of Psalms that talks about nature, as well as nearly half of the Old Testament and sundry parts of the New Testament where the texts talk of earth history and human history. "So you are talking about 20 to 25 percent of the Bible you have to toss out immediately. But it doesn't stop there. The very doctrines we hold true become threatened."[2]

Wise is addressing the "Fourth International Conference on Creationism," held near Pittsburgh in August 1998. His topic is not only belief in an unerring

Bible but also the theoretical presuppositions used in science. The term "neocreationist" suggests a next generation, a group of Christian sons who, unlike their fathers, spent many arduous years earning advanced degrees in natural history. For two years as an undergraduate student, Wise managed fossil invertebrates at the Field Museum of Natural History in Chicago. Then he went on to Harvard, earning his master's degree and then a doctorate under the formidable Gould.

Wise gives credit to Gould, an Alexander Agassiz Professor at Harvard, for being a fair, even noble, doctoral adviser, and includes him in a personal pantheon of admired Harvard naturalists. The pantheon also includes Alexander Agassiz's father, Louis, who when he died in 1873 was the last special creationist in America with standing as a preeminent scientist. Louis Agassiz did some of the best scientific work of his day, and Wise has ambitions to do likewise. That has put him at the helm of the ICC conferences, running the peer review on papers about geology and paleontology. He and a small group of peers are trying to upgrade creationist science, first by screening out slipshod materials, and they are ready for cries of "Elitism!" from kitchen-table activists or ministries. This is not new to him. The charge comes not only because of his Harvard credentials but also because he defends scientific "model building" over flailing at the evolutionist hydra.

"Create a whole different system," he says. "I am about building a replacement model, one that will do a better job than the other. I'm not about to waste my time showing people who know it already what the weaknesses of their theory are. What I'm fighting for in neocreationism is a completely different game."

Wise became enamored of nature during his childhood in the farming community of Rochelle, Illinois. At age nine he accepted Christ at the First Baptist Church. By college age, when he was trying to reconcile his beliefs with his undergraduate major in science, the early creationist literature that had given him an initial hope began to fail badly. Even Henry Morris's *Genesis Flood* lost its persuasive power. This time in his life, when he earned a bachelor's degree in science at the University of Chicago, was "not a fun period," Wise says. "It had embittered me against creationists. That's why all my contributions to creationism, up until my graduation from Harvard, were negative contributions."

In the long hallway that connects classrooms outside Wise's office at Bryan College, glass cases brim with specimens—fossils, pickled embryos in sealed jars, stuffed birds, and shelves of beetles and seashells. He has been on the faculty at this tiny, nondenominational Christian college since 1989, teaching natural science half-time and conducting research in the remaining hours. He does not place much hope in the God-as-designer "intelligent design" argument. He says the key to a durable creationist model is natural history, the geological life of the planet.

Wise and his geologist colleague Steve Austin have been investigating the Grand Canyon and Death Valley. In a side branch of the Grand Canyon, they have found long, conical shell fossils of a creature called the nautiloid. Wise moves his finger across a distribution graph of the fossils. "They show water flowing," he says. "Now that's interesting because the traditional interpretation would not suggest that limestone is formed under moving water." He is not talking about just any water: "The Grand Canyon, by our account," he says, "would have been formed and deposited by Noah's flood."

He did not say this in presenting his initial findings at a poster session at the 1995 meeting of the Geological Society of America. More modestly, the display argued that "a school of nautiloids was killed suddenly by high-temperature toxic waters from a hydrothermal mound, then transported to the flanks of the mound and buried rapidly." Passersby at the exhibit "liked the argument, but they didn't like the conclusion," he recalls. He and Austin returned to the society meeting in 1999 with a paper updating the canyon research but with minds changed on the mound theory and emphasizing instead evidence that a "gigantic population" of nautiloids "was overcome by a canyon-length catastrophic event impacting an area exceeding one thousand square kilometers."

The Geological Society had drawn more than six thousand people to the Denver assembly. Ten weeks earlier the Kansas State Board of Education had stripped large-scale evolution from testing standards, and the geologists naturally took sides. "There was a lot of anticreationism at the meeting," says Wise; it dominated an entire Sunday, a luncheon, and a speakers forum. In all the uproar, he says, someone speaking from the floor cited the "good geology" done by some creationists, despite cases like Kansas. Wise has frowned on such school board politics, preferring other ways to teach God and science, such as in the home or in private schools.

In creationist circles, there is hardly one way that Genesis and a global flood are approached, for expertise varies wildly, and doctrinal emphases do also. On occasion, Wise might rather spend time with a hard-nosed atheist doing good science than with an old-earth theistic evolutionist or an old-earth intelligent design proponent. For either of these groups "to associate with me is professional suicide, as I've committed professional suicide being a young-age creationist," he says. "Their position, the respect that they hold, is too important for them to associate with me."

While in Denver, he and Austin had also reported on their mapping project in the Kingston Range near Death Valley, California. In his office a few years earlier, Wise has pulled out a very large topological map, unfolding it to show its weaving lines for elevation. The Kingston Range is made up primarily of granite, metamorphic rocks, conglomerates, sandstone, and dolomite, a common rock-forming mineral. The Kingston Peak Formation is a conglomerate of rocks, and the region surrounding it is cut by earthquake faults. Included in this formation at a strange orientation are older rocks up to a mile long and a half-mile wide. "It's askew. It's weird," Wise says. "This rock is oriented this way, when everything else in the region is oriented this way." He moves his pointer across the map. "Otherwise, how would you recognize a mile-long rock when you're standing on it?"

To get wedged there, he says, the rock had to have fallen from a higher point and with a tremendous force. The force would have peeled off one hundred meters of the mountaintop, a massive flap of earth that then would have settled back down on the intruding rock. "Spectacular stuff," says Wise. He rejects the conventional theory that glaciers were powerful enough to create the Kingston Peak patchwork or to sling a rock a half mile in diameter into the peak's side. "Plus, there's limestone in here, which you wouldn't expect in the cold environment of a glacier. And the list goes on and on," says Wise. In Denver in 1999 he made the case for "a catastrophic sedimentary-tectonic model"—in effect, an underwater landslide triggered by a super-earthquake.

Now he takes out a map showing the floor of the Pacific Ocean and the mountainous terrain of western North America. Giant wedged-in rocks are found all along the former coastline of the continent. "Same age rocks," he says. "We're going to argue that this whole sequence, all of these, are avalanche deposits off the edge of a continent. Didn't include a California or anything out here. The edge of the continent was inland, and the entire edge of that continent collapsed in an enormous earthquake."

As with the nautiloids, here is a reasonable local theory with a big-picture implication that is off the map for conventional geology. "What we're going to argue ultimately is that this is the beginning of the flood. This was an enormous earthquake which cracked the surface of the earth and propagated across the earth's surface," he says. In this model the flood results not from rain but from an upheaval of the ocean floor. Driven by "gravitational energy," the ocean crust may have sunk, being denser and cooler than the mantle below, and this in turn pushed molten earth up into the ocean—the engine that geologists now believe is what spreads ocean floors and makes continental plates move. If such an upheaval raises the ocean floor a kilometer and a half, the ocean would cover the earth. This drama parallels orthodox plate tectonics but postulates a different time line from its millions of years. Wise's model needs plate movement in a few biblical days. "You accelerate the plate tectonics, the result is what you see described in Genesis," he says.

Questions about the rate of plate movement and dating of the rocks, Wise concedes, are the main sticking points for the young-earth approach. Radiometric dates for the Grand Canyon or tectonic movements show them to be incredibly ancient. In 1986 Wise told the first ICC that it was useless to question the methods of radiocarbon dating. If it was a problem of inaccuracies, he says, the dates would average around a mean. But they all come out very old. He believes something else is happening. "It's an area that right now I feel that's a weakness for the creation model, and a strength for the evolutionary model, the old-age model. However, I believe there's plenty of evidence to indicate that we're right on the edge of understanding it. It's very exciting."

He pulls from his desk drawer a CD that contains the National Geophysical Data Center's list of all rock ages dated radiometrically in private and public laboratories. The arduous task ahead, Wise says, is to look for dating patterns that suggest alternate interpretations. "I think radiometric decay is valid. It's indicating age, but that's not a one-to-one correspondence between radiometric years and chronological years." Wise's position is downright heretical among professional geologists, who, at least in public, unequivocally swear by the integrity of their science. To argue otherwise, Wise needs their kind of federal funding: in the mid-1990s, dating a rock cost $500 a pop.

A scientific alternative is to focus on theoretical science rather than take the dating head-on. Scientists have been driven by all kinds of weird ideas, so a literal Genesis is hardly shocking, Wise says. "In the history of science, who cared where your idea came from? Who cares how irrational, how metaphysical a particular theory happens to be? I guess sometimes I get frustrated with those who are out to destroy me and invalidate anything I do and say creationists don't have a right to do science and shouldn't be allowed to present scientific papers and all that." Fellow scientists have rejected Wise's requests to share data, a normal gentlemen's protocol in the selfless pursuit of truth.

Having been active in science research and teaching for a few decades, however, Wise says he has a leg up on the U.S. science education establishment's number one problem: boredom. He can excite students, he claims, while secondary school science scores plummet and colleges see more science dropouts. "By the time students get into college, they hate science," he says. "Science is not memorizing the names of a bunch of bones. Science is dynamic. Science is a verb. It's something we do. Scientists get out there."

It is generally agreed, moreover, that every scientist, being human, has a philosophical bent that motivates him or her to go out in the field to pick up a few more pieces in a billion-piece puzzle called reality. So what's the problem? Wise asks. "Evolutionists have been doing this with the data for a hundred, two hundred years in some cases. They're still struggling with data, after putting that much money, man-hours, and the rest into that data. We as creationists, in my opinion, have only been doing this for about ten years, at least people trained in the science."

His claim is one the National Academy of Sciences would not buy, but Wise says the neocreationists are finding as much evidence to corroborate their model as evolutionists are finding to back theirs. Both sides, moreover, struggle with data that contradict their bigger view. "But if you consider all the fields as a whole," he says, "we're having about as much success as the evolutionists. In some fields, more success, in other fields less success."

GOD OF THE OLD EARTH

Evolution means too many things to please John Wiester (rhymes with "Easter," he says). It can mean change over time, ancestral relatedness of life, an undirected mechanism that creates new things, and how humans came to be. "We have got to define terms so that we can have a rational discussion," says Wiester, a short man with round, cheerful features, white hair, and a taste for cowboy hats. "The key is to unpack the language."[3]

Wiester does his semantic unpacking a little north of Santa Barbara, where the California mountains plunge into the Pacific. He is a small-time rancher trained in geology at Stanford University who also teaches geology and earth science at two Christian liberal arts schools, Biola University and Westmont College, near Los Angeles. In his rustic ranch-house office, where I visited in 1996, he has built a fifteen-year repository of the words and phrases flung around in the evolution-creation debate. In the process, he has witnessed a culture war that he says makes the search for knowledge secondary to advocacy. "Ninety percent of this is not science at all," he says. "It's all rhetorical maneuver."

Given his opinions about terminology, he prefers not to talk simply about evolution in general. "I'll talk to you about microevolution," he says. "I'll talk to you about macroevolution. Talk to you about common ancestors ['common descent']. And I'll talk with you most of all about Darwinism." Those are the four terms on which he has focused most intensely in his efforts to make the scientific language more precise.

He begins with microevolution, which Mayr defines as "evolution at and below the species level." Wiester has read *The Beak of the Finch*, a Pulitzer Prize–winning book by Jonathan Weiner about a Herculean study of how drought, death, abundance, and breeding changed the beak sizes of Galapagos Island finches over twenty years. If that's evolution in the flesh, as the book claims, then he is an evolutionist, Wiester says. "We have 5 percent variation in the size of finch beaks. Impressive. Then the media says, 'How can anyone doubt evolution if we've got a 5 percent variation.' Yeah, right!"

The problem with the book's thesis is that it makes a gigantic leap from an observation of microevolution to the claims of macroevolution, that new body plans develop over time from established ones, Wiester says. While beak variation has been explained, he says, macroevolution is one of evolutionary biology's "unsolved problems and unanswered questions." The great Dobzhansky said that the "working hypothesis" of the New Synthesis was that genetic variation has "equality" with change in larger complex creatures, called the "higher taxa."[4] In other words, finch-beak change is equal to amoeba-to-human change—except for the time involved.

Wiester calls this one of the great "shell games" of Darwinist educators. He says they recite the classic textbook definition of evolution—which is essentially about genetic heredity, explained as "any change in the relative frequencies of alleles [sets of genes] in the gene pool of a population"—and then go on to the next page, where they recite from an illustration that shows how amoebas evolved into fish, and apes into Homo sapiens.[5]

The claim is that microevolution leads to macroevolution, and from here the next leap is that all organisms on earth share a common ancient forebear. They have "common descent." Wiester respects the idea of common descent but only when it is presented as a hypothesis in the larger theory of evolution. He acknowledges that organisms have similar shapes and parts, shared genetic material, and, in vertebrates, a likeness as embryos. Yet on each count, he says, there is contrary evidence. The fossil record is notably lacking in transitional stages, and researchers are faced with two other enigmas: wildly different genetic programs between organisms, and embryo dissimilarity at their earliest stages when they should be similar. Indeed, it is not controversial in evolutionary science to say there were several ancestors—Darwin said "one or a few."

Wiester's word watch is aimed primarily at educational powers. One favorite target is the *Science Framework* set down in 1990 for California schools.[6] "The framework says that to be scientific, the word [evolution] must be defined and used with consistency of meaning," he says. "They have violated that in every case. It's not defined. It's used inconsistently." And abundantly. "The word 'evolution' is used 230 times."

Wiester's focus on education hints at his concern that science not lead young people to disbelief, an issue that comes from his own experience. It was only in adulthood, with many regrets, Wiester says, that he found a valuable faith. In his earlier life in Southern California, he presided over Astro Industries, a manufacturer of ultra-high-temperature instruments for nuclear and aerospace firms. He had enjoyed "power, prestige, and the playboy philosophy of life." He was a male hominid amid abundant females of the species, though the theory of evolution was far from his mind. Three failed marriages later, he met his Christian wife— and his God, with a passion. "Is there a creator God or not?" asks the convert. "Did God create, or didn't he? Or was it done by Darwin's mechanism of natural selection without any outside input?"

To answer that for the next generation, including his own grandchildren, he set out to reconcile his knowledge of geology—which included an ancient earth and universe—with the Genesis account of Creation. "This was strictly as a brand-new Christian," he says. "Completely untutored." At the time, Wiester did not know about a precedent for his enterprise, the work of a theologian who had taught at Biola in the late 1940s. That Baptist scholar, Bernard Ramm, accepted the ancient earth of modern geology and eventually espoused progressive creation. He only grudgingly used the required young-earth materials at the school and set about to research his own material. In 1954 he produced *The Christian View of Science and Scripture*, the first book of its kind. Wiester did not know of the text or realize that he had embarked on a similar kind of apologetics.

He got some evolutionists' help along the way. One summer, Wiester met the noted geologist Preston Cloud, author of *Cosmos, Earth and Man* and professor emeritus at the University of California at Santa Barbara. Cloud had debated young-earth creationists on that campus in 1976 and had never forgotten how an audience poll had given them the victory. "So he had an account to settle, of which I had no idea," Wiester says. "But anyway, he loved the book I was writing and went through it. 'That is right. This is wrong.' He said, 'It's a great book!'" In 1983 the mainline evangelical publisher Thomas Nelson released the book in question, *The Genesis Connection*. It gave Wiester a calling card and an old-earth reputation. "At this time there was nothing coming out but young-earth literature," he says. "So this was a major breakthrough."

Walter Hearn, a longtime newsletter editor and secretary for the American Scientific Affiliation, read the book and invited Wiester to attend the ASA's annual meeting in 1985. "Oh, my gosh!" Wiester recalls thinking. "There are other scientists that do this stuff." Hearn pegged the new enthusiast from the start: "Wiester's an activist."

That inclination soon led Wiester to a second project. The previous year, in 1984, *Science and Creationism: A View from the National Academy of Sciences* had been released to battle creationism and had been sent to all the nation's secondary schools.[7] Even though geologist Preston Cloud was one of its eleven authors, Wiester did not like its tone. His urge to fire off a rebuttal, however, evolved into something more: an alternative booklet—"a view from the American Scientific Affiliation." He and Hearn solicited the ASA board's support to form the Committee for Integrity in Science Education. Next they spent a year writing *Teaching Science in a Climate of Controversy*.[8]

Wiester pulled together the funding, and forty thousand copies were sent to public schools. He had stepped onto some well-established turf. The National Association of Biology Teachers attacked his science guide as dangerous. The *Science Teacher*, a journal, collected a series of scathing essays: "Scientists Decry a Slick New Packaging of Creationism."[9] The backlash surprised Wiester. He was, after all, an old-earth geologist, and his coauthor, Hearn, was an evolutionist. "They not only bashed the booklet, they distorted it," Wiester says. "When asked to correct their distortions, they wouldn't. That's when I realized that this isn't about science. This is about a political culture war."

The ASA has a membership of a few thousand. An average annual meeting may draw 150 attendees. Fewer still are activists for the group, so Wiester's advocacy stood out when he became chairman of the ASA's regular Education Commission. With his prompting, the ASA's executive council issued a resolution focusing on terminology in public school science. It stated that "the terms 'evolution' and 'theory of evolution' should be carefully defined and used in a consistently scientific manner." It also urged "forceful presentation" of scientific data, clear distinctions between evidence and inference, and candid discussion of "unsolved problems and open questions."

After doing his research for a few years, Wiester was convinced that the nation's educational materials had been touting a consistent two-pronged Darwinian message. It seemed to say first that Darwinism means "you are the result of a purposeless accident," and second that religion is subordinate to science. To illustrate his argument, Wiester has picked out three of the most salient examples in mainstream texts. One is the high school textbook *Biology*, by Miller and Levine, which says that "evolution works without either plan or purpose."[10]

As a second example, he points to the NABT's 1995 "Statement on Teaching Evolution," which had a preamble saying that evolution was an "unsupervised, impersonal" process. Noting that the NABT disclaimer also says that science "neither refutes nor supports the existence of a deity or deities," he winks.[11] "So, they're saying, 'We've nothing to say about a deity. We are completely neutral. Except that your creator is Darwin's unsupervised and impersonal mechanism of natural selection.'" From 1996 onward, Wiester frequently cited the "unsupervised, impersonal" phrase in his writings and public talks. His argument showed up in the letters sections of journals such as *Commentary, Christianity Today*, and *First Things*. The drumbeat was hard to ignore. "Help! I need a copy of your Feb. 3 letter to the editor," signaled the NCSE's Eugenie Scott.[12] What followed was "l'affaire NABT," in which the organization agreed to drop the two offending words. Scott says she agrees with Wiester that science should not make theology-like statements.

Wiester's third example of how Darwinian language works is the National Academy of Sciences' mass-distributed *Science and Creationism*. It reiterates the Academy's assertion, issued in 1981, that science and religion are "mutually exclusive" kinds of knowledge. What grates more on Wiester is the conclusion: "Scientists, like many others, are touched with awe at the order and complexity of nature. Religion provides one way for human beings to be comfortable with these marvels."[13] Wiester winks. "If you can't deal with reality, it's a comfort," he says. "You get a little pat on the head."

What impact his linguistic policing has had may never be known. The Academy changed its wording when it updated *Science and Creationism* in 1999. Gone are the "mutually exclusive" statement and allusions to religion making people "comfortable." Instead, the colorful thirty-three-page publication says: "Scientists, like many others, are touched with awe at the order and complexity of nature. Indeed, many scientists are deeply religious. But science and religion occupy two separate realms of human thought. Demanding that they be combined detracts from the glory of each."[14]

The text war may be a big part of Wiester's legacy, but he started out trying to show young-earth people a better way. He thought their "impossible" geological stand detoured from a more basic question: God as Creator. After some culture-war experiences of his own, though, his impatience cooled. "Now, I feel like I have more in common with our young-earth friends than I do with the left wing of theistic evolution," he says. Young-earth people at least challenged evolution as atheism. "They held a very strong line. Kept the issue alive," he says. He hopes against hope, perhaps, that creationists will put the age question on the back shelf and concentrate instead on what he calls "intelligent design versus Darwinism." In 2000, he provided a vehicle on this front for high school students, a cartoon book called *What's Darwin Got to Do with It?*[15]

Meanwhile, the content of his geology classes is drastically different from what he learned as a college student in the 1960s. Now geology is tied to a new paradigm—informed by plate tectonics, mass extinction, and dinosaur death by asteroid. "Many of the major formations are laid down in a very catastrophic way, whether a volcanic eruption, a thousand-year flood, or a ten-thousand-year flood that lays down thick layers of sediments," says Wiester. "Maybe 50 percent of what we see in the geological column has been laid down in a very short period of geological time, by a catastrophic event. These kinds of events, in fact, define what are called the eras and minor periods in geologic time."

Scientists today will say they are not blind to these terrestrial eruptions and that the laws of gradualism and upheaval both are heeded. If they are quick to distance geological upheavals from young-earth catastrophism, Wiester is slower. Young-earth creationists "were right to emphasize catastrophic events in geology," he says. "Publicly, on radio programs, I give them credit for that."

GOING BACK TO GENESIS

Civil engineering professor Henry Morris, whose texts on the physics of bridges, roads, and water are staples of college courses, is uneasy about the moniker "leader of the young-earth movement." Thousands of believing scientists, he says, helped play that role. But it was no accident that many of them were engineers. "I think there is a connection," he says in his office at the Institute for Creation Research, nestled in Santee, California, a semidesert town turned affluent San Diego suburb. "The reason is that engineers are more practical. Things have to work. You can't just speculate about something, like a biologist can, or a philosopher.... The hard, pragmatic approach of the engineer tends to make him look at the Bible in terms of evidence, and the same way with evolution." He pauses. "When you design a bridge, you can't just speculate on how big to make the girders."[16]

That is his answer to one anticreationist salvo that goes thus: not being pure scientists, engineers have easily fallen for creationism. The criticism has roots in a hundred-year class division between theoretical and applied science in America, but now it has gotten a new gloss. Sociologist Dorothy Nelkin, for example, theorized that the mobility of engineers, from one bridge or NASA project to the next, creates an uncertainty that turns many to fundamentalist churches—where they become creationists. "A lot of the early creationists came from fundamentalist families; they were the bright kids who went off to study science," she says from her New York University offices. "If they were in aerospace, they moved almost annually, and they found churches in those areas."[17]

Morris listens to this theory in his upstairs office. "That's all nonsense," he says. Professionals in the applied sciences, he says, are skeptical of evolution because of the harsh light that practical science puts on reality. So what the studies show makes sense to him: scientists on the mathematics, chemistry, and engineering end of the spectrum tend to hold more religious belief, while disbelief is highest down toward the biologists. Still less belief is found in Nelkin's discipline of sociology, followed closely by psychology.[18] Engineers like Morris have long suffered under the class pride of theoretical scientists. By presiding over the National Academy of Sciences, the theoreticians have defined science in America, at least until Congress gave "coequal" status to the National Academy of Engineering in 1982. The two academies were characterized as "one composed largely of academic members, the other of practicing engineers," which makes sense to Morris. Creationists have even argued for two kinds of science: "singularity science" for academic speculation on onetime origin events and "operation science" for working with phenomena driven by present physical laws.[19]

For our interview, Morris was more than happy to have me over, but he wondered why anyone would want to write yet another book on creationism. Among the fifty or so by Morris is his *History of Modern Creationism*, a classic overview. He credits another history, Ronald Numbers's *The Creationists*, as the only critique of creationism to approach objectivity. Still today, Numbers tells that story. "Henry Morris's success was precisely in getting large numbers of conservative Christians to agree on a particular interpretation of Genesis," he told a science forum in 2001. "And then in the nineties along come the design theorists who say, 'Oh, let's just set aside the relatively unimportant matter of the best way, or the right way, to interpret Genesis, and create this big tent of people who oppose evolution.'" So books have been written, but the story still unfolds with what Numbers calls "a tremendous amount of tension between" the design and young-earth camps.[20]

Morris had arrived at his San Diego home in 1970, late in a career that had included earning his hydraulic engineering doctorate and teaching at the University of Minnesota, and heading up the departments of civil engineering at Southwestern Louisiana Institute (now the University of Louisiana at Lafayette) and at Virginia Tech. Having arrived at the Virginia school in 1957, Morris soon gave his department a name for producing "the highest research activity of any department on the campus, except maybe statistics," he says. He had a national audience as well through his five books on his specialty, the most popular a textbook on hydraulic engineering. He truly wore down his pen, however, writing about Bible belief and science.

The interest began in his own student days, when he was studying engineering at Rice Institute (now Rice University) in Houston. Reared a nominal Southern Baptist, Morris had slid from a progressive creation belief to theistic evolution. Then, one day in nearby El Paso, he caught an "impressive" "Sermon on Science" by evangelist Irwin Moon. He also learned of Harry Rimmer, a Presbyterian evangelist and self-made scientist who was America's best-known antievolutionist. In time, Morris dissented from Rimmer's belief in an old earth and local flood, but he would be galvanized by Rimmer's book *Modern Science and the Genesis Record* (1940).

The year before, Morris had graduated from Rice. It was the "Lord's doing," he says, that the navy assigned him to return and teach science to naval personnel going off to war overseas. By then he had joined the Gideons International, a lay fraternity that distributes Bibles, and had become an advocate of young-earth creationism, or "what [Baptist theologian] Bernard Ramm called me: a holder of 'naive literalism.'"

Teaching servicemen, Morris saw firsthand how science was causing faith to stumble. "So I went into a real program of study in the evolution literature," he says. "There wasn't much creationist literature at the time." In fact, he says, old-earth works, much like those produced by theologian Bernard Ramm in the 1950s or John Wiester in the 1990s, were not uncommon. "It is 'young-earth creationism' that was practically ignored in the evangelical world until recent years," Morris says. Back at Rice in the war years, he says, "I made a detailed study of the Bible, verse by verse, to see everything it had to say about origins or science and the flood and the early patriarchs. I found that there was no way, if I was going to say that I believed the Bible to be the word of God, that I could also believe in evolution. Finally I became convinced I couldn't believe in an old earth either."

As a teaching tool, in 1946 he wrote *That You Might Believe*, which has a chapter on Creation, one on the flood, explanations of fulfilled prophecy, and other evidence for the Bible's authenticity. The shadow of the Scopes trial still hung over education, Morris recalls. "It made everybody in the academic world afraid to teach creationism."

During his four years of teaching at Rice—a science-oriented school where Julian Huxley had founded the biology department—"I never did find another faculty member who was at all sympathetic to what I was doing," he says. He asked the president of Rice to invite Harry Rimmer, the evangelist, to the campus. The president refused, so Morris presented Rimmer to his Bible class.

After he joined the Virginia Tech faculty in 1957, Morris not only remained active with the Gideons but in 1962 founded College Baptist Church (now Harvest Baptist), affiliated with the Independent Fundamental Churches of America. At the time, Baptist preacher Jerry Falwell was becoming known. Though Falwell's affiliation was different—the Baptist Bible Fellowship—Morris invited him to College Baptist for a series of revivals and sent his own children to Falwell's Bible summer camps in Lynchburg, Virginia. "So we were good friends," Morris says.

This was all around the time when Morris cowrote *The Genesis Flood* with theologian John Whitcomb. Its publication in 1961 had earned him many speaking engagements, especially at Christian colleges. "I found that most all of these, especially the liberal arts colleges, had felt they had to accommodate evolutionary thinking to some degree, or at least the old-earth concept, in their program," he says. Morris's book and speaking schedule alerted Virginia Tech's evolutionists to his stance, and he says he knows of two times when someone on the biology faculty urged that the president fire him, to no avail. Still, he was shopping around:

Was there a Christian college that would let him establish a full-blown creationist curriculum?

He would not try such a venture until 1970, but in 1963 he was helping to pioneer the Creation Research Society. At the time, he and his colleagues had no idea how political their intellectual efforts would become. "All of them unanimously felt that we should not get involved in the political approach at all," he recalls. "Not even hold conventions, but just to concentrate on doing research and writing. And speaking whenever invited, but not in a political context." Today, the society continues as a fellowship of members with no professional meetings.

Then a trip to Los Angeles changed everything. It was 1970, and Morris went to Biola University to share a panel in a lecture series with fundamentalist Baptist pastor Tim LaHaye. In San Diego, LaHaye's downtown church was opening the fledgling Christian Heritage College, which in two years would move to a suburban campus, and the minister offered Morris a role. "I told him, no, I wouldn't be interested in a Bible college, or a science department. But if he made it a Christian liberal arts college built around the concept of creation in all the disciplines instead of evolution, science as well as humanities, and have an associated creation studies center, yeah, I might be interested," Morris recalls. LaHaye was willing to give it a try. "So that's why I came out here."

While the college did emerge as an unabashedly creationist institution, enrolling six hundred students in 2000, the hub of Morris's work became the ICR, which he founded in 1970. And the best was yet to come. Morris had lectured around the country, but by 1973 the debate format was drawing thousands of students. "And at that time, there was almost nobody else," he says, noting the subsequent increase in the number of activists and the quantity of books published. "So it has grown. We think the scientific evidence is strong, and we think the biblical evidence is compelling."

If not head of a movement, Morris was president of its most prestigious institute, and in October 1995 he passed that mantle to his son John. Duly credentialed with a doctorate in geological engineering from the University of Oklahoma, John pledged to broaden the institute's outreach, particularly onto the Internet frontier. The succession, part of a ceremony honoring Morris the elder, was part of a weekend "Back to Genesis" seminar in the northern Virginia suburb of Woodbridge. During the day, the event was geared to dinosaurs, Bible stories, and a few thousand children, and then it shifted to adults at night. The gathering illustrated that the movement's strongest constituency still was among homeschoolers, church schools, and church Bible classes.

The evening ceremony also featured the unveiling of *The Defenders Bible*, in which Morris defended the faith in sixty-four hundred footnotes and eight major appendices. In some ways, it was a modern alternative to the vaunted *Schofield Reference Bible*. Produced in 1909 by American C. I. Schofield, that Bible had been a mainstay of dispensationalist believers—akin even to the theology of Morris or LaHaye. But on the creation topic, Schofield's notes supported an ancient earth with a gap of eons from Creation to Adam. Now, *The Defenders Bible* provided notes for a young earth and flood geology.

A quarter century earlier, a Noah's Ark craze had first drawn the media spotlight to creationism. John Morris had helped stoke this media fire with his book *Adventure on Ararat*, written after an expedition to Turkey in 1972 in search of Noah's ship. Hollywood filmmakers also caught on. In 1979 the film *In Search of Noah's Ark*, narrated by Leonard Nimoy of *Star Trek* fame, aired on nationwide television. During this decade, however, the mass media overlooked the journal diatribes and textbook battles that as early as 1972 pitted creationists against organizations representing biology teachers. Those scales fell from media eyes around 1980—when creationism got political.

Henry Morris says he never urged a legislative initiative, but once local activists got into the game, it was hard for his ICR to deny them help. "They have felt that the only way to do this is to get the legislature to pass a law that you have to have a two-model approach," giving evidence for both evolution and creation, he says. "And some of them even wanted to ban evolution altogether, which, of course, is never going to happen."

The two-model idea grew out of Yale law student Wendell Bird's thesis from 1978. The constitutional requirement for government neutrality in religion, Bird said, requires presentation of both sides of the human origins story in public schools because evolution teaches atheism.[21] Bird became a lawyer for Morris's institute and in one of its tracts offered a two-model resolution that school boards might adopt without making it statutory. But South Carolina conservative Catholic Paul Ellwanger, founder of Citizens for Fairness in Education, carried Bird's proposal into the legislative arena. He developed a model bill that was pushed by politically active preachers and churchgoers in more than twenty states. Their efforts succeeded in Arkansas and Louisiana, but district courts and the U.S. Supreme Court stopped those and other attempts cold.

Looking back on his forty years of creationist advocacy, Henry Morris feels his warning about politics has been proved correct. "That's self-defeating to try to compel it under penalty of law or a fine," he says. But others went ahead anyway. Once the legislative activism began, Morris says, "then we thought, well, we should help them to word the legislation in such a way that it would be more scientifically oriented than religiously."

It is debatable how much political strategy the creationists had in mind, but clearly they were up against the old fundamentalist dictum of separatism, to be "in the world but not of it." The new Christian right shattered the dictum by 1980. Some early creationists were even harbingers of such activism, but there was also a degree of Bible-believer disapproval. Morris's coauthor for *The Genesis Flood*, Whitcomb, put his warning this way: the risk of undermining faith in the Bible got "exceedingly high" when Bible belief was justified by arguments from science.[22]

Such creationist efforts in the secular arena did face many setbacks, from negative court rulings to cultural derision. Social movements tend to have a cycle, and once they reach a high point of energy, something begins to dissipate. Morris says he has seen no such slump. But it may be as inevitable as the second law of thermodynamics—the law that order moves toward disorder—so central to the creationist argument that God must be pumping order into a natural world.

A major infusion of energy came in 1992, when the ICR opened its new complex. Its plain business-park exterior belies a bright cavalcade of creationism inside, particularly a museum that even gets plaudits from evolutionists. It drew twenty thousand visitors in its opening year.

Downstairs by the classrooms, outsize photographs are mounted on hallway walls. They show the large canyons created in a matter of months by water coursing through the hardening ash of Mount St. Helens in Washington State. They inspire the roughly forty students, mostly in summer courses, seeking the institute's four kinds of master's degrees. "The degree programs are all nationally accredited," Morris says. True, but only by the Transnational Association of Christian Colleges and Schools. This association, which was specially recognized by the U.S. Department of Education in the first Reagan term, comprises only schools with doctrinal statements that uphold "the six literal days of the creation week."

The contest is now for a younger generation, says the white-haired godfather of creationism. Perhaps they will have an easier time, but with new challenges as well. "I don't think necessarily that Darwinism will grow," Morris says. "I think some kind of pantheistic evolution will grow, the New Age type of thing." With "so many holes" in Darwinism, he says there will be more attempts to account for jumps in nature, self-organizing principles, or some substitute for intelligence that make sense of nature's wonders.

"They are not willing to accept a biblical God, so they go back to pantheism. Nature is God. Gaia. Some very eminent evolutionists are going that direction now, and I think that's the trend of the future. They can have design there that has intelligence to it, but at the same time avoid a commitment to the biblical God of righteousness and holiness."

THE CREATOR WHO HAS NO GAPS

A hearty race of Dutch Calvinists has populated central and upper Michigan since the early 1800s, tilling farms, building churches and denominations—and coping with the occasional heresy. The more intellectual among them founded two colleges, Hope in 1851 and Calvin in 1876, and Dutch entrepreneurs erected a cluster of publishing houses to transmit the piety and research of their scholars. Today, Baker, Zondervan, and Eerdmans are among the most respected publishers in American Protestantism.

In 1960, Calvin graduated a bright student named Howard Van Till, whose love of science led him to earn a doctorate in physics at secular Michigan State University and then return to Calvin as a professor. He entered the science-versus-Genesis debate and wrote a book, *The Fourth Day*, that Eerdmans released in 1986. His life in Christian Reformed Church circles would never be the same.

"If I had simply been another member of the denomination teaching at Michigan State University or something like that, they"—his creationist critics—"would have paid absolutely no attention," Van Till says several years after surviving a kind of heresy trial, which ended in 1991. His notoriety, in fact, came about because of his effectiveness as a teacher. The college had sent him out to thirty locations to give his basic lecture on science and Genesis. Its short version was published prominently in an "occasional paper" by the college. When he expanded its thesis into a whole book, *The Fourth Day*, his ideas were declared to be no longer acceptable. "I represented a threat to a certain way of picturing the biblical text," Van Till, a tall, willowy man, says years later over lunch on a snowy day at Calvin.[23]

Immigrants from the Netherlands, his parents had moved from Michigan to central California, where his father reared him on faith and openness to "the world"—including science. He was not the first among Calvin's science professors to wander from an older defense of a literal Bible. The well-known geology professor Davis Young wrote in 1977 that "the biblical data favor an essentially global flood," but by the mid-1990s he had changed his stance, saying that physical evidence indicated a local flood was more likely.[24]

The two main concerns of Van Till's book were not new; what made them noteworthy was that he wrote as a Christian physicist. He looked at how Near Eastern culture shaped the Old Testament description of natural events and how modern cosmological science can inform the Bible story of origins. No one at Calvin College was reading Genesis 1 as an account of a literal six-day creation, but he had the bad luck to stand out. "I chose, wisely or not, to use the common vocabulary of the sciences, which, of course, includes the e-word, evolution," says Van Till. "I chose to use that word because it's a fine word within the sciences, though it can be abused by what I will call preachers of a philosophical naturalism."

His book was mostly about the development of elements, stars, galaxies, and planets. Then, in "just a couple sentences," he says, he endorsed biological evolution. He could cite the offending paragraph in his sleep. "Top of page 258," he says. It reads: "I see no reason whatsoever to deny that the Creation might have an evolutionary history or that morally responsible creatures might have been formed through the process of evolutionary development."

Printed statements such as the preceding gained Van Till a formidable enemy: John Hultink, editor of the *Christian Renewal* newspaper in nearby Canada. The paper had begun a crusade to root out liberalism in the Reformed churches, and for a year Hultink wrote diatribes against Van Till's work. Then another critic, Grand Rapids businessman Leo Peters, weighed in. For three years, Peters used his considerable wealth to buy thirty full-page or double-page advertisements in the *Grand Rapids Press* lambasting the book and the Christian Reformed Church's failure to do anything about it. "He saw the occasion of the publication of my book as the occasion, I think, to publicly punish the denomination and Calvin College," says Van Till, who adds that he thinks Peters was embittered with the church. "The vast majority of the denomination just saw the ads as trash and were embarrassed."

The tensions went back at least to Henry Morris's *Genesis Flood* in 1961. When Morris spoke at the Calvin College's seminary chapel, "the faculty seemed quite displeased with my book," Morris recalls. He also spoke to the Reformed Fellowship, a group of conservative pastors and laymen, and to them he later "submitted a report" on his reception at Calvin. Morris does not think it resulted in "any significant action," but he was later keen to say the investigation of Van Till "resulted in a whitewash."[25]

Whatever its origins, a small, vocal group focused on the physics professor and his book. About twenty letters of complaint reached Calvin's trustees, and before the next annual church synod, others made overtures, or resolutions, for action— implicitly, to fire the heretic physicist. "In the minds of many, that was basically it," he says. "Within the first year after publication, Calvin's board of trustees had appointed a subcommittee to meet not only with me but with two colleagues in geology, Clarence Menninga and Davis Young, who had expressed similar ideas." In the thick of this controversy, in 1988 the three coauthored a criticism of both fundamentalist creationism and Sagan-like scientism, *Science Held Hostage*.[26]

At Calvin, a four-member subcommittee met monthly with the three scholars, hoping to settle the unrest before it became a denominational issue. Young and Menninga were excused after a year of discussion, but Van Till was kept under scrutiny for four years. In early 1989 he took a sabbatical to England. "I wanted to get away from all the noise," he says. The inquiry continued by letter, and then came a series of questions on doctrinal matters. "By that time, the breadth of the questioning had grown considerably," he recalls. "Now it was questions of faithfulness to the creeds." He responded to the subcommittee in a nine-page letter.

Because there was no formal charge against him, for which there would be procedures, the subcommittee was ad hoc—and thus could go on indeterminately. "It was terrible," Van Till recalls. "There comes a point when your faithfulness to the community is being sorely questioned, and then your whole career as a Christian scholar."

By the time of his reply from England, his responses had become "so candid" that the trustees wanted the subcommittee to make a final recommendation, but the panel split. "Two said no action is necessary. The other two said, 'We should institute formal charges,'" Van Till says. The charge would have been that he did not teach in conformity with the creeds specified in "The Form of Subscription," written at the Synod of Dort in 1618–19.[27]

"Well, in my letter to the board of trustees, I did raise fairly substantial criticism to that Form, that subscription." Yet few believers could know with surety all that the form required, he says, or correlate all its contents with the Bible and the modern world.

Van Till returned to Calvin at the end of his sabbatical and resumed teaching. Turnover on the subcommittee brought a new chairman, who seemed to be a peacemaker—but he decided the entire process should begin again with a meeting with Van Till. "I stormed into the president's office the next morning and said, 'I am not going to meet with this committee again,'" he says. "They will have to do what they have to do.'"

"It wore me down. . . . I basically couldn't go any longer." Finally, over six months the subcommittee wrote what Van Till calls "a very fine document," to which he responded in writing, leading to a final "negotiated document" stating concerns but not pointing fingers; all parties signed in May 1991.

This dispute over academic freedom did not catch national headlines like clashes at other church-related universities: clashes, for example, in which scholars at Southern Baptist schools in the early 1960s downgraded Adam and Eve, miracles, or Jesus' divinity to symbolic stories, or in which Catholic theologians challenged papal teaching authority in the 1960s and 1980s. Among evangelical liberal arts colleges, however, the Van Till case was considered significant. "There were lots of other related Christian communities, educational communities, watching us," he says. "Because they said, 'If you people at Calvin College can't find a way to resolve this, we won't either.'" No clear precedent seemed to have been established when Van Till's ordeal was over. The stream of events had been messy, beginning with public protest and going into ad hoc resolution without formal steps through faculty or academic committees. Further, the dispute was not about doctrine, ecclesiology, or morals—it was about how Christian intellectuals teach science.

Van Till went on to refine his concept of how the biblical God is sovereign in physical and biological evolution. He introduced the twofold concept of the "developmental and operational economies of the created world." By these "economies," or systems of resources and capabilities, a transcendent God is also imminent because "the only capacities for action that created matter has are the capacities that God gave it." He insists this is not an impersonal deism.

There is precedent for his claim in mainstream Reformed theology. James Mc-Cosh, a Scottish theologian who was president of Princeton University in the 1870s, had no problem with the idea of a sovereign God allowing nature to go on its own, for God still would have the glory: "If man could construct . . . a watch which should produce other watches telling the hour through all time, our admiration of the skill of the artist would not be diminished."[28]

Van Till's remaining critics contend that he seems to be giving Darwinians all the credit for understanding nature, even as Darwinism faces its own share of modern scientific puzzles. This occurs despite the fact that many Darwinians are atheists and some are anti-Christian. Van Till responds to fellow Christians this way: only a Creator who is always present in the powers of nature deserves admiration, a Creator who in fact endowed the human mind to conduct science. Special creationists, who see God's occasional intervention in nature, are eager "to find gaps in Creation's developmental economy." The same goes for the more sophisticated "intelligent design" approach. He calls it "Paleyan-design," after William Paley, and faults it for its "glib use of intervention by God in the formational history of the Creation."

Van Till spent his last few years at Calvin in relative calm and retired around 1998, gaining the title professor emeritus. The editor of the Canadian newspaper split from the Christian Reformed Church, Peters died, and some foes on the faculty moved on. "Most of the critics are simply out of the picture, and I am still teaching," Van Till said in 1997.

On this day he is in Seattle, participating in an event for the American Association for the Advancement of Science. He has moved into moderate to liberal ecumenical forums. He is a member of the American Scientific Affiliation, where evolutionists such as he are not rare, but he has also done some missionary work to the anticreationist National Center for Science Education—where in a talk he advised against science proffering its own religion of "scientism."

"In the eyes of some in my own smaller denomination, I'm off on the left," he says. In his widening circles, however, he clearly represents a more conservative stance.

He wishes young dissenters Godspeed, because he knows church authorities on a college board can thwart a nascent career. "I'm wrapping up," says Van Till. "I can be a little more candid now than I was earlier. I will probably be finding myself more and more candid as time goes on."

Well-formed science changes one's ideas, he says. "I think that through all these things, one's concept of God and of divine action does change."

Van Till has been offered evangelical forums for presenting his ideas. He wrote a chapter on "theistic evolution" in a collection of three viewpoints in the evolution-creation debate, published by Zondervan. He also was the keynote speaker at the InterVarsity Christian Fellowship's student conference, titled "Following Christ—Shaping the World" and held in Chicago at the end of 1998. Such student fellowships and evangelical publishing always have allowed leeway for debate on science issues.

Since 1953, when the InterVarsity student magazine *His* endorsed theologian Bernard Ramm's attack on young-earth fundamentalism, the student group has been open to progressive creation and theistic evolution.[29] To this student conference, Van Till presented his divine-economy idea under new terminology. He called natural science "creaturely science," implying the proper standing of the object of investigation. He warned against "folk science" such as that proffered by six-day creationists or atheistic naturalists. A generous God, he told the students, gave the Creation a "robust formational economy." It has all the "richness of being" needed to give matter the capacity to bring about new forms; thus, God is not forced to intervene, as creationists require, to impose a missing "form." Yet God still is present in sustaining nature, guiding providence in humanity, and acting by way of miracles—the most central presumably being Christ's bodily resurrection.

Van Till, as a critic of the intelligent design and "theistic science" movement, has offered a counterproposal called "optimally gifted creation" (OGC).[30] He is trying to get away from terms such as "theistic evolution." He wants to emphasize God's creativity and generosity in evolution, not nature's apparent gaps. "Fully aware that there are, and always will be, gaps in our knowledge of the Creation's formational capabilities, the OGC perspective sees no need to posit the absence of certain key capabilities that would necessitate occasional episodes of irruptive divine action," Van Till told the students.

Having been a teacher, he went on to work with the John Templeton Foundation, dispensing grants for science and religion projects by scholars and grants for teachers—at secular schools, Christian liberal arts colleges, and even conservative seminaries—to develop courses on science and religion topics. But what if this strikes the conservative Christian schools as liberalism and a scientific affront to theology?

"Then, the tragedy is going to be that when they have a very vigorous conservative evangelical community in North America, it will be completely isolated from the intellectual development of that same North America. It will have lost contact with a very important portion of that larger community. A lot of that has already happened. The university crowd doesn't think it has to pay attention to Christian scholarship. They think Christian scholarship is an oxymoron. There's a history to that. And I think it's a tragedy that has to be undone."

CHRISTIAN ANTICREATIONISM

If things had gone differently, one of public television's most spectacular science documentaries might have featured two characters, a reserved Harvard-Smithsonian astronomer—and God. Lacking the necessary $2.2 million, however, Owen Gingerich could not transport his six-part series from the drawing board to the airwaves.

Besides a few other lengthy PBS science treatments—from the nine-part *The Mind* in 1988 to the seven-part *Evolution* in 2001—the mightiest of all was the thirteen-part *Cosmos,* a "personal journey" hosted by Carl Sagan in 1980. It cost $8 million to produce, spun off a *New York Times* best-seller, and sold television rights to forty countries, but as far as Gingerich was concerned, the upshot was its memorable godless claim: "The cosmos is all that is, or ever was, or ever will be."[31]

Episode six of Gingerich's proposed series, *Space, Time, and God,* would have been "Is the Cosmos All There Is?" The series would have presented a medley of science, a look at an ancient universe, and a new view of the Galileo affair. "And so on," Gingerich says wistfully.[32] It would have been a personal story, "subtle in my philosophical approach." Yet in a polarized evolution-creation debate, it was a no-win proposition. To mention God was to alienate corporate and science-programming funders; to discuss the fact of evolution would offend many Christians. To be subtle and nuanced, moreover, was not to live up to television's show business requirements, as Sagan's young, beatnik enthusiasm had done so well, riding a "spaceship of the imagination" to synthesizer music.

"Christians who have lots of money they'd like to put into the media just disappear into the woodwork as soon as the word 'evolution' is mentioned," Gingerich says of such large film projects. Working with public television's top science program, *NOVA*, would also throw up obstacles. "*NOVA* is committed to reductionism," he says. "If one were taking money from the National Science Foundation or from *NOVA*, there would be sufficient censorship to denature what I would want to say."

Gingerich did not always have much to say on the evolution-creation debate. He was reared a Mennonite, a pacifist church he called "the quiet in the land." In 1953 and 1954, he wrote articles for the journal of the American Scientific Affiliation, which served evangelicals in science, one of them arguing that "the evidence seemed against" astrophysicist Fred Hoyle's "continuous creation," which explained that an infinite, steady-state universe expanded because new hydrogen atoms appeared continuously: it rivaled the big bang. But it was more in the field of astrophysical computing, and of astronomical record keeping, and finally scholarship on Copernicus, who first theorized a sun-centered solar system, that Gingerich's reputation began to speak for itself. That career trajectory would take him through the question of God in the university, the politics of historic commemorations in world science, and finally the evolution-creation debate.

After earning his doctorate at Harvard, Gingerich got on the faculty and soon learned the rule of thumb that professors should avoid topics touching on religion. "Atheist professors can and frequently do voice their personal views in their courses," Gingerich says, "but to mention one's Christian beliefs in a comparably forcible way would no doubt bring charges of proselytizing." It reminded him of his own youthful struggle over science, when he nearly turned away from his natural interest because it had too little to do with the Mennonite call to God's service. His Mennonite math teacher set him on course again, saying, "We shouldn't let the atheists take over any field."

So at Harvard he found an academic way to keep his science students' minds open. For an essay requirement in astronomy, he offered as one topic option, "Is there a conflict between science and the idea of a personal God?" He announced that the topic was experimental and that he, rather than the teaching fellows, would grade that one. To his overworked chagrin, nearly half the class chose that topic. No complaints of "sectarianism" arose, but he could raise this case of academic excitement when Harvard debated offering courses on religion, a case that helped swing a positive vote.

Near to Harvard, the Smithsonian Astrophysical Observatory housed New England's most powerful electronic computer, used to track satellites. Gingerich's work at the observatory put him in the vanguard of high-speed astronomical computing. Eventually he was appointed as head of the Central Bureau for Astronomical Telegrams, the arm of the International Astronomical Union that names comets and disseminates information about transient astronomical phenomena. After pulsars were discovered in 1967, and their sightings became more frequently reported, telegrams bearing the Gingerich signature went out more widely than ever.

By the time the five hundredth anniversary of the birth of Copernicus approached in 1973, Gingerich had a natural seat on the international organizing committee. Poland, the land of the astronomer's birth, hosted one of four major celebrations around the globe, but only in this communist country did it bear "the heavy burden of political overtones." In hindsight, Gingerich recalls the event's "dark current of European ethnic rivalries about who could claim Copernicus as a cultural son."[33] For the 1943 anniversary of Copernicus's death, the Nazis took title of him as German stock, and now Poland was hitching itself to the Copernican star for his Polish birth. The Russian and German scholars identified as possible keynote speakers for 1973 only opened these old political wounds, so Gingerich was urged to fill the gap: he gave the keynote address at Poland's celebration.

He had meanwhile become an expert on the existence of 598 first editions and second editions of Copernicus's seminal work from 1543, *On the Revolutions of the Heavenly Spheres.* The Gingerich catalogue came in especially handy in the 1990s, when a "rash of thefts" struck the collection, and his listing helped identify the stolen artifacts whenever they surfaced.[34] The Copernicus celebration held in the United States was under National Academy of Sciences and Smithsonian Institution auspices, and the theme was "the nature of scientific discovery"—a topic that Gingerich considers central to the entire science and religion debate.

Years later, he traveled to England to lecture on Galileo and there told an interviewer that all great science, by its nature, seeks not proofs but coherence. "Newton had no proof the earth moved or that the sun was the center of the planetary system," he said during that 1999 visit to Oxford University. "Science doesn't march ahead by proofs," he says. "Science advances by coherency." And for a Christian, he says, "a belief in design can also have a legitimate place in human understanding, even if it falls short of proof"—but gives coherence. By 1999, of course, design had become a highly charged code word for antievolution and the intelligent design movement, both of which Gingerich had distanced himself from. Natural theology—trying to prove God by nature—had severe limits, he believed, especially when you want to jump to a Christian tenet such as accepting Jesus as the incarnation of God. In that sense, Gingerich says, "You can only get so far with these arguments from natural design." By a "step of faith," and by an understanding of science, God's design may be apprehended in the cosmos, he says. "Not a leap, but a step."[35]

A Gingerich synonym for design and coherence has been "tapestry," an image he draws from astrophysical science, where mathematics and real stars are not dead-on matches. The data are like threads, and science assembles "little threads woven together to form the grand tapestry of understanding." Here in this fabric Gingerich sees a "grand design."

Gingerich first took his "subtle" approach to the wider public in 1981, when he gave the first Dwight Lecture, turning it into his manifesto on God and nature. The lecture, given at the University of Pennsylvania, came with a profession of faith to be signed by the speaker because it was underwritten by Protestant benefactors. Rather than sign it, however, Gingerich offered his own description of the faith, which the patron accepted. "I have sometimes described it as my pro-Christianity, anticreationism lecture," Gingerich says ruefully. The "long and close-knit lecture" was reprinted in anthologies in the long run.[36] In the short run, it was re-orated by Gingerich more than forty times, often on a lecture circuit arranged by the American Scientific Affiliation.

The early scandal of a Harvard-Smithsonian scientist professing Christian salvation would quickly wane, and Gingerich stood comfortably amid the swirl of the evolution-creation debate. "When somebody wants to have a conference and they say, 'Round up the usual suspects,' I am very often one of them," he says.

His coming forward as a Christian evolutionist, he says, was required by the force of a pincer movement, with young-earth creationism charging on one side and Sagan's godless *Cosmos* advancing on the other. "I was hoping to indicate that there is a middle way, that you can be a serious Christian and take Scripture very seriously without, however, going flat out on a literalist approach that runs you into head-on conflict with modern science," he says. He also wanted to show atheists that faith had intellectual credibility. Sagan took him seriously enough to consult him about a bibliography for his Cornell students.

Yet the middle path was not necessarily smooth. The central problem for a Christian in science, where Scripture says God is almighty and science says the universe is all chance, is to reconcile the two. That often is a theme of Gingerich's talks. Contingency in nature is certainly more interesting for scientists, he says, and perhaps it is more fascinating to God as well. "It's a universe of uncertainty and chaos," Gingerich says at yet another conference lecture.[37] "God may very well have intended self-conscious, contemplative beings, quite possibly in many different latent possibilities fully known to the Creator, without specifying which of these possibilities would be actually realized. The more we look at the universe, the more it seems to be constructed of the innate freedom, with a fundamental role for contingency."

That idea forces him to speak of God as being surprised by what happens. Such an idea is logical, but it does present a quagmire for orthodox Christian belief and can provoke "shocked looks of heresy," Gingerich says. "I just think about it a lot, but I don't have an answer." If there is a God, he continues, God must be responsible for all of natural history, including the quirks. "You have the tremendous development of dinosaurs, which runs into a complete dead end, as far as developing promising sentient life is concerned. And you look at the tremendous number of species that have gone extinct. And it is clear whatever God is doing in making it possible for us to be here, it is not just a straight-line march to us. There's a kind of tinkering and contingency."

Another taboo in science is to claim that the universe hinges on human life. Indeed, Sagan was famous for reminding humans of their obscurity in a backwater of the cosmos. Cautiously, Gingerich suggests that perhaps man does have pride of place. "Think of it like a giant plant there to bring forth one small, exquisite flower—man. This is an arrogant statement, I know, but at the same time exceedingly humble because of the burden of responsibility that it places on us all."

And that gets back to design, an idea that all creationists evoke, but in a variety of ways. The intelligent design push, which mainly criticizes macroevolution, does not speak to Gingerich, though its questioning of atheism does. But the cause of atheism remains debatable to him. The "claim that evolution is tantamount to atheism is incorrect," he says.

Evolution can say plenty about the course of life, but that still leaves space for God, he says. "If you are just going to shake up atoms in a box, you're not going to get anywhere. The probability of the DNA forming is incredibly low. So it is the whole game plan of science to find these clusters, these hierarchies, these pathways, catalytic ways in which life can form, in which these mutational jumps can occur and can build up to macroevolution. But the fact that those pathways are there is what I consider to be part of God's design, if you will."

Truth be told, moreover, "I am even prepared to give serious consideration to the idea that God is continually at work at the lowest quantum level, where it's totally and forever concealed to physics." In an age of code words in the evolution debate, there's another truth Gingerich tells. "Some years ago, before the word 'creationist' got so terribly hardened with a capital 'C,' I would have freely said I was a creationist with a small 'c' because I believe that God is creator of the universe and creator of the plan that allows the creation of life to go forward." Despite such confidence in the divine, he sees an age of disbelief ahead. Higher learning and science will sow doubt and more atheism. "What one is hoping to do is to get the young people who are still open-minded on this and who haven't decided," he says.

When NOVA garnered funding from a Microsoft empire affiliate to produce the seven-part Evolution series for the fall of 2001, the producers did not turn to Gingerich for the last segment, "What About God?" What is more, intelligent design scientists had refused to participate, saying their appearance in part seven simply pigeonholed them in a PBS religion ghetto, and so that concluding segment ended up featuring mostly evangelistic young-earth creationists.[38] Still, Gingerich's Dwight Lecture, if not a television series, was around for next generations seeking a subtle, middle path. In one anthology, the lecture was cheek by jowl with two other selections that have had staying power.[39] "I was sandwiched right between Pope John Paul and Asa Gray, which I thought was interesting," Gingerich says. Both are theistic evolutionists and, of course, Christians in good standing.

The road winds upward and around the rocky clefts of central Pennsylvania until the vista opens on a grassy mountaintop. At its center stands a modern structure that looks like a setting for a James Bond movie, circa 1960. The fieldstone edifice, with its large windows and steel girders reminiscent of a Frank Lloyd Wright design, used to be a research facility for Iacocca Hall, the top-flight laboratory owned by Bethlehem Steel in nearby Bethlehem.

Now it is the work space for biochemistry professor Michael Behe (rhymes with "kiwi"). When the American price of steel was depressed in the 1980s, Lehigh University bought the complex for $10 million and made it the home of chemical engineering, the College of Education—and the biology department. Behe is not a likely cast member for a Bond movie, except perhaps as the backroom scientist who supplies Agent 007 with crafty spy devices. Though the short, bearded professor may look the part, he is no laboratory recluse. He grew up in a large Catholic family, has a large one of his own, and while teaching undergraduates also advises two graduate students in his laboratory.

The professor's real-life drama hinges on the assertion in his 1996 book, *Darwin's Black Box*, that Darwinian natural selection loses its power to explain organisms at "irreducibly complex" levels, down among the cells and molecules. At these levels, Behe says, "a class of biochemical systems"—those controlling blood clotting, vision, and various cell functions—look designed, because all the pieces have to be there for the system to function. By making this claim, he earned a place in the National Academy of Sciences' rogues' gallery of creationism. He feels that the federally chartered and funded science agency abused its authority when it published an anticreationist booklet warning the public against such theories as "irreducible complexity" and "intelligent design," though it did not mention Behe by name. "By intelligent design I mean to imply design beyond nature's laws," Behe says.[40]

He asserts this as a tenured scientist who both teaches and pursues original research, and has the respect of those who dismiss creationists who just write and speak. "Many so-called scientists among creation science are not real scientists," says evolutionist Kenneth Miller. "Michael is a real scientist."[41]

Philosopher Daniel Dennett, the great promoter of materialism and Darwinism, downplays the impact of Behe's book but also credits its author for being a working scientist. "He has not put the heebie-jeebies into his colleagues," Dennett said at a forum with Behe. The biochemist has merely enlarged the sense of unfinished business in research on evolution. To his credit, Dennett says, Behe "disassociates himself from the wilder brands of creationism."[42]

Behe does not accept the moniker "creationist," even as it is frequently offered. Darwinian evolution works fine, he says, in explaining common ancestry among the flora and fauna of the earth, and it proposes a lawlike mechanism, natural selection, that produces novelty and fitness in organisms over millions of years. His specialty, though, is biochemical systems, which could only be a "black box" of mystery to Darwin. If gradual evolution by natural selection is universally true in all of nature, Behe asks, where is the evidence? Where are the ancestral pathways that pieced together so much cellular machinery—biochemical machines that, absent even one part, could not operate?

"Evolution and Darwinism worked just fine until you got to the bottom," says Behe, standing in his two-room laboratory. The history of biology, he says, has been a quest to move from the complexity of organisms down to their simplest moving parts—but nature takes the opposite approach. "Things actually become more complex the lower down you go. Just like a machine," he says. Airplanes have a simple aerodynamic exterior by which they adapt perfectly to flying, but behind the frame and skin are mechanisms of bewildering complexity. "And that's what has happened to biology," he concludes, "although people still keep a stiff upper lip."

Behe's book set out a challenge for biochemistry: to explain, in even one case, how a gradualistic accumulation of essential parts could produce a self-replicating cell, a molecular machine, or a chemical process as complex as vision. In such cases, he said, all the key parts had to fall into place at once to survive the eliminative power of natural selection. Perhaps his greatest offense was to point out a paralysis in the field of so-called evolutionary biochemistry, which he summarized in one of his public talks. "Darwinian theory has not published on how a complex biochemical system might have been produced," Behe says. He reached this conclusion after scanning every issue of *Journal of Molecular Evolution* and finding nothing about that topic. Most of the articles—80 percent—compared the molecular sequences of two different organisms; another 15 percent reported on research into the origins of life; and the rest were odds and ends. "In its twenty-year history, the journal has never published a paper, not one, on how complex systems such as the ones I have just described might have come about. There's a complete absence of investigations in this area." This defies logic, he claims, because under the "publish or perish" dictum that drives all academic research, someone should have taken up the topic.

When the dust had settled, eighty reviews had been written of Behe's book. Lay readers said it was fascinating and provocative, but science reviewers attacked it from all possible directions, including efforts to discredit Behe's point about journal articles. Amid this uncoordinated criticism, however, Behe found contradictions that threw the burden back on his critics. Some reviewers argued that there were "hundreds, even thousands," of science papers explaining the molecular pathways of evolution, but others agreed with Behe that the subject remained a frontier. The latter group was bothered most by Behe's defaulting to "intelligence"—everyone knows that means God—to explain a tough problem. Behe had given up, they said; he was a quitter. Still, Behe has his wager on the table: either there are many papers, or there are none. He suggests that his critics should put their heads together on this one. Well, he asks, "Which is it?"

That question also could apply to another contradiction among his critics. Some said his theory is not scientific because, alluding as it does to divine intelligence, it cannot be tested and disproved; other reviewers, however, cited examples in the literature that they said disprove Behe's thesis.[43]

His colleagues at Lehigh were not as acerbic as the book reviewers. In the biology department, "half of the faculty here don't care," Behe says. Biology has become specialized. It has become harder to keep up with the Darwinian implications in every area, so biologists muddle through. "Most people can see that the explanations are mostly stories, and they may be true, and they may not be. And it's hard to know all this stuff, and they say, 'Ah, the heck with it.'"

Behe has spent his career looking at molecules, and he is not shy in his partisanship. Biologists only explain the surfaces of biological life, not its guts, he says, but he works at the level that counts. "I'm a biochemist, so I teach biochemistry," he adds. He teaches the first part of a two-semester biochemistry course and also instructs graduate students on the structure of proteins and nucleic acids.

He learned how to do this in postdoctoral work at the National Institutes of Health (NIH), where he investigated "weird DNA"—Z-DNA that has three or four strands (instead of two) in its helix. He recalls working in the NIH lab with a colleague, a woman who was also a Catholic, and they analyzed the supposed miracle of the first living cell coming into being by historical accident. "What would you need?" they asked each other. "You need a membrane, a power supply, and you need some genetic information. You need a replication system. And we kind of stopped and looked at each other. We said, 'Nah.'"

The smaller of Behe's two lab rooms contains a bench for using radioactive material and where DNA molecules can be tagged. Nearby is an electrophoresis machine, in which electron currents separate DNA molecules. The larger room has two spectrometers, which reveal molecular structures by seeing how the molecules absorb or reflect light. This represents the current state of his field, in which researchers spend most of their time looking at the DNA codes in proteins and organisms to say how they are similar and how they differ. Such "molecular sequencing" has exploded in recent years, becoming the work of countless researchers and the topic of many dissertations—and of the articles that fill the *Journal of Molecular Evolution*. Molecular sequences can illustrate the Darwinian principle of "common descent"—that their common DNA is proof positive that insects, reptiles, birds, and humans grew out of a common gene pool. They had a common biological ancestor.

That's well and good, says Behe, but common ancestry is not the "black box" at the center of all the protest his book has sparked. "Evidence of common descent is not evidence of natural selection," Behe tells his critics. In fact, the failure to explain in detail how molecules join in a chemical cascade that produces vision or how all the moving parts of the flagellum, the hairlike oar of a bacterium, were engineered is a major dead end. He thinks the "reasonable" tack for science is to stop beating that horse and turn to new approaches that are "ultimately more fruitful" than the Darwinian framework.

That framework satisfied Behe most of his life. His doubts arose only in 1987, when as a professor he read Michael Denton's *Evolution: A Theory in Crisis*. The book held great surprises for him, and he went on to read Dawkins's *The Blind Watchmaker*. "I was geared to think of people like this, who objected to evolution, as being cranks," Behe says of Denton. A reputable paleontologist at the University of Chicago had written a positive blurb for the jacket of Denton's book, prompting Behe to think, "Well, maybe I should read it." The book made no allusions to God or the Bible, and it was driven by good old skepticism. Evidence that Denton knew his biochemistry was an added enticement for Behe. "I remember becoming rather angry after having read the book," he says, "because I felt that I had been duped." Duped in his bygone college days: "I was taking university courses in evolutionary biochemistry, and I was assuming this was the way the world happened. Nobody ever said this theory has difficulties that may or may not be solved."

Catholics have not faced the problems with evolution that evangelical Protestants have. Indeed, Behe's "frontal assault on Darwinism"—Dennett's phrase—was credited with being part of a Darwin criticism that finally had "exploded out of its evangelical Protestant ghetto."[44] Behe had learned evolution in Catholic schools and his university education, and it never struck him that it led to atheism, though debates about God's existence had come up in college dormitory bull sessions. Looking back on his graduate studies, he is amazed this never occurred to him before: evolution was always presumed, though the details were so often avoided. For example, when he took a top-flight course on "evolutionary biochemistry," it was really just biochemistry. Having decided his mentors jilted him, Behe limited his reprisals to "rude remarks" when someone extolled Darwinism. He hinted at problems in his biochemistry class, especially when it looked at molecular functions in a cell.

Then, in 1988, Lehigh University invited its professors to develop seminar courses for freshmen. It was a call for volunteers, so Behe and other willing faculty members got free rein. "My course was titled 'Arguments on Evolution.' So you are expecting to hear arguments against evolution," he says. The semester course typically drew about fifteen students. Soon enough, however, some professors expressed resentment over Behe's volunteerism. "When the book came out, there were a few grumbles about presenting this course to naive freshmen who didn't have the background," he recalls.

Students, however, worried less about what they believed and more about the workload Behe gave them. "Almost to a person," Behe says, "students coming in believed in evolution. They watch programs like *NOVA* and TV shows in general where they take in evolution as a background assumption." The heart of the course is reading Denton's and Dawkins's books. Behe believes that this argumentation breeds confidence in young students, who often doubt their own brains when something doesn't make sense. "This Oxford professor says this is the way it happened, so they think the fault is their own if they can't make the connections," he explains. At semester's end, some students were no longer unequivocal about evolution; the committed evolutionists were cognizant of the research challenges ahead.

For three years Behe marshaled his modest dissent on the mountaintop, and then Johnson's *Darwin on Trial* came out in 1991. He read it and then one day came across an index citation for a review in *Science*.[45] He found the review but says, "It wasn't an analysis of his arguments at all; it was a warning against this book: 'Scientists be aware that this dangerous book is out there. It's confusing the public.'" Besides calling Johnson a creationist, it chided him as a lawyer writing outside his field. "I got mad at this, too, so I wrote a letter to the editor. And they published it."

When Johnson read the letter, he was recruiting a team to debate scientific materialists at Southern Methodist University. So he invited Behe, who accepted. In the 1992 debate, says Behe, "The odd thing was that all of the scientists gave philosophical arguments. They didn't give scientific arguments. In fact, I think I was the only one there who was giving a scientific argument. So, we thought to ourselves, yeah, we could do this." The book idea soon followed.

Behe modestly has set out to correct what William Paley, in his *Natural Theology* of 1802, and Darwin, in his *Origin* of 1859, got wrong. He says Paley erred by seeming to argue that God designed every irrelevant detail underfoot, and Darwin seized on that to prove the silliness of the design argument. He scoffed at the idea that God had taken any part in designing his nose. To others he said: "I cannot persuade myself that a beneficent and omnipotent God would have designedly created the Ichneumonidae with the express intention of their feeding within the living bodies of caterpillars, or that a cat should play with mice. Not believing this, I see no necessity in the belief that the eye was expressly designed."[46]

For Behe, a modern scientist who can see inside the black box, it all depends. The insect and cat well could have evolved their symbiotic behaviors. But how could the interlocking chemical cascade of human vision just fall into place? Nature, then, is an ensemble of some designed things and many more things that evolved willy-nilly, he posits.

When Paley said "design," he argued that God was the artificer. Behe insists that he can stop short of theology. He'll argue this on two precedents set in modern science. First, science already has looked for intelligence. The National Aeronautics and Space Administration has searched the heavens for extraterrestrial signals through its SETI—"search for extraterrestrial intelligence"—program. And no less a scientist than Francis Crick, who discovered the DNA helix, had so tired of "ineffective theories" for the origin of life that he proposed it might have been "deliberately transmitted to earth by intelligent beings," a process he called "directed panspermia."[47] Behe feels quite orthodox in this context. "All these weird ideas," he says. "Why are prominent scientists proposing weird ideas, if it is obvious that evolution is true?"

The second precedent in his favor, Behe says, is that scientific theories should be accepted on their merits for solving limited problems. His problem is biochemical machinery, not a philosophical proposal. Still, subjective biases creep in. Einstein disliked facts that were not aesthetic, so he fudged his data. Sir John Maddox, editor emeritus of *Nature*, once editorialized, "Down with the big bang."[48] He disliked the theory because it gave aid and comfort to theists.

If Einstein and Maddox swung too far one way, science should not err in the other direction by censuring scientific approaches just because people read philosophy into them, Behe says. "There are some scientists who simply don't want there to be design," he concludes. "If you don't want the biological world to be a result of intelligent activity, then it has to be a result of unintelligent activity and natural law." Everyone comes at the big question from some angle, but so what? That's not his concern as a biochemist. "It will be interesting to find out who the designer is, especially for those folks who already have a candidate in mind."

By the fall of 1998, *Darwin's Black Box* had run its cycle of national debate. Behe had crisscrossed the country probably fifty times to speak, and then a major speech at Virginia Tech "pretty much closed out that era with the book." In the process, he became witness to the fiery politics of the evolution-creation debate. It was not irreducibly complex, but it did seem to have no end.

7. POLITICS

Before *Origin of Species* appeared, British social classes and professions were already locked in rancorous political debate over the idea of organic evolution. Its materialism offended public order, and its novelty enticed social radicals. Darwin felt the intense heat of this social cauldron. In the view of historian Adrian Desmond, "Darwin was frightened for his respectability," so he held back publication of his ideas for twenty years.[1]

There has also been a political dimension to the evolution debate in America, revealing some of the culture's deepest divisions. The fractures run between cosmopolitan and nonurban culture, between state and federal authority, and then cross into the realm of funding science by the national purse. Many of these national dynamics of region, religion, and social class are compressed into the illustrative state of Virginia.

Since the Second World War, Virginia has seen a population boom in its northern counties. Under the expansion of federal government, the tax revenues and workers that flowed into the District of Columbia turned northernmost Virginia into a federal suburb. By the 1990s, a quarter of Virginia's population lived on 3 percent of its territory—four northern jurisdictions with Fairfax County as the high-density cornerstone. The 2000 census found continued growth, but the direction of Fairfax County was already typical in 1994: half of its residents worked for federal agencies or their suppliers, the so-called beltway bandits that ringed the capital city. The county's highly educated workforce averaged fewer than three persons per household; at $64,000, median incomes were twice the national average; and more than seven in ten women worked outside the home.[2]

The cultural picture is different downstate. Aside from urban and culturally rich Richmond, the Old Dominion's other population centers abut naval installations, which disseminate the conservative values of the military, or spread across rural and small-town Virginia, where conservatism stems from Bible-based faith and local politics and where Jerry Falwell and Pat Robertson could turn local church work into ministry empires.

Somewhere between these two extremes on Virginia's cultural map lies Charlottesville, the rural yet academically liberal city where Jefferson built his university. In 1989 the city played host to a seminal education summit of the nation's governors. Though the forum was national in scope, it also addressed the delicate balance between states' rights and federal control. Speaking to the group, President George Bush called for a decade-long project—America 2000—to boost student performance by establishing uniform national standards. The federal role would be to provide funding and consultants, and in deference to the states, Bush swore off any new laws or regulations to enforce the education reforms.

The federal government was not so retiring in 1994. At the urging of President Bill Clinton, the Democratic Congress enacted the Educate America Act, or Goals 2000. The law added federal oversight to grants for "standards-based" reform, which began with states writing new standards in history, English, math, and science. For this last subject, the National Academy of Sciences urged states to copy its *National Science Education Standards* (1996), provided free and with no copyright restrictions. Each year of the 1990s, the National Science Foundation, with offices in northern Virginia, disbursed between $3 and $4 billion to research and education, and some of it went to creating the national science standards. The NSF funding history included the $13,000 it gave to the Darwin Centennial in Chicago in 1959, used to bring one high school biology teacher from each state to attend the celebration. So it was no surprise when the NSF also helped pay for the science standards, which presented "Evolution and Equilibrium" as one of five "unifying processes and concepts in science." The document recommended an inquiry approach to learning—problem-solving projects rather than memorized facts—and it contrasted the scientific "way of knowing" with unreliable methods such as mysticism, myth, and superstition.[3]

When the millions of dollars in Goals 2000 grants began to flow from the Department of Education in 1994, however, Virginia was one of only a few states to decline. Governor George Allen, a Republican, argued that the funding would usurp states' rights and allow federal intervention in the local schools. Most of Virginia's 134 school districts (organized by county and independent city) agreed with the governor. Forty-five districts—Fairfax County among them—wanted the earmarked federal cash, a total of about $7 million each year. Virginia finally accepted the Goals 2000 funding in 1996—feasible, the governor said, because the Republican-controlled Congress had deleted some of the federal oversight provisions. A balance had been struck, and the longest sustained education reform movement in memory stayed on track in all fifty states.

. . .

Though federally funded, science does not provoke as many public controversies as, for example, how taxes are used in public schools. Big science does touch on local politics, especially in allocations for local research or industry. But in Washington it is less vulnerable to partisan squabbles, perhaps because few Americans know how the wheels turn in the federal science machine.

Roughly five thousand science professionals shape the annual U.S. science budget. They range from industry CEOs to university researchers and are augmented by another three to four hundred science or industry groups whose lobbyists prowl Washington corridors. Science issues may catch occasional head-lines, but they do not swing elections, says Jon Miller, a science educator at North-western University. "Nobody has ever been elected to Congress on a science issue, or primarily for that," he says. Beyond Washington, about 15 percent of the pub-lic—twenty-seven million Americans—is "attentive" to science and technology, he reported in the late 1990s, but no more than 8 percent could make sense of the technical and policy issues. Miller believes an attentive segment of a quarter or a third of the public would be ideal. Then, he says, "You can be sure there'd be a lot more policy debate. More democracy at work for science."[4]

Such political participation could bode well for science, as it had in the 1950s, when popular support for America's space race with the Soviet Union pried open U.S. coffers for the largest spending on science in human history. That same democratization, however, is a tool for critics of science, including creationists who resent that their taxes underwrite public instruction in evolution.

Emblematic was a 1975 protest over NSF funding of a social science curricu-lum called *Man: A Course of Study* (*MACOS*). With a grant of $4.8 million in 1970, a group of New England intellectuals produced a curriculum that studied animal behavior to decipher human conduct. The Netsilik Eskimos were also studied to show how an icebound environment can determine human ethics, such as a be-nign euthanasia for the elderly. With independent funding, the *MACOS* designers also went over the heads of school boards by providing it free, along with train-ing, to interested teachers. Parents, editorial writers, and social conservatives who would eventually staff the Reagan White House allied in protest. Though science reform had started up again in the 1990s, back in the 1970s it was dead in the water after the *MACOS* conflict, recounts sociologist Dorothy Nelkin. "This controver-sial course based on evolutionary assumptions provoked an attack on the NSF that was to end the period of science curriculum reforms."[5]

Mistrusting the NSF's judgment after *MACOS*, some members of Congress tried to legislate a new rule: the House Committee on Science and Technology would review all NSF grant decisions. The effort failed, partly because the NSF had quickly defunded the controversial social science project. Then the NSF began a process of putting each "program area," of which there are now eleven, under a separate directorate—so a controversy in biology or sociology, for ex-ample, would not affect the others. None of those programs includes the word "evolution," but the subject does show up under the Division of Environmental Biology.

The NSF's public treatment of the topic is mixed. Scientists report that syn-onyms for the e-word are encouraged in grant proposals. Still, the NSF trumpets some evolution studies in press releases and on the Internet. So it was with biolo-gist David Reznick, whose NSF grant led to a 1997 report in *Science* that West In-dies guppies, when free of predators, evolve incredibly faster than fossil records suggest. "Our work is part of a growing body of studies that clearly demonstrates that it is possible to evaluate evolution with experiments in natural populations,"

Reznick said in the press release. Short-time guppy evolution, in other words, makes a case for large-scale evolution over vast time: "It is the scarcity of experiments that is the source of some of the criticism of the theory." Biologist Michael Zimmerman, who gave an anticreationist public talk at the NSF in 1989, was unhappy with the agency's equivocation. "Unlike other talks scheduled at NSF headquarters, mine had not been advertised," said Zimmerman, known for his surveys of students and teachers. He presumed it was a measure to avoid provoking "misinformed" congressional aides who walk the NSF halls, and he noted the "fear that now pervades" the NSF. He urged Capitol Hill to run a science budget process "that insulates scientists and educators from the winds of religious doctrine."[6]

Each year in Washington, a presidential budget reflects something about the philosophy at the White House, but rarely does the sliver of money allocated for science signal anything particularly ideological. The Reagan administration, for example, boosted science spending once the 1970s recession had ended, though its shift to defense research in the first budget cycle eviscerated science education. Spending on the environment has not been a Republican priority, a point made clear when the GOP took over Congress in 1994 for the first time in forty-five years. But that was a time when all of Congress faced a deficit scare, so a mostly bipartisan budget ax fell on the National Biological Services and enforcement of the Endangered Species Act, programs favored by evolutionists. Science spending grew during the surplus-producing Clinton era. And Texas governor George Bush entered the White House vowing to spend mightily on biomedical research, a windfall for many evolutionists, and to make the more abstract basic research—such as the study of West Indies guppies—a federal priority.[7]

On the Bush watch, moreover, a debate arose in Congress on whether Washington should be telling states how to teach evolution. Already, this top-down approach was seen in the federal *National Science Education Standards* for states, to which Congress attached financial incentives. For the first Bush education bill, Senator Rick Santorum of Pennsylvania, a Republican, used a top-down strategy with funding-related legislation for an opposite reason. He introduced a two-sentence "sense of the Senate" amendment saying that public school science teachers should inform students of the controversy over evolution and urge them to think critically about its claims. With accolades from key Democrats such as Edward Kennedy and Robert Byrd, the amendment passed 91 to 8. The eight dissenters, in fact, were Republicans mainly averse to Capitol Hill trumping state control.[8] By the time Bush signed the bill in January 2002, however, the evolution language had been revised and moved to an "explanatory" appendix—raising new questions about its legal clout.

■ ■ ■

Presidents have also invested the evolution-creation debate with their own words. In 1962, the yachtsman John F. Kennedy pondered on why everyone at an Americas Cup banquet felt so committed to the ocean. "I think it is because we all came from the sea," he said. "We are going back from whence we came."[9] There had been Protestant opposition to the election of the first Catholic president, but there was no hue and cry over the evolutionist remarks.

The opposite happened in 1980, when candidate Reagan answered a question at a Dallas assembly of evangelical ministers; he said that evolution "is a scientific theory only" and that "the biblical theory of creation" had a place in public schools. The presidential contest was peaking, so President Jimmy Carter replied through White House science adviser Frank Press that he found evolution "convincing." On creationism, Carter said schools must uphold the separation of church and state. The debate continued to hit home for Carter in the 1980s, when his own Southern Baptist Convention was torn apart over Bible inerrancy. Asked about this, Carter is still quick to offer, "I don't see anything wrong with God having created the evolutionary process." In fact, he had worked on persuading Harvard evolutionist Gould that there was a God at all.[10]

As Gould tells it, meanwhile, Reagan was amenable to evolution, a claim that Reagan staff neither confirmed nor denied. When anthropologist Richard Leakey lunched with Reagan, he came away reporting that the president "had a strong and genuine interest in evolution," recalls Gould. "Reagan wasn't against evolution personally. Those creationist noises he made were for political purposes." The deeply religious Reagan left only one specific legacy for the young-earth creation science movement. During his presidency the Department of Education recognized the Transnational Association of Christian Colleges and Schools, which fundamentalists had organized in 1979 to accredit colleges that taught origins only as "the six literal days of the creation week."[11]

Democratic presidential contender Al Gore, however, excelled on the topic in his best-seller *Earth in the Balance* (1992). Mixing Bible story with the science of ecological doom, Gore explained that "human evolution itself was shaped by dramatic transitions in global climate patterns during the last 6 million years." Evolution had given humans a "very long period of childhood," with its divinely ordained dependence on parents. He warned that hubris about genetic technology could "take from God and nature the selection of genetic variety and robustness that gives our species its resilience and aligns us with the natural rhythms in the web of life." The evils of war and terrorism revealed "primal fears and passions" of the evolutionary past, but the future could be bright by adding a new divine commandment: "Thou shalt preserve biodiversity."[12]

The origins topic would explode for presidential contenders in the sultry days of the 1999 campaign trail, thanks to the Kansas State Board of Education's vote in August to drop some aspects of evolution instruction from its science standards. Democratic rival Bill Bradley sided firmly with evolution. Through spokesmen, Gore and Bush were more delicate, supporting evolution, but also creation, and favoring local school control. By the next news cycle, however, Gore backpedaled on creationism. "On the second day, the media created a new campaign issue—evolution v. creation," said the *Kansas City Star*. "And the presidential candidates said, 'This is not good.'"[13]

.   .   .

Elections come and go, but through the 1990s the U.S. government funded three gigantic projects that touched on the question of origins. Their goals were to master the human genes, map the evolutionary relationship of all living things, and discover life beyond earth.

Americans are most dazzled by space flight. But the Human Genome Project momentarily rivaled that glamour on February 15, 2001, when the secrets of human DNA were announced by an international consortium and a private firm, Celera Genomics, which had purportedly raced to get there first. The project, begun in 1990 with an estimated overall $3 billion budget, testified to the coming "century of biology," especially as the dreams of physicists were mothballed along with a multibillion-dollar atom-smasher, the Superconducting Supercollider, which Congress cut from the budget in 1993.

The genome finale came four years ahead of the original schedule, but for all the planning, science was caught by surprise: humans had fewer genes than believed. The three billion chemical codes on human DNA apparently bunched into 25,000 to 40,000 protein-producing genes, whereas the number was previously believed to be 60,000 to 100,000. In sum, humans are produced by only twice the number of genes that make a fruit fly, worm, or plant, and humans have only a few hundred genes that are not in a mouse. To illustrate how contentious this debate is likely to be for another decade or more, however, a rival laboratory at Ohio State University in mid-2001 disputed the low count, arguing from its mathematical study that 65,000 to 75,000 genes make up the human genome. Meanwhile, betting scientists have been putting their money on a range from 27,402 to 153,478 genes.[14]

Despite the ambiguity, scientists have prepared to tackle this so-called human blueprint more as a mass of probabilities than a precise map, and computers and bioinformatics are the tools of choice. "We are turning our understanding of human biology into something that is mathematical," says Glen A. Evans of the University of Texas's Southwestern Medical Center lab, which helped bring in the "post-genome era" and its promise for medical science. If the official 2001 findings are accurate, then evolutionary theory must adjust to a demise of the one-gene approach, as in one gene produces one evolutionary trait. "It appears that humans make more out of their genes than do more simple organisms," genome expert Mark Bloom of the Biological Sciences Curriculum Study said at the time of the announcement. Science must now look for "alternate expression from the same gene, gene regulation, gene interactions, and protein modifications."[15]

Single-gene comparisons had been standard practice for evolutionists who compared, for example, the relatedness of humans to chimps on the molecular tree of life. But if the genes are so much fewer in number than once expected, a new method must now be developed, especially for human evolution, said David Baltimore of the California Institute of Technology. Now, he wrote in *Nature*, "I wonder if we will learn much about the origin of speech, the elaboration of the frontal lobes and the opposable thumb, the advent of upright posture, or the sources of abstract reasoning ability from a simple genomic comparison of human and chimp." Clusters of interacting genes are apparently what evolved worms to mice and chimps to humans, he suggested. "Another half-century of work by armies of biologists may be needed before this key step of evolution is fully elucidated."[16]

The molecular tree also deals with evolutionary time, so use of the so-called molecular clock will also face new theoretical challenges. The clock—which is a mathematical calculation on computers—assumes that more genetic diversity (also called mutations) between two species means they branched from a common ancestor longer ago. What is tortuous, however, is making "assumptions about the rate at which mutations occurred," says John Maddox, former editor of *Nature*. Because scientists do not agree on rates, there is no consensus on the results produced by various investigators. Said Maddox in 1998, "The molecular clocks now being built may be little better than the timepieces in use 2,000 years ago, based on dripping of water through a fixed orifice into a measuring vessel."[17] Only three years after that assessment, the Human Genome Project forced the clocks to be even more complex in their interpretation. Undaunted evolutionists, however, still hold to these timepieces as the only way to write a true genealogy of life.

The 1990s also spawned a great project called Systematics Agenda 2000, with an equally fantastic goal: to identify all of earth's fifteen to thirty million species. "A grand inventory of the entire world is a pretty incredible prospect," says James Rodman of the NSF's biology division. "The idea is that it's not an impossible task."[18] The agenda began as a global vision after the 1992 Earth Summit in Rio de Janeiro, where presidents, premiers, and foreign ministers met, and where, amid the receding rain forests, E. O. Wilson declared a right to life for biodiversity. The project's U.S. arm works under the quasi-governmental Panel on Biodiversity and Ecosystems (PBE), part of a wider science advisory committee to the president. To accomplish the inventory, researchers aim to collect a specimen of each species and then enter its image, DNA sample, and environmental notes onto a global Internet registry. This dragnet of nature is more than "stamp collecting," for its goal is to find the next plant, fungus, insect, or bacterium that will revolutionize medicine or agriculture. This is the work of systematists, dedicated evolutionists who say their profession is the oldest, dating from the earliest time of picking up plants and deciding which could be eaten. Modern classification was invented by creationists in Europe, but after Darwin, systematics focused on the evolutionary lineage among organisms. When the NSF made its first ninety-seven grants in the early 1950s, systematists got eleven of them.

More recent appeals for funds to upgrade PBE favorites like the National Biological Information Infrastructure, a database and Internet link to world systematics, can lack political capital, however. Taxpayers have been moved to save majestic bald eagles and mountain lions, but fungi, molds, and insects don't have the same appeal. "It's the hardest thing in the world to try to sell," says ornithologist Joel Cracraft of the American Museum of Natural History.[19]

Salesmanship in American science has reached its apogee in the NASA space program, which in the 1990s revived its occasional focus on extraterrestrial life as its contribution to the federally funded search for origins. That quest shook the nation awake in the summer of 1996, when NASA said it had evidence of life on Mars. Martian rock that fell to Earth in Antarctica contained chemicals and bacteria-like shapes suggestive of living microbes. After NASA's press conference, stacked with top officials, President Clinton held his own news event to laud the search for life's beginning. The skeptical field of science was immediately divided over the evidence, but theologians called by reporters were not half as reluctant.

If most religion scholars said that God could put life wherever he pleased, the most conservative ministers questioned the scientists' motives. "What they're really after is not age-old life but a way to show that God had nothing to do with Creation," said Jerry Falwell. "They want to show there are other forms of intelligent life somewhere." Indeed, when a budget surplus finally appeared in 1998, NASA made good on its earlier enthusiasm: it formed the Astrobiology Institute, declaring that the Mars evidence, a possibility of water on one of Jupiter's moons, and discovery of new planets made it a cutting-edge field. With a $15 million budget in its second year (of a five-year funding period), NASA's Ames Research Center in Northern California—in conjunction with ten other institutes and universities—set out to probe the origins of terrestrial life, but also life in space.[20]

The Mars rock episode of 1996 had shown a new side of U.S. science. Gone were the days when scientists could be detached, because the era of automatic prestige and guaranteed funding was over. The call went out in 1994, for example, for "civic scientists" to engage the public. "The science community requires a much more public and civic persona," NSF director Neal Lane had told assembled scientists. "The public likes science, but do scientists like the public?" Lane is a physicist—a practitioner in a field whose funding heyday has passed. Yet the backing for the Human Genome Project, a new field of astrobiology, and the Bush administration's zeal for biomedical research obscure what some call the general penury of evolutionists. Historian Ruse says that even premier evolutionists like George Gaylord Simpson always had to scrape for funds: "By about 1940, Simpson as a paleontologist was finding it particularly difficult to get grant money." More than forty years later, by another account, cutting-edge geneticist George C. Williams of the State University of New York also "was having a hard time getting grant support for his research."[21]

Still, if government funding must do today what great private fortunes had done before—paying evolutionists to build natural history collections—it is certainly not a dismal record, especially at universities. They have become great centers of basic research, housing a third of it in 1950 and about half at the end of the century. Through the 1990s, universities received nearly a fifth (18 percent) of all science dollars, a third of which went into biological research on the environment, husbandry, or microbiology.[22] Quite naturally, then, academia became the home of America's foremost evolutionists as the universities in turn cheered their efforts to bring in grants.

For every grant that paleontologist Richard Bambach can win, his Virginia Tech employer takes a cut. "So if I get a $60,000 grant, I really get $40,000 to spend," he says. "The rest is overhead." Compared with the cost of supercolliders or human genome computers, Bambach adds quickly, the price of studying evolution is low. Important work can be done by reading technical papers and science journals, taking the occasional field trip, and visiting other collections. "So it doesn't cost me much money to do a lot of fairly high-visibility, large-scale theorizing on the history of life," he says. "Somebody like me, who works on the history of life, the whole thing, comes cheap." Harvard geneticist Richard Lewontin, who has been a university scientist for forty years, explains with candor how matching government money for genetic research far exceeds "every penny I spend in this lab." An ample cut of the funding also "goes into the general coffers of the university." What society gets for that investment is debatable, he says. "I wouldn't have the chutzpah to say to the public, 'You ought to do this for me because of the wonderful things I'm going to do for you.' It's a lie. We're living a life which is a good life, and to solve problems intellectually. It's a lot of fun."[23]

Not a few creationists have written enviously of the evolutionists' funding. During depositions for the Arkansas creation science trial of 1981, the state's attorney grilled Francisco Ayala on the issue of financial gain. "He was suggesting that by being an evolutionist I was getting all these big grants and making lots of money," Ayala says, "suggesting I would become a great figure because I have a $1.5 million grant. I was amused in a way." One critic of the evolution establishment puts its taxpayer endowment in the billions, and in a democratic spirit urges that Congress might want to take a look.[24]

But as in any research profession, the funding is never enough. To stay vigilant on the funding front, evolutionists have increasingly sold their work on the practical benefits. Evolution research, they emphasize, protects the planet's food supply, fights disease, and manages fragile ecosystems buffeted by modern industrial societies. A benchmark for the practical benefits to "societal needs" argument was a white paper worked on in the late 1990s by eight natural-science organizations and published as *Evolution, Science, and Society*. The document pulled no punches on where funding problems might arise: "Creationist opposition to evolution is so vocal in the United States that it has threatened federal funding of evolutionary research."[25]

The eight groups also pointed to federal largesse to medicine and defense, however, arguing that research in evolutionary biology should have the same federal overview. That will take some political will, especially in environmental policy, "since there are economic, political, and informational limits on the number of species we can save," the naturalists said. In the shadow of the "large-scale projects," such as the Human Genome initiative, they asked Washington not to forget the small and diverse projects, for evolutionary biology "has progressed due to the interplay" of subdisciplines. "We therefore affirm the preeminent value of individual-based research programs." To make the appeal still more persuasive, they asserted that naturalists come in all philosophical stripes, among them a plenitude of religious believers, as well as some atheists and many agnostics. Anyway, personal beliefs are beside the point because science "cannot provide answers to ultimate philosophical or ethical questions," the naturalists added. "Antievolutionists have charged that evolution robs society of any foundation for morality and ethics," they noted, countering firmly, "Evolutionary science has never taught any such thing."[26]

. . .

Nevertheless, the politics of evolution and creation are never far from a cultural debate on values and power. This can hurt even the creationist side because it muddies the intellectual issues, says historian Mark Noll. After exploring the evolution-creation debates of Darwin's century, Noll concluded that the issues have been captured by political agendas. "Creation-evolution debates today are often an entrance point to talk about, really, other things," he says. "Like financing of public schools. Like the role of the government. People use the language of creation-evolution. They use the language of 'scientific research' or 'faithfulness to the Bible,' but they're really not talking about these things." They are talking about social and political power. From another vantage point, David Raup would not disagree. The clue for him is the deep animus that has shown up on both sides in the debate. "I think it's largely political," he concludes. "It is that the creationist movement, if there is such, is seen as an attack on secular humanism, which it probably is. And the whole liberal establishment would come tumbling down if we gave in on that point. It's an incredible paranoia." Because Raup is an evolutionist and secular humanist, he chuckles as he says maybe the paranoia is warranted.[27]

In an electoral season, evolution versus creation rarely stands out as a single issue. The two stances, however, predictably ally with other values-laden topics, such as parental rights and environmentalism, true hot buttons for Democrats and Republicans. According to past and present exit polls, the 1996 election solidified those alignments, say political scientist James Guth and his research colleagues. It was a turning point in modern politics: a voter's "worldview" was now the strongest predictor of party choice, twice as strong as income or sex. The old alliances of immigrant-Democrat and white Protestant–Republican had given way to a new party base, now determined by secular or religious affinity. Most of the 28 percent of Americans who are "functionally secular" voted Democratic, reported Guth, making the political heirs of Roosevelt, Kennedy, and Carter "a party of secularists, religious liberals and modernists." On the other side, "an increased number of religious traditionalists have become Republican."[28]

Republicans are not of one mind on evolution, of course, not even conservative Republicans. A great divide falls between pragmatists and social conservatives, illustrated by John O. McGinnis's argument in *National Review* that "a Darwinian politics is largely conservative politics." He said that "the new biological learning holds the potential for providing stronger support for conservatism than any other new body of knowledge." Biology and conservatism both recognize self-interest, sexual differences, and "natural inequality." And the evolutionary view of humans as competing for survival justifies the state's supervisory role, which he deemed more realistic than the "pure libertarianism" in the GOP's conservative wing. Despite such theoretical detours, evolution is a Republican issue mostly because of the religious conservatives. The Democratic Party cares because of its backbone of teachers' unions and allies such as Boston public television, producer of pro-evolutionist documentaries and the *NOVA* series. The Republican connection is plain to see. It was no surprise when, in 1999, the Boston station was caught exchanging membership fund-raising lists with the Democrats.[29]

. . .

The waning years of the twentieth century provided two superlative examples of the politics of evolutionism versus creationism, one at the state level in Kansas and the other at the local level in Fairfax County, Virginia. Differing in local color, the two stories share elements that speak of America in general.

Kansas was only a bit player on the national and international stage for most of 1999. That changed in August, however, when its state board of education voted 6 to 4 to strip ancient time and large-scale evolution from proposed science standards to be used in assessment tests of students. Since 1994, the state had received about $4 million annually in Goals 2000 grants, and in 1998 the time came to review the Kansas science standards. In June the state commissioner of education summoned the Science Writing Committee to draft a new vision and list of what students should know when they graduate from science class each year.

During the 1980s, conservatives in many states began to complain about a trend in top-down public education, which included centralizing of school funding and a new pedagogy called "process" or "inquiry-based" learning. Like other conservatives, those in Kansas favored local control of schools and traditional learning. So in 1996 they successfully fielded four candidates in the state board of education race. That split the board five to five between conservatives and moderates, a division that was in place when the Science Writing Committee began its task. The committee adopted the *National Science Education Standards* as its model and for almost a year of meetings and hearings wrote four drafts of the Kansas standards. When the school board's conservatives realized that their objections to too much evolution would not be heeded, they sprang a political surprise: they offered a less evolutionist alternative, written by an ad hoc panel. The state's Science Writing Committee, hoping for a compromise that would garner a majority vote, diluted its product. Still unsatisfied, the conservatives stripped from that document all mention of an ancient universe or large-scale evolution. Small-scale evolution, which covered genetic mutation and Darwin's law of natural selection, was left intact, and the document was adopted by a 6-to-4 vote.

The drama disclosed the split between moralists and pragmatists in the Republican Party, which dominates Kansas. The governor had led moderates back to power in the state GOP in 1998, but in the 1999 board vote, social conservatives "reasserted themselves with a flourish" and gained a "huge symbolic victory," said one Kansas political reporter.[30] The conservatives said they had merely given local control to the 304 school districts. By excluding some evolution themes from the assessment tests, which can be a basis for penalizing slacker schools, each locale could consider its constituents in deciding how extensively it taught evolution. When the dust had settled, in fact, the Kansas science standards had increased the number of evolution citations from 70 in the 1995 document to 390 in the new version. That fell short of the 640 evolution references put down by the Science Writing Committee, however, so the media thundered about a rising tide of antievolution.

The Kansas story finally showed that a creationist putsch, whatever its motivation, cannot succeed in modern America. From international media attention to the GOP infighting, the outside pressure was enormous. The National Academy of Sciences and two other science groups denied Kansas the copyright for material it copied from the national standards. By eliminating ancient time and large-scale evolution from the ninety-four-page document, they said jointly, Kansas had "adopted a position that is contrary to modern science and to the basic visions and goals" of science learning. The board of education voted 7 to 3 to reword the 130 passages that had potential copyright problems, but by the fall of 2000, the conservative majority had been ousted in a board of education primary. When the new board met in 2001, its majority vote restored the Science Writing Committee's last version.[31]

The prairie storm in Kansas took on remarkable proportions, but as is shown by the Fairfax County case, most local evolution disputes stay parochial. Virginia voters had decided to end their distinction as the only state where local school boards were appointed, so in the fall of 1995 they held the first elections. In Fairfax County, thirty-five candidates vied for twelve seats with four-year terms. The contenders were "nonpartisan," but parties tacitly backed their candidates, and a cast of usual suspects jumped into the drama. One was the local American Family Association (AFA), a morality watchdog group that years earlier had formed at Capital Baptist Church, whose pastor was a graduate of Falwell's Liberty University.

With a newsletter that reached 3,600 voters, the association asked every candidate eight policy questions to go into a voters' guide. Only question number 5 had to do with evolution: "Would you support the teaching of both the creation and evolution theories of origin, as allowed by law?" Three of the remaining seven questions concerned the sex-education curriculum, which had traditionally been the most contentious education issue in Fairfax. One other asked about student-led prayer at graduation. Voters' guides had become a common staple of labor unions, environmentalists, and finally the new Christian right, but the handouts all shared one characteristic: candidates who disagreed with the group usually did not answer the questions. When the AFA voters' guide was printed as a simple tally in July, only one Democratic-backed candidate, Mark Emery, had replied—with a negative. Twelve of the conservative candidates had answered yes on question number 5.

The other usual suspects in county politics, the Democratic Party and the National Education Association chapter, were about to come forward as well. The AFA newsletter had been out for two months, but it sparked attention only after a candidates' forum at a Catholic church in mid-October. Someone, perhaps a provocateur, asked whether "biblical creationism should be taught as part of the science core curriculum." Three Republican-backed candidates seemingly agreed but later said they had recommended it only for social studies or in voluntary discussion, as the Supreme Court had allowed. "The creationism issue sort of mushroomed from that forum," says Emery, a Roman Catholic, former schoolteacher, and physicist who worked for the federal government. The bloody shirt had surfaced, and Democrats waved it. Emery built a media campaign on the topic. "It is possible," he wrote, "that all twelve new school board members could implement policy that would introduce biblical creationism as a factual part of the science curriculum." He quoted Gould as his science authority. And he compared creationism to belief in a flat earth.

"It was a defining issue," exulted Fairfax County Democratic Committee chairman Mark Sickles. "That was the alarm bell. It alarmed people, and they said, 'Who are these people?'" This was the first school board story sexy enough to make the front page at the *Washington Post*, and an anticreationism editorial followed. Sickles held a Halloween Day press conference to expose the voters' guide. He warned, "If we elect these kinds of candidates who have social issues as their number one priority, we will become a laughingstock," and then he sent the alert to ten thousand Fairfax swing voters. AFA director William Nowers, a Baptist creationist and retired navy engineer, was puzzled by how question number 5, which was recommended by a conservative Catholic, had languished in oblivion and then suddenly exploded late in the campaign. "Personally, I'm delighted at all the attention," he said. "The polls show all the time that most of the public supports alternative views to strict evolutionism."

Election Day 1995 was cold and rainy, but the turnout of four in ten registered voters was high for an off-year election. By fairly narrow margins, the Democratic-backed candidates won eight of the twelve county school board seats, and the party's majority hold on county government was not remarkable in a Democratic-run federal suburb. Republican-backed Gary L. Jones, a former U.S. Department of Education official, lost in his run for board of supervisors and suggested that question number 5 had been the political torpedo even for him.

Reelected board of supervisors chair Katherine K. Hanley analyzed the victory in rhyme: "Fairfax Countians again have chosen the mainstream, not the extreme." The *Fairfax Journal* said about the same with its headline, "School Board Vote Backs Mainstream," and its opening paragraph, "Fairfax County voters elected a School Board that . . . opposes teaching creationism in science classes and restricting sex education in schools." Witty columnist Les Fettig unmasked the sham of this "nonpartisan" election of citizen educators. "We now have a complete set of 12 political party people sitting there . . . all backed [by] and captive to the psychotic ideologies of their special interests."

Across the Potomac River, Georgetown University political scientist Clyde Wilcox watched the races closely. He suspects that in some parts of rural Virginia creationism is taught in science class. But when it comes to cultural settings like northern Virginia, Democrats can never lose by attacking Pat Robertson or alleging "stealth" politics by Christian Coalition voters' guides, though the guides have never contained a creationism question. "There's a long history in Virginia of Democrats running against Pat Robertson. It always works to a certain extent," Wilcox says, adding that in the Fairfax election, "It's interesting that the media focused most on creationism. It became the hot button."[32]

8. SCHOOLS AND TEXTBOOKS

The U.S. Constitution makes no mention of a Creator, but neither does it require that Americans be taught biology. This has left the evolution debate in public schools to balance on the First Amendment, which has two prescriptions for religion: government shall not establish it or curtail its free exercise. Some have called this the "neutrality" doctrine.

But when teachers, parents, and students become interested in the idea that, in the Supreme Court's words, "a supreme being created humankind," the neutrality of public school science can be sorely tested.[1] "The devil isn't quite so much in the details," one educator told a first-ever U.S. Commission on Civil Rights hearing on religion in schools. "The devil is in the concept: What does it mean to be neutral?"[2]

The three-part hearings, which opened in May 1998, subpoenaed fifteen witnesses to the upper reaches of the YWCA building in downtown Washington, D.C. Though witnesses testified under oath, this was no hardball probe. Since the 1980s, the legal pendulum had swung toward an emphasis on individual religious freedom, and Commissioner Robert George, a Republican appointee and professor of politics at Princeton University, had called the hearings to assess the "free exercise" rights of students and teachers under the 1984 Equal Access Act (which gave Bible clubs the same rights as chess clubs) and to inquire into the status of religion as an academic subject in public education. When creationism came up, it was treated as the strictly religious topic that two Supreme Court rulings have deemed it to be.

"The Supreme Court has made clear that religion-based views of origins certainly can be talked about," testified Elliot M. Mincberg, legal director of People for the American Way. "They can be talked about in history, in comparative religion courses. . . . But I think the courts have made clear that it isn't appropriate to teach creationism in the same way that one teaches evolution."[3] Humanities professor Warren Nord, who had cited the "devil in the concept" of neutrality, testified that teaching "about" religion as a relevant academic subject was not going well. If religion was treated as an artifact of history, said the University of North Carolina professor, the discussion on evolution and human origins was also stuck in the past. "As things are now," he said, "students will only learn the conventional scientific account [or] fundamentalist creationist accounts [that] get bootlegged in by some biology teachers."[4]

Historically, the search for neutrality on the evolution topic has been approached in three ways. In the 1920s, belief in the evolution of man from apes was seen by some legislatures as quasi-religious. They barred schools from teaching evolution just as they had outlawed Bible instruction. In 1968 the Supreme Court took a different approach. While evolution was neutral, "an entire system of respected thought" in science, it said, the ban on teaching it had the "sole reason" of defending "fundamentalist sectarian conviction." So the ban itself was a religious force that destroyed neutrality.[5] The third approach was an attempt to achieve neutrality by exposing students to a "balanced treatment" of facts about both creation and evolution. The state of Louisiana argued that it legislated the mandate for the sake of academic freedom for students, and thus it had a secular purpose. In 1987 the court held the law to be an "establishment of religion" because the state had a religious purpose in mind.[6]

Nord did not cite the 1987 ruling, but he testified that something akin to balance—a "different points of view" approach—was still an option for school curricula on the grounds of academic freedom and a liberal education. Anything, he said, would be better than the existing, hackneyed Bible-versus-science dichotomy. "What students should learn is the controversy," he said. He suggested that a biology teacher might begin the unit on evolution with a session on the naturalistic assumptions of science. "One could have some discussion of creationist alternatives, or a variety of religious alternatives to . . . neo-Darwinian accounts of evolution, in a science course," he explained. Such a tactic requires trained instructors, he said, but is worthwhile because students "should understand religious as well as scientific ways of making sense of nature."[7]

Commissioner George, agreeing with the diverse-viewpoint tack, cited Dawkins's influential work *The Blind Watchmaker*, which openly refutes religious belief in its discussion of evolution. If that was a popular "science" book, George said, science books indeed could harbor antireligious messages that would undermine classroom neutrality. "We should soberly consider whether the quest for neutrality might lead us in the other direction, to actually narrow what counts as scientific in these curricula," he said, thinking out loud.[8]

When the hearing report was delivered to the president and Congress in December 1999, its executive summary noted a "significant decrease in the number of claims of [religious] discrimination in public schools."[9] It was a season when the Department of Education issued expanded guidelines on religious freedom in schools, and President Clinton's weekly radio addresses said schools should not be "religion-free zones." Nowhere, however, was there proposed a solution for the evolution-creation debate in schools.

The public's solution has typically been "fairness," or teaching both sides. That is what a 1999 Gallup poll found: nearly seven in ten Americans favored "teaching creationism *along with* evolution in public schools." A similar public opinion came in the same year in a People for the American Way poll: while 83 percent of Americans wanted evolution taught in public schools, a majority also wanted students exposed to creationism somewhere in the curriculum. Indeed, when the poll asked why teach a creationist viewpoint at all, the majority sided with the purely secular rationale of having an "open mind" and with students getting "both points of view."[10]

When the Supreme Court in 1987 struck down the Louisiana mandate to give creationism equal time to evolution in biology class, it left the door open for quite a few alternatives. The ruling, a concurring opinion said, did not usurp "the traditionally broad discretion" of state and local officials to set curriculum. Neither did it say "a legislature could never require that scientific critiques of prevailing scientific theories be taught." What is more, said the majority, "teaching a variety of scientific theories about the origins of humankind to schoolchildren might be validly done with the clear secular intent of enhancing the effectiveness of science instruction."[11]

Despite such leeway, neither the public nor educators are any clearer on where religious origins might be taught. In the People for the American Way poll, 17 percent of respondents said "religious explanations" for origins should be taught in "another class," which the poll called philosophy. Another 29 percent allowed creationism to be mentioned as a "belief" in a science class but not as a "scientific theory."[12] None of these falls under the Supreme Court's idea of a "variety of scientific theories about the origins of humankind"—except perhaps the new proposal being called "intelligent design." Its advocates present it as a scientific theory, with a secular purpose, that considers how nature was formed by an unnamed intelligence.

This proposal came up in the third hearing by the U.S. Commission on Civil Rights, held in Seattle. Eugenie Scott testifed. "I disagree on intelligent design theory," she said. "I see it as a synonym for creation science." Having heard the possible sources of the intelligence, Commissioner Constance Horner pushed it further. "Can you define intelligent designer, or intelligent design," she asked, "without reference to a conscious deity or without reference to an alien being?" The advocate for the theory was Stephen Meyer, a philosophy professor with the Discovery Institute. "You can define it by reference to a conscious mind without stipulating identity of the same," Meyer answered. "Design theory, for obvious reasons, fits nicely in a theistic worldview, but it doesn't entail—it's not a proof of God's existence." The debate between design and evolution distracts from science learning, said Scott, who was rejoined by commission chairwoman Mary Frances Berry with the opinion that "any debate is worth it" for students. Scott held fast. "There are many valid controversies in science that students could debate; whether evolution took place is not one of them," she said. "We can debate how it took place."[13]

The next year, Meyer was coauthor of a legal guidebook for educators on how and why intelligent design—Darwinism's "chief scientific rival"—was legal in biology class. "Design theory does not fit the dictionary definition of religion, or the specific [legal] test for religion," says the guide, written with two law professors. "Design theory and scientific creationism differ in propositional content, method of inquiry, and, thus, in legal status," the lawyers argued. "Design theory is not based on a religious text or doctrine." Its exclusion from a science classroom, therefore, is a form of "viewpoint discrimination" that the Supreme Court deems unconstitutional; it injures the First Amendment guarantee of free speech.[14]

When the *National Science Education Standards* officially came out in 1996, they did not comment on constitutional issues but did comment on religion four times. While saying that class instruction may relate science to "larger ideas, other domains, and the world beyond," the standards propose no particular mode of relations between science and religion, but they do draw a rather resolute line: "Explanations on how the natural world changes based on myths, personal beliefs, religious values, mystical inspiration, superstition, or authority may be personally useful and socially relevant, but they are not scientific."[15] So, naturally, evolutionists recoil at the propositions that design theory is science and that "different points of view" should be presented to students on the first day of biology. "Why should you single out my evolution course as the place to do that?" protests Gould. "Put it in philosophy class."[16]

That is exactly what advocates of the legal "teaching about religion"—and thus about origins—would like to do in high school social studies or humanities. These advocates see two formidable obstacles, however. First, high schools usually lack humanities electives. High school students on average must complete three units of social studies (civics and history) to graduate, a tight space for even the slightest discussion of the evolution-creation debate. The average allowance for electives is twice that, or six and a half units.[17] But few high schools have electives in philosophy, comparative religion, or current events. Teachers must be trained for a specialty elective. And scarce indeed is the teacher who, taking a science-and-religion approach, can shepherd a course from Newton to Darwin, from Genesis to the *Origin*, from DNA and the big bang on to the Darwinian synthesis, creation science, spiritual ecology, and intelligent design.

"The topic is usually handled in high school history with the Scopes trial, a stereotype," says Charles Haynes, a scholar with the First Amendment Center and a pioneer in the "teaching about" approach to religion in schools. In social studies, he says, religious views of nature would invariably be presented as inferior to scientific views. On this, Haynes is on the same wavelength as Warren Nord, the humanities professor with whom he was coauthor of *Taking Religion Seriously Across the Curriculum*. "Why should science and social science get to define the ground rules by which religious claims are to be interpreted and assessed?" asks Nord, noting that this is how it works in history, economics, literature, psychology, sexuality, and morality. Haynes concludes that schools lack the political will to present students a "neutral" comparison of the "two kinds" of knowledge. "The vast majority of science teachers think that science is winning," he says, "so why give any ground?"[18]

. . .

In public education, blue-ribbon panels on how to teach evolution have been plentiful. The challenge has been to move from findings to actions. In 1986 biologist William Mayer, a former university professor and curriculum designer, wrote, "The recommendations of the 1890s are remarkably parallel to those of the 1980s," so what is needed is an "acceptable mechanism for inducing change."[19]

In the twentieth century, evolutionists have seen at least two successful pushes to teach the topic in public schools—the Biological Sciences Curriculum Study, which had its heyday in the 1960s, and the *National Science Education Standards*, issued in 1996. Biologist and educator Arnold B. Grobman said the BSCS project worked because it trained teachers, produced textbooks, and got the books into half the nation's classrooms. "The appearance of evolution in biology classrooms did not come about as a result of numerous recommendations by prestigious national and regional committees," he said, "but rather, because of an effective program of implementation" made possible by large-scale federal funding.[20]

In the 1990s as well, funding played no small part in the states' implementation of the *National Science Education Standards*, which present evolution as one of five "unifying concepts and processes in science." By 2000 that theme was echoed to varying degrees in the science standards of every state but Iowa (which didn't have standards). The new state standards, which are the basis of curriculum, teacher training, and assessment tests to gauge how schools are doing, managed to "cover evolution in a generally comprehensive manner." That was the assessment given by Texas Tech University professor Gerald Skoog to the 2000 meeting of the National Science Teachers Association. More than thirty of the states tied the science standards to assessment tests. "What is tested," Skoog told the assembly, "tends to be taught."

Skoog evaluated how much the various concepts of evolution showed up in the state standards. The bare-minimum concept of biological diversity was found in the standards of 46 states, for example, and it went on: species change (44); natural selection (43); evidence of evolution (33); speciation (19); descent from a common ancestor or descent with modification (17); pace and direction of evolution (10). The last concept was human evolution, adopted by the standards of only four states—Connecticut, Indiana, Michigan, and Utah. Not a big number, Skoog surmised, calling it "obvious evidence that policy makers and educators continue to be intimidated by anti-evolutionists." Upon these state trends, the Bible Belt had stamped a traditional pattern. While South Carolina adopted evolutionist standards without an expected fight, and Arkansas put them down more firmly than expected, a "very brief and restricted" allusion to evolution characterized the documents in Louisiana, Mississippi, Georgia, Florida, and Alabama. "Optimism is warranted," Skoog told some of the seventeen thousand teachers and professionals who flocked to Orlando, even though standards "do not alone possess the leverage to force major changes."[21]

Efforts to teach evolution can be stymied by the teachers in America's sixteen thousand diverse school districts. In the Ohio public schools of 1987, 90 percent of biology course teachers surveyed had included a "substantial evolutionary component," but nearly one in five also viewed creationism as scientific enough to include. At about the same time in Texas, all biology teachers seemed to touch on evolution, but more than half in Biology I spent only five classes or fewer on the topic, while nearly a third (28 percent) said they also "teach creationism"—a third being roughly the number arrived at by all studies of biology teachers who want equal time for creationism. A decade later in Louisiana, just 15 percent of biology teachers dedicated seven and a half days or more to evolution. Biology instructors seemed less reluctant to teach evolution than in the 1960s or 1970s, says Skoog, but many still lacked a minor or major degree in biology. "It may be presumptuous to assume," he says, that "evolution will soon be emphasized in the manner envisioned by the *National Science Education Standards*."[22]

Teachers of science, meanwhile, have also asked for sympathy. They must face students from all kinds of families, pupils with "a host of misconceptions concerning biological evolution," reports one teacher. "Many are secretively or openly hostile toward the topic." Researchers in one study of high schoolers concluded that "highly religious students" with a belief in Creation "are less likely to give it up during instruction." The commonsense hypothesis, they added, is that "acquiring a new belief is easier when you do not have to give up a prior belief." They concluded that religious students might need a class debate "to explore the alternatives," thus provoking their critical thinking and instructing them in scientific inquiry.[23]

To calm both parents and educators on this process of teaching around or through student beliefs, the National Academy of Sciences offered some helpful distinctions. "Students are not under a compulsion to accept evolution," it says in *Teaching Evolution and the Nature of Science*, a 1998 publication. Still, the academy says, "If a child does not understand the basic ideas of evolution, a grade could and should reflect that lack of understanding, because it is quite possible to comprehend things that are not believed." Finally, all national and state programs, guidelines, and texts are at the mercy of the private classroom. Scott states the mystery succinctly: "When that classroom door is shut, nobody knows for sure what goes on—unless Johnny comes home and says, 'Guess what I learned in school today, Mom?'"[24]

. . .

What Johnny learns about evolution usually is found in his school's biology and earth science textbooks. They come in many versions but have had enough in common to add some unity of thought to the nation's vast archipelago of schools.

Before 1900, natural science was taught in three parts: zoology, botany, and geology. Textbooks by geologist James Dwight Dana, zoologist Louis Agassiz, and botanist Asa Gray "dominated the limited high-school life-science textbook market during the first post-Darwinian generation," says historian Edward Larson. After public education became compulsory in the early 1900s, the three natural sciences were collapsed into one biology course, and the second generation of post-Darwinian biologists took over the writing of textbooks. They presented Darwin's theory with great certainty, though field scientists still were fiercely debating its particulars. "Public high schools were teaching evolution decades before the anti-evolution crusade, with the presentation seeming to grow more dogmatically Darwinian over time," says Larson, who has leafed through all the early texts.[25]

The most widely used life science textbook in that era was *A Civic Biology*, the book from which Dayton high school teacher John Scopes purportedly taught human evolution. His prosecution in 1925 gave publishers their first lesson in the politics of state textbook markets. A year later, a *New Civic Biology* came out with the word "evolution" expunged. Charles Darwin's visage, which had graced the frontispiece of the previous text, had disappeared, and in its place had been printed an image of the digestive tract. The most popular text in the 1930s, *Dynamic Biology*, mentioned evolution, but only at the back of the book, where students read, "Darwin's theory, like that of Lamarck, is no longer generally accepted."[26]

So, between 1900 and 1930, the presentation of Darwinism in biology textbooks ranged from dogmatic certainty to near omission. With every new decade, moreover, textbook treatments might fluctuate, as seen in the long-surviving text *Modern Biology*. First published in 1921, *Modern Biology* treated evolution and human origins. In 1926, after the Scopes trial, the publishers deleted the human evolution reference but kept up a gradual increase in evolution coverage until 1950. In the charged atmosphere of the Cold War era, when the public easily associated communist atheism with evolution, *Modern Biology* diluted its coverage of Darwinism. About a decade later, when the federally backed BSCS biology texts exploded on the scene, *Modern Biology* followed suit with more emphasis on evolution. The 1965 edition revived the statement that "many anthropologists" believed humans shared a common ancestor with primates. By the late 1970s, the publishers had again rolled back evolution coverage in the face of school board fights and creationist activism. A compromise position in the 1990s reached a sort of plateau: full coverage of evolution, but with disclaimers. "Scientific data," the 1990 edition proclaimed, "have been used to present this material as theory rather than fact." It also said that "every effort has been made to present this material in a nondogmatic manner."

In all of this, the BSCS's tenth-grade *Biological Science: Molecules to Men*, which came out in 1963, set the commercial standard for a decade. Because of it,

"Competing textbooks also began to give unprecedented emphasis to evolution."[27] *Biological Science* came under attack from creationists, resulting in its deletion in 1969 from the list of science texts approved by the Texas Board of Education, a major textbook buyer.

Textbook sales are no small business in the United States. Elementary and secondary school textbooks generated $3.3 billion in 1998. This symbiosis between the crass marketplace and elevated truths on the textbook page has long been noted, most pointedly by Hermann J. Muller in his speech to the Central Association of Science and Math Teachers in 1958. He called the soft-pedaling of evolution to sell textbooks in various states a "vicious cycle." The publishers had chased the almighty dollar, he said, and science teachers needed to band together to resist. "Can we let the profit system destroy us?"[28] Muller cited the South in particular, but at the end of the century, the textbook market was driven primarily by the three largest states, California, Texas, and Florida, which consumed 22 percent of the textbooks sold each year. The two largest merit a closer look to assess Muller's scenario of evolution versus capitalism.

The Texas Education Act of 1974 bore one of the first fruits of creationist activism there. For any textbook that discusses evolution to be adopted, the act said, it "shall identify it as only one theory of the origins of mankind." For a decade, publishers seeking to place their textbooks in the Texas market added a statement to that effect, what is now called the "theory, not fact" disclaimer. Ten years later, the Texas attorney general declared that the clause was an unconstitutional establishment of religion, and so it was rescinded. Textbook publishers got a stronger signal than ever in 1991. For the first time in thirty years, the Texas Board of Education said it is "required" that biology textbooks adopted by the state's text selection committee include evolution. Later in the 1990s, Texas joined the national science standards movement, firming up the status of evolution despite persistent creationist opposition.[29]

In California, politically appointed education officials in the 1960s also were responsive to creationist complaints. In 1970, the state's science framework allowed creationism into the classroom, stating that "creation in scientific terms is not a religious or philosophical belief."[30] This remarkable concession to creationism, which took center stage in a textbook adoption war in 1972, in the end never prodded a secular publisher to mix evolution and creation in a textbook—though creationists had put out a book they hoped California would adopt. *Biology: A Search for Order in Complexity*, first published in 1971, was re-released in an improved second edition in 1974. In that year, however, the California Board of Education dropped its recognition of creation in the science framework, replacing it with an antidogmatism rule. It dictated that in textbooks and class instruction, "dogmatism be changed to conditional statements" in cases "where speculation is offered as explanation for origins."[31] The policy shaped textbooks that were vying for the California market. Before the antidogmatism rule, one textbook had read, "As reptiles evolved from fishlike ancestors, they developed a thicker scaly surface." After the rule, the text read, "*If* reptiles evolved from fishlike ancestors, *as proposed in the theory of evolution*, they must have developed a thick scaly surface."[32]

The California science framework of 1978 asserted more boldly that "all living organisms on earth have a common ancestor from which they have diverged by evolution during about 3 billion years." This provoked a creationist complaint in 1979 and then a lawsuit in 1981, but all that resulted was a judge's order to recirculate the antidogmatism clause. The several years of "recognition" and leniency toward creationism in California, one study found, produced high school graduates who were far more exposed to both creation *and* evolution than were their peers in Texas and Connecticut.[33] The latest California science framework, adopted in 1990, is among the strongest nationally on evolution, saying it "is an accepted scientific explanation and therefore no more controversial in scientific circles than the theories of gravitation and electron flow." That was one reason that 2000 reviews of state science standards ranked California, with a few other states, at the top—a point not missed by market-wise textbook publishers.[34]

. . .

The Commonwealth of Virginia is one of twenty-two states that adopt textbooks at the state level, and its process mirrors the basic ways of all school systems. With state selection, however, there is a carrot-and-stick power: school districts are denied some funding, or free textbooks, if they select books not on the state list.

Textbook review in Virginia comes every four years, and when it did for science in 1995, the board of education in Richmond summoned four committees on natural sciences. They perused the market offerings, judging textbooks by content accuracy, topics, bias, assessment tools, and laboratory helps. In biology, eleven textbooks qualified for state financing, and that list was sent to the districts. They were chosen as the best for implementing state-level guidelines, as well as guidelines in districts such as the Fairfax County Public Schools. "The textbook is a tool to help the kids learn our program of studies," Fairfax's science coordinator Jack Greene said in 1996, a year after the list was drawn. Biology is the most common science course and is so "oriented to hands-on" instruction that students spend half of all class time in the laboratory. To get just the right textbook, Greene said, "There's heated debate." It is mostly about utility, not evolution.

To pick its books, Fairfax County assembled its own sixty-four science specialists. The team on biology spent three months poring over the eleven approved texts, heard pitches from publishers, and then in January selected four. One was for advanced classes, and the fourth augmented the two primary texts. They were displayed for public scrutiny in a regional library and the district area offices, and then in April the county school board took an up-or-down vote. The four texts were approved with no protests. "Evolution is one of the strands, and it will be taught as the 'theory of evolution,'" said Greene. The BSCS biology textbook was not adopted in the 1995 cycle, but its latest edition was in the library at the county's selective science magnet school, the Thomas Jefferson High School for Science and Technology.

In all the evolution-creation wars of the 1970s and 1980s, Virginia faced neither a "balanced treatment" bill nor a motion for "only a theory" disclaimers or anti-dogmatism rules. Virginia was a conservative state, but the demographic mix did not lend to creationist initiatives, let alone victories.

The pattern of such crusades was clear, however: a few lone activists instigated and stirred all the action. In Fairfax, seventy-two-year-old William Nowers, a tall, courtly former U.S. Navy engineer with an abiding interest in the origins debate, took on the role of David against the Goliath of anticreationist forces. He had played a catalytic role in the 1995 county school board races when he had published the Christian newsletter with its question number 5, and though there were other creationists around, advocacy had fallen on his shoulders alone. "If I hadn't pushed it myself, it wouldn't have gotten as far as it did," he says. "It's not a popular cause."

Nowers pushed again in 1997. The parents of a ninth grader at Jefferson High School complained that the "blue version" BSCS textbook, *Biology: A Molecular Approach*, derided creationists. Most biology textbooks have an early section on the nature of science. Some mention the Galileo affair. Others say tribal myths differ starkly from scientific thinking. The BSCS "blue version" featured an entry on page 15 that contrasted science and pseudoscience and put creationism, astrology, and miracle cures in the latter category. It said creationism could be studied as a religious belief but not as science.

The parents were riled not only by the pseudoscience label but also by the linking of creationism with astrology. "We don't send our children to public school to have their faith ridiculed," said their complaint. They charged that the book violated school board policy number 1460: that the county "shall not engage in any activity that either disparages or advocates religion." They took their protest first to their local school board member, and he discussed it with Jefferson's principal. They quietly asked Nowers for advice: Could a disclaimer satisfy the complaint? Nowers came up with wording to neutralize page 15, but when the entire school board caught wind of it, a formal process was demanded. Nowers became the advocate for the parents.

The Fairfax County School Board could have ordered the biology textbook off the shelf, as it had done with *Jump Ship to Freedom*, a book that offended blacks, and *Sex, Lies, and the Truth*, a film that offended conservative parents. Instead it went to Jefferson's Challenged Materials Committee. (Such a committee had been set up in every school in the county in the wake of the firestorm over sex education in the previous decade.) The committee rejected Nowers's plea to use a disclaimer or dispose of the books, and so the issue moved up to the county's eleven-member review committee, made up of principals, parents, and students. Nowers was by now contesting more than the offensive entry. "The textbook, in presenting evolution as fact, teaches only one side of a controversial area while denying all evidence in opposition," he testified. "This does not meet the intent of the [Virginia] Standards of Learning." He said the Fairfax standards on evolution were worded similarly to the Humanist Manifesto II, a document that espoused evolutionism as an atheistic religion. By this stage in the complaint, the proposed disclaimer had been winnowed down to only sixty-three words: the county school system "disclaims and regrets the disparagement and ridicule of persons having faith in the Creator of the Bible."

The Jefferson High School principal voiced legal worries to the review committee. "It's not clear that we have the right to affix labels to copyrighted material," he warned. He was followed by a stoic biology professor from Jefferson High School who read a booster statement for the science program and addressed the issue only in the last sentence. The school's six biology teachers recommended that "no disclaimer be affixed to the inside cover or any other part of the textbook."

The review committee agreed, and when the parents appealed its decision, the issue truly arrived at a political level, the elected Fairfax County School Board. When the question came before the board, ten of the twelve members were present, and though they usually sat anywhere around the table, this time they sat on opposing sides. "If you actually look at the sentence, it does not say that religious belief in creationism is a pseudoscience," said chairwoman Kristen Amundson. Board member Mychele Brickner had charged in writing that "equating a belief in a Creator to astrology is undermining and disparaging religious beliefs." Board member Carter S. Thomas, who first heard the complaint from the parents, said at the table that the textbook entry was a "gratuitous shot by the publisher." The board voted 6 to 4 to take no action.

The parents ended their fight, worried that their son would be stigmatized. A month later, three taxpayer mothers of students in county schools other than Jefferson sued the board in circuit court for an "arbitrary, capricious" vote and "an abuse of discretion." Policy number 1460 was invoked again. The mothers asked the court to mandate the disclaimer and provide "further relief as this court deems appropriate." The court issued a "final order" in October 1999; because there had been no action on the lawsuit, it was deemed dormant and would be expunged from the docket. At the time of the court order, the "blue version" biology textbooks, seventh edition, were on the library shelf at Thomas Jefferson High School.[35]

. . .

In his home state of Nebraska, biology teacher Gerald Skoog had reached his high school classroom each day by driving past the home and statue of William Jennings Bryan. The great commoner, in his antievolution crusades of the 1920s, sought neutrality in public school instruction on evolution. One thing Bryan accomplished by promoting antievolution laws in a number of states was to weaken the treatment of the subject in the textbooks of the century.

To Skoog, a Lutheran layman and Sunday School teacher who taught Darwin's theory in biology class, Bryan's legacy was like a dark shadow cast by the statue. Yet it was Bryan in bronze and Muller in print—the "One Hundred Years Is Enough" speech—that had finally cinched for Skoog the topic of his doctoral thesis: How did American biology texts treat evolution from 1900 to 1968? "So I completed my dissertation and moved to Texas in 1969," joining the College of Education at Texas Tech University. "In the 1970s, I published my first article on the subject."[36] He once was president of the National Science Teachers Association but perennially is a public voice on biology textbook wars. "What the public tends to argue is, 'Out of a sense of fairness creationism should be included in science and in textbooks,'" he says. "Well, I want to tell them fairness is not a good criterion."

Writing in 1984, he said biology textbooks had suffered a bad decade. "Anti-evolutionist activity and economic pressures of the marketplace are the main forces that have caused the decrease in evolution's coverage." To avoid the market pressures, textbook writers invented synonyms for evolution. The texts became filled with such terms as "change with time," "variation through time," "organic variation," "evidence of change," "theories of change," and "species formation." By 1990, the term "evolution" was back in style, revived on state and publisher resolve, and allowed to resurrect where creationist protests had waned. Word came of a "new generation" of biology textbooks that treated evolution in "an unabashed and uncompromising way."[37]

Creationists had always complained of censorship, Skoog says, but he contests it wasn't so. To censure is to suppress materials published and in circulation, but his study of biology texts from 1900 to 1968 found that they lacked mention of creationism because it had no explanatory power. In all, he found three references to either special creation or catastrophism, but they were brief and derogatory, and the few mentions of creation science came only after the 1981 court ruling as a note on current events.

One licensed biology instructor in Alabama is close to Skoog's heart, but only in having shared with him the tedium of leafing through the pages of innumerable biology textbooks, looking for the telling sentence, word, claim, or anecdote. Norris Anderson had twice served on the Alabama State Textbook Committee by the time, in 1995, he was the chief author of the state's famed disclaimer about evolution, "a controversial theory," with advice to students and parents that "any statement about life's origin should be considered as theory, not fact," for no one was around to see when and how life appeared on earth. Anderson, formerly an "ardent evolutionist," did not mind the profession's heckling, but he took it on the chin when the editorial page of the *Mobile Press Register* said, "Must we now portray Alabama as a state that mocks science, confuses children, insults its teachers and trivializes religious faith?"[38]

The editorial typified a "highly emotional" reaction by someone unfamiliar with the data, Anderson wrote in a lengthy response paper.[39] The plumb line of the Alabama textbook committee was the Alabama Course of Study: Science, which had called for "an inquiry-based" approach, "understanding of fundamental assumptions" in science, and "critical thinking, openness to new ideas, skepticism." The biology textbooks available to the state did not meet these criteria on evolutionary biology, Anderson says. So the disclaimer was designed to fill those gaps. In sum, the textbooks failed to distinguish between genetic diversity (microevolution) and the origin of major phyla (macroevolution). Only one text mentioned the Cambrian explosion and its Darwin-shaking fossil record. All had failed to discuss, he says, gaps in the fossil record and rarity of new major species in recent earth history. Equally absent from even the best biology book was any puzzling over DNA, its origin, and its achievement of holding the "instructions" for biological life itself.

"Study hard and keep an open mind," the disclaimer's last lines say. "Someday you may contribute to the theories of how living things appeared on earth." To his critics, Anderson says, "I am unaware of any attempt to use the insert to bring creationism into the classroom." He says its purpose is to chase indoctrination out of the classroom and "promote full disclosure of both the strengths and weaknesses" of scientific theories.[40] Anderson had earned his bachelor's degree in science education at the University of Michigan and his master's in natural science from Oklahoma State University before going to California to teach high school and college biology. In a contest for writing science texts, his talent was noticed, and he landed a job as a textbook and film writer for BSCS in Colorado, the front line of evolutionary textbooks. He finally settled in Alabama, founding an educational project called Cornerstone, a Christian ministry.

Anderson—whose beliefs adhere to an ancient earth, and somewhere in its history God's special acts of creation—agrees with Skoog that the biology textbooks of the 1990s did become far more evolutionist than in the past. But for Anderson, it came with a wrinkle: the texts of that decade seemed driven by reaction to creationism. "Textbook writers feel so besieged and under attack, and that may be true, but the problem is that they reacted in a dogmatic way." He points to trade textbook statements such as "One thing is certain—birds evolved from ancient reptiles," or "Theories are the solid ground of science, that of which scientists are most certain," and "One of the great wonders of our existence and life itself is that it has all arisen through a combination of evolutionary processes and chance events."[41]

The textbooks have increased their reliance on authority, not the persuasiveness of honest assumptions and the inexorable building up of facts, he says; they read like catechisms and legal briefs, not scientific literature. "These materials will say, 'Scientists know this to be true.' The authorities know this, and therefore we should accept it." No fan of creationism in the science classroom, Anderson is proud of the earliest BSCS biology textbooks for their "intellectual honesty." He can hark back to a BSCS text from 1968, which says, "You might believe without any shadow of a doubt that all frogs are made up of cells, but this still does not make your belief a fact." From other texts he admires: "The next unit of this book is organized around the assumptions of the heterotroph hypothesis," and after a different text's unit on the origins of the cosmos, the author asks, "Was there a Big Bang? No one knows, but you now know some of the evidence."[42]

Anderson says he quit the textbook business because of its slide into an imperious tone. The question he is asked today by religious parents and by educators is: Which textbook has the most honest treatment of evolution? "Frankly," he says, "I don't have a real good one."

The creationists have tried to fill this market gap. The closest they came was *Biology: A Search for Order in Complexity*, written by members of the young-earth Creation Research Society. It was first published in 1970 by Zondervan, an evangelical house, and went into a second edition four years later. The volume had the impressive, glossy look of any fat textbook, and its pages mostly bore the facts of conventional biology. But in key areas the Creator was evoked. "Flowers and roots do not have a mind to have purpose of their own," says the first edition. "[T]herefore, this planning must have been done for them by the Creator." Where the evolutionist would claim common descent by the evidence of common features in species, *Biology* credited "an all-wise Creator employing a single efficient pattern in His creation."[43]

Textbook panels in Alabama, Georgia, Oklahoma, and Oregon had adopted *Biology*, but it was the Indiana adoption that made legal history. The ACLU swiftly sued, and in 1977 the state supreme court said purchase and endorsement of the textbook was an establishment of religion. Marked with legal defeat, the textbook faded in all states. The Institute for Creation Research also issued a "public schools edition" of its paperback *Scientific Creationism*, amended to remove a last chapter titled "Creation According to Scripture." This book, available since 1974, was "written primarily with teachers in mind."[44]

Teachers also were targeted by another well-known text, *Of Pandas and People* (1989), which the ACLU also loved to hate. Described as a supplementary text for general biology or biology teachers, *Pandas* focuses on "the central question of biological origins." It is the work of intelligent design theorists who are not interested in the age of the earth but do question the idea of common descent of all species. Whenever dead ends seem to appear in biochemical evolution, comparisons of DNA in molecular biology, or yawning gaps in the fossil record, the text proposes the alternative theory that "intelligence" might have been at work.

Pandas was pummeled by evolutionist journals and the leaders of science teachers' associations, yet it is still selling to teachers for course preparation, and by the time it was in its tenth year, school districts in at least eighteen states had made bulk orders. The ACLU spotted the title in the library of a Ohio school district in 1999 and was ready to swoop. The book's most theological term is "intelligent design," but the ACLU winked: everyone knows that is code language for God. So vulturelike was this watch on *Pandas*, however, that before the year was out even the *New York Times* chided this great defender of free speech for bordering on censorship: "Despite the book's measured tone, its mere presence in a school so outrages anticreationists that some of the most vocal advocates of free speech are trying to have it banned."[45]

Then came the book *Icons of Evolution*, written by Jonathan Wells, a Berkeley-educated biologist and intelligent design fellow at the Discovery Institute. Using standard textbook review methods, Wells focused on how the seven most popular examples of evolution were treated in ten big-selling textbooks. His review gave an A to evolution examples that showed "full disclosure of truth [and] discussion of relevant scientific controversies." An F was given when the textbook entry "dogmatically treats a theory as unquestionable fact, or blatantly misrepresents published scientific evidence." The highest cumulative grade for any of the ten textbooks was D+, which earned Wells the opprobrium of the evolutionist authors. One textbook writer, however—Bruce Alberts, president of the National Academy of Sciences—told the *New York Times* he would drop what Wells had called faked images of embryos, by Ernst Haeckel, from the next edition of his *Molecular Biology of the Cell*. The reporter conveyed Alberts's view that "Haeckel's drawings were 'overinterpreted,' or highly idealized, rather than outright fakes."[46]

If the 1998 hearings of the U.S. Commission on Civil Rights probed religion in schools, the *National Science Education Standards* had its own galvanizing effect in school science, most excitedly in Kansas in 1999.

In reaction to the strong evolution emphasis in the standards, the Kansas State Board of Education downgraded evolution by not making all its concepts part of the assessment tests. Students, the critics said, would be ill prepared for college, modern life, and America's economic race with the world. "Who will explain to our children why A's in Kansas science classes do not lead to success on national tests?" queried biology professor Helen Alexander of the University of Kansas. "Why would businesses locate in Kansas if education is not a priority? Wake up, Kansans. This is your children's future." It was a winning concern for some partisans, but social critics like Yale Law School's Stephen Carter wondered if one really had to master evolution to succeed. "I am not sure this proposition is true—plenty of people somehow manage to have rich and productive lives with little understanding of science."[47]

But that understanding factor is what people like Wayne Carley, executive director of the National Association of Biology Teachers, fight for. He, too, worried that Kansas high school graduates might stumble in "the scientific and technological world around them." Their young intellects would miss rigorous training, for "teaching about evolution develops in students powerful skills of analysis and evaluation."[48] Clearly, high school is only the prelude of the evolution-creation debate. It is in higher education where the maturing minds of many young Americans will arrive at a personal, free, and reasoned conclusion.

9. HIGHER EDUCATION

The Darwin Centennial of 1959 had echoed across the United States in museums, parlors, and newsrooms, but its hub was a university. The University of Chicago celebration, held over a wintry Thanksgiving week, amounted to five days of "scientific discussions, pageantry, ritual, and theatrical spectacle." Having drawn 250 delegates from 189 institutions and 14 countries, and having packed in 2,500 other registrants, the event put the midwestern university "on the intellectual map of the world."[1]

A century after *On the Origin of Species*, universities had clearly become the main repository of Darwin's legacy in America and key purveyors of the current consensus in evolutionary thought. Higher education could no longer imagine an era like the 1920s, when some college presidents soft-pedaled evolution to avoid fights with a state legislature or a powerful trustee. Only once in its history did the American Association of University Professors (AAUP) have to stand vigilant for the academic freedom of science instructors. Formed in 1922, its Committee on Freedom of Teaching Science fielded reports "that presidents and professors indicate widespread anxiety" about the era's antievolution crusade. But it was a campus battle never fought, and the committee disbanded in 1929.[2]

Though university scientists today have bemoaned the new cultural criticism of science, there are no strictures on teaching. The greatest complaint from natural scientists might be that, still, some students reject the theory of evolution. One campus study found that six in ten university freshmen and sophomores knew about and accepted evolution. Among upperclassmen, the proportion rose to eight in ten—an increase but not a clean sweep. The students most exposed to the science of evolution are the million or so who, in average years in the 1990s, major in science of all kinds. A tenth of this number go on for advanced degrees. If there is a vanguard for evolution being trained, it may be the roughly one hundred thousand students each year who receive degrees in natural science.[3]

Back in 1959, when the New Synthesis in evolutionary theory was at a high point, the University of Chicago was in the forefront of training students to advance Darwin's work.[4] One of them was Dean Kenyon, a junior and a physics major when the Darwin Centennial hit the campus. He not only felt grateful to be at the very center of the Darwin celebration, but it swayed him to pursue biology. "I went to many of the meetings there, saw many of the great figures in evolutionary biology," he recalls decades later. He heard a keynote address by Julian Huxley and a talk by Sir Charles Galton Darwin, a grandson of the great naturalist. "I became very excited about devoting my life to further research on the evolution theory."

After graduation, Kenyon headed for Stanford University to study for a doctorate in biophysics. Then he crossed the San Francisco Bay for postdoctoral work at the University of California at Berkeley, where he toiled in the laboratory of Nobel laureate Melvin Calvin, the man who had explained photosynthesis. By the mid-1960s, Kenyon was teaching biology at San Francisco State University, where for thirteen years he taught the main evolution course for science majors. His coauthorship of a provocative book, *Biochemical Predestination* (1969), gained him rank in the field, followed by tenure in 1970 and full professorship only four years later. He taught mostly in the newer Hensill Hall, which abutted the older Science Building, where the basement hallway was enlivened by a red-stripe time line of the earth's evolution.

Kenyon taught in the era when television audiences across America watched the 1980 *Cosmos* series. In one episode, Carl Sagan stood by a large, spherical Pyrex flask that was coated inside with a mudlike substance. By Sagan's account, this muck bore amino acids, the very building blocks of proteins and life. He used the prop to reenact a famous experiment conducted by Stanley Miller, a graduate student at the University of Chicago, in 1952. Miller had filled a sphere with gases presumed to have covered the primitive earth—methane, ammonia, and hydrogen—and then, as they swirled about in the vapor of boiling water, he pierced them with electricity, as if ancient lightning. The gases precipitated organic compounds, including simple amino acids. Thus, said Sagan, did life originate.[5]

Kenyon had investigated just such questions on the origin of life in Calvin's lab, but he ended up dissenting from Sagan. "In fact, what he was pointing to was what I would say is the strongest evidence against the case of chemical evolution. That material is nonbiological, so it ties up all the carbon in a huge macromolecular network of carbon atoms that exists in no living organisms, and it sort of traps the material. I don't know how you can argue that that's the basis for further development. It stops the thing in its tracks."

Kenyon was no stranger to a university world in which a Sagan-like outlook—scientific naturalism—allowed little room for things religious. As *Biochemical Predestination* was published, Kenyon spent the 1969–70 academic year on a fellowship at the Graduate Theological Union in Berkeley, where he reviewed the contemporary literature on the relationship of science and religion. As an Episcopalian, he was not inclined to see any conflict between God and Darwinism. Yet this was for him a season of intellectual doubt. In the mid-1970s, a student gave him a book that challenged the idea of purely chemical origins of life: *The Creation of Life: A Cybernetic Approach to Evolution*, by European creationist A. E. Wilder Smith. Kenyon made time during the summer for what he thought would be a handy refutation of the work. "I found out, in fact, I could not answer the arguments," he says. Thus began a period of "serious personal rearrangement of thought and anguish of soul" that took him up to the 1980 fall term. He was a tenured professor, and he had to make a decision.

"Just go public with my doubts? Take my chances?" He asked himself that question, then proceeded to do just that, perhaps naive about the consequences that would follow when a few students complained about his comments in class. The story would make the *San Francisco Examiner.* "Well," Kenyon says, "I had no idea of the fallout." He was summoned to three faculty hearings to testify on what he taught in his courses. Department chairman William Wu responded by laying down the "5 percent doctrine": no more than 5 percent of a course could include criticism of or doubts about Darwinian theory, and that was how Kenyon proceeded through the 1980s.[6]

Kenyon showed his doubts about evolution in other ways. During the December 1981 creation science trial in Arkansas, he booked a hotel room in Little Rock, ready to be a witness for the creationist defense. The trial tide turned quickly under an unfavorable judge, so Kenyon packed and left, an episode reported in the journal *Science.* Back at San Francisco State, he was gradually stripped of his courses. The new department chairman was Crellin Pauling, son of the Nobel laureate and a staunch evolutionist. Nearly thirty years down his career path, Kenyon found himself, as the 1980s closed, largely restricted to teaching Biology 100, or human biology, to 160 nonscience majors, as well as his senior seminar in biomedical ethics. "Going by the 5 percent doctrine," he says, "I could justify raising counterarguments in that course, devoting about a lecture and a half out of thirty lectures."

In the meantime, he had been putting his name on public documents that were hardly friendly to evolution. An expert's affidavit that he wrote on the problems with evolutionary theory was used in an appeal before Louisiana courts after the state's "balanced treatment" law was struck down in 1981. He also wrote the foreword to a 1984 book on the origins of life that introduced the earliest "intelligent design" concepts, mainly by showing the mathematical and environmental improbability of life's having arisen by random forces of chemicals on the primordial earth. Then, in 1989, the intelligent design high school textbook *Of Pandas and People* came out under his coauthorship. As was the custom at San Francisco State, his publishing success—anti-Darwinian as it was—was posted in the biology department's faculty lounge.

Finally, in 1992, sparks flew in Kenyon's Biology 100 class, where four students complained about his citing a Gallup poll that showed how creationists outnumbered naturalistic evolutionists in the United States. He also mentioned alternatives to orthodox evolution theory. When the department chairman and dean of the School of Science told Kenyon to stop, he asked for specific guidance on what it was about his teaching content that did not present evolution properly, and what issues were prohibited. For his part, Kenyon believed he was fulfilling his professorial obligation and teaching the essentials. "I never would come straight out and say, 'Darwin was dead wrong,'" he says. "I remained impressed. Darwin was one of the greatest naturalists that ever lived, and a lot of the material in *Origin of Species* is wonderful biology, accurate. In that sense I'm not an anti-Darwin person."[7]

Whatever his attitude, Kenyon's superiors charged him with injecting religion—the idea of intelligence in nature's design—into a science course. "I order you not to discuss creationism in your class," said the chairman. The dean wrote, "Advocacy of these views has no place in the science classroom." Kenyon was removed from Biology 100 for the 1993 spring term under the chairman's ruling that he at times "presents an essentially religious view" in science class, which "challenges the academic freedom of students who challenge his view."[8]

This time Kenyon was in a mood to fight. He lodged a complaint with the university's Academic Freedom Committee, which ruled in mid-1993 that his "academic freedom was abridged" and his removal lacked due process. The committee said he had the "right to present even objectionable viewpoints" on his subject, but it paused over creationism case law—even citing the five criteria for science from the Arkansas trial—and wondered whether presentations of intelligent design created a church-state problem at a publicly funded university. Yet in the balance, the committee concluded, the censure of Kenyon came "perilously close" to an "official suppression of ideas." Neither the department chair nor the dean would reinstate the dissenting biology professor, so he appealed to the Faculty Senate. The AAUP now became Kenyon's defender, just as it had prepared to defend biology professors in the days of antievolution laws. The AAUP backed the Academic Freedom Committee's report with the "hope and expectation" that Kenyon would be reinstated. The Faculty Senate, voting 25 to 8 for Kenyon, revealed a remarkable split: the eight science professors united in opposition; the rest of the faculty representatives took Kenyon's side. James Kelley, the science dean and an oceanographer, said Kenyon's attempted ouster was still justified after "years of student complaints."[9]

For critics of Darwinism, Kenyon became a cause célebrè. His case was a factor behind the Discovery Institute's establishment of the Center for the Renewal of Science and Culture, at which Kenyon became a research fellow.[10] When the 2001 spring term ended, Kenyon retired, leaving behind his long, windowed, and somewhat barren office on the fourth floor of Hensill Hall. Twenty years after the first dispute at San Francisco State, he had wrapped up his career with a teaching load of graduate seminars. His wife, who was reared a Catholic, returned to her tradition in 1988, and he converted to Catholicism. Though leaving university work, Kenyon continued his science as a Discovery fellow researching the statistical and linguistic "texture" of coding and noncoding DNA.

When I talked with Kenyon at his offices, he looked back on the academic freedom clash with some amusement. Science had no pride of place in other university departments, and that may have been his salvation in 1993. He tried to stand in the shoes of the twenty-five faculty members who voted for his reinstatement. When they saw him in the dock, he wondered, did they think, "This crazy creationist guy over there, he's an educational entrepreneur like we are. And we don't like the science area anyway. And maybe he's going to break out of that, and he'll become one of us. So let's vote for him." Kenyon hoped that was not their sole logic, but at avant-garde San Francisco State it would not be a surprise.

. . .

Such philosophical conflicts may be rare in science departments, but they can be plentiful in schools of law or political science or humanities departments. Due process can protect the innocent, and institutions can always bend a little, as with the "5 percent doctrine." They also can try to intimidate an untenured—or even tenured—faculty member to quit.

A different remedy is to create different schools for different outlooks and let matters such as Darwinism, feminism, or a legal philosophy ride on that. That is what Virginia Tech, a state school, and Liberty University, a private school not sixty miles away, have done in regard to the evolution debate in higher education. Because evolutionary biology is part of established science education, Virginia Tech is like all secular schools—not mandated to instruct, even for knowledge's sake, about creationism or other theories inimical to textbook evolution. Yet such topics invariably come up, it seems, and so there is an evolution and creation element at each of the institutions. The schools' modes of dealing with them differ in the following ways.

Virginia Tech requires life science majors to take some kind of introductory biology course, which is where evolution comes up. Biology majors must take the in-depth "Evolutionary Biology." But as life science students move on to special fields, they may never study evolution again. If a student wants to hear criticism of the evolution theory, it may arise in class—and, as was seen in the Kenyon case, it is very often a question of teacher-student relations. But the usual source for a secular university student will be the antievolution speakers who, relishing campus free speech, frequent the nation's universities. No secular school, meanwhile, is inclined to give a credited course on antievolution, though critical courses may indeed uphold academic standards: take Lehigh University, where Michael Behe taught a freshman seminar, "Arguments on Evolution."

Liberty University, in contrast, is constituted to teach creationism. It also wants to produce biology graduates with bona fide credentials, so early on it was forced to take the Bible and theology out of the science curriculum. Now students learn creationism in a required humanities course, "History of Life." The creationist convictions are strong. A Virginia Tech student whose father teaches at Liberty would not accept the "Evolutionary Biology" course requirement; this student switched to a physics major instead.

The stories of the two schools and some of their teachers illustrate how the evolution-creation marketplace of ideas can work at a university.

. . .

Virginia Tech is a vast campus of blockish cement buildings that grew up around the ivy-draped Victorian architecture of the old central commons. The biology department enrolled about fourteen hundred majors in the mid-1990s. By the time some of them arrived in the classroom of botany professor Duncan Porter for giant doses of ecology and biogeography, they had digested the basics of evolution in introductory biology or the required "Evolutionary Biology" course and moved on to prepare for careers in fisheries and wildlife management. Many more biology majors go into the medical fields. As they specialize, evolution increasingly is bypassed, except for those needing the analytical skill. "Some of them may go into comparative physiology and need to know something about relationships," says Porter.

Few, if any, are going into the evolutionary field in which Porter has worked. "That's starting to be a big problem because taxonomists, and I was trained to be a very classic taxonomist, are not being trained much these days," he says. The research funds and careers have all detoured into molecular biology, and so even botany majors must keep up with that trend. "I tell my students, 'If you don't know these [molecular biology] methods, you're going to have a hard time getting a job.'"

The biology students also should be aware, if it is important to them, that in America there are perennial arguments over evolutionary theory. "You don't want them to think there are no problems," Porter says. "They should know about the creationist approach to explaining these things." He wouldn't go this route with high school students, who are not yet critical thinkers, but higher learning is a new game. "When you get college students, you assume you're getting the best students coming out of high school, and you can teach them to start thinking."[11]

Down a long corridor from the botany laboratories in the Life Sciences Building, associate professor of zoology David West finishes discussing a test with an undergraduate student. In his genetics class, students acquire a framework on which to hang knowledge they will gather in life. "I look upon my students ultimately as being informed citizens," says West, a tall man with curly brown hair. "Others say, 'Oh, we must train them. Make them all geneticists.' That's nonsense. Or even all scientists!" The world does not need more geneticists, he says after thirty-four years teaching the subject. What it needs is a citizenry that one day can deal with genetic screeners who will attempt to predict each person's genetically driven fate in health or even temperament.

West is optimistic that the future will resolve, one by one, the mysteries of genes and the history of life. One mystery for which he has no use, however, is religion, and he wonders how Christians in evolutionary science keep it all together. "They must be compartmentalizing their minds like crazy," he says. "It's what I tell my Unitarian friends. They are walking on the brink, but I always add, 'and lack the courage to step off.'"

West earned his doctorate in bird taxonomy, but his interests became wide-ranging, extending to ecological genetics. He is fond of butterflies for these studies. They adapt to environments in well-understood ways, and this is what his students learn when they repair to a nature site on a nearby mountainside. In whatever venue, West exudes confidence that science will crack the mysteries, even though much about genetics, for the time being, is under a shroud.

"We don't understand very much about the genetic foundations of development," West says. "We don't understand how you get from one major body plan to another. So you have people creating ad hoc hypotheses to explain how things have changed, and saying genetics won't tell us. I would suggest that genetics *will* tell us if we find out more about it." Meanwhile, he holds that the genetic similarities in organisms are the best evidence of a common evolutionary ancestor, better evidence even than fossils. "The only demonstrable connection is a genetic one. That's what provides the thread," he says. "I am very candid with students that we really don't have the fossil record to make those connections and to make strong statements about it. But there's no question we have new sources of information." There's nothing wrong with speculation, but he believes protracted arguments about evolution are for inside the field, not his introductory courses.

West is no fan of "big science," with its bureaucracies and giant budgets, and perhaps because of this aversion to big and powerful science, he is not looking to overthrow or discover a grand principle. "I march to a different drum," he says. "I get interested in a lot of different things, looking for connections. But that's not the way one gets promotions." As he predicted to me that day in 1996, he retired without having been made a full professor.[12]

Elsewhere in the Life Sciences Building is the laboratory office of paleontologist Richard Bambach, one Virginia Tech professor who is searching for the grand principle. He has become a player in the field of big-time evolutionary thinking, and that was evident again in 1999. That summer, the Kansas State Board of Education had voted to limit evolution in its science standards, and so the Geological Society of America's subsequent meeting became the venue for a Paleontological Society "short course" on the evolution-creation wars. Bambach was commissioned to present "Teaching Evolution Convincingly and with Clarity," as he seeks to do in his own classroom.[13]

His career has dovetailed with a period of great ferment in evolution debate among specialists, not to mention the gadfly of creationism. But that doesn't exhaust similar levels of future thrill for his students, even if science is mostly routine. "I don't think they have a less exciting future," he says. "The times of great excitement are self-made, when people make breakthroughs." He is proud of what some of his own graduates have done: become editor of the journal *Paleobiology*, get appointed to the faculty at the University of California at Los Angeles. A few have bagged important apprenticeships with senior evolutionists. "We get students who are good enough to turn into players," he says of Virginia Tech. "We don't get the people who are naturally going to be leaders."[14] That, of course, falls to a Harvard, Berkeley, or University of Chicago.

Even a university is no final rampart against the forces that stir confusion on evolution, says Bambach, a burly man who favors a close haircut, short-sleeved shirts, and bolo ties. He cites "The Mysterious Origins of Man," aired on NBC one Sunday evening in 1996. Hosted by Charlton Heston, the program suggested that humans lived with dinosaurs and even cited claims that human and dinosaur footprints had been found side by side in the bed of the Paluxy River in Texas— an assertion that even creationists have abandoned.

"My graduate students just thought it was horrible," says Bambach, who sat at a long laboratory table gesturing. Undergraduates in oceanography, however, were full of curiosity. "We took two whole class periods to discuss this," he recalls. The weird science of TV had gotten his students' attention, so he used that as a hook for explaining the geologic column and its layers dating back billions of years. From that he moved on to explain how science works. "There is this fable that we are suppressing things," he says, adding that it doesn't work that way in science. Scientists want fame, and an authentic discovery of human and dinosaur fossils cheek by jowl in the same historic rock would certainly be the ticket to achieving it.

Reared in the District of Columbia, Bambach as a youth wandered through the Smithsonian's National Museum of Natural History, and a curator labeling a meteorite gave him some broken-off chips, as if oracles of his future. "And I've still got them." He returned as a student volunteer in 1952 and then, four years later, as an intern in geology. For his doctorate, he worked on four-hundred-million-year-old Silurian clam fossils in Nova Scotia.

During his training, the great debate in paleontology focused on the amount of diversity in early life and its rate of expansion on earth. To Dean Kenyon, in his "5 percent" sessions, this question posed a severe problem for Darwinism, which postulates a few simple forms at the beginning, and then widening diversity. The Cambrian period, Kenyon argues, actually saw the rapid appearance of many of the body plans that still dominate today.

The paleontologists under whom Bambach learned looked at the same fossil record and defined the problem differently. One solution was the "equilibrium" theory of David Raup: the Cambrian produced a fundamental range of life-forms, which then evolved through countless variations. More fossils are found toward the present, Raup argued, because older ones were worn away. Paleontologist James Valentine had a rival theory: basic life-forms continued to arise after the Cambrian's biological explosion. As to fossils, says Bambach, "Jim read the record more literally. We don't find many things early, and now we find lots. And this might indicate that we had low diversity which increased later."

The two views finally were merged and explained in a consensus paper that Bambach wrote with Raup, Valentine, and paleontologist J. John Sepkoski.[15] This was Bambach's entrance into the very-large-scale aspects of evolution. Today he calls himself a "paleoecologist," someone who follows a fairly new approach in evolution that tries to discover the natural surroundings of the past: "How did organisms adapt to the conditions of an environment?"

In 1996, he applied the environmental question to the great Permian extinction, which 250 million years ago witnessed the death of 90 percent of life on the planet, life that moved mostly in the oceans. Bambach and his two colleagues proposed that the ocean floor had become layered, capturing carbon dioxide. When the ocean floor experienced an earthquake or shift, "too-rich water came up to the surface," and its carbonation destroyed primitive life. "We've put together a very complex scenario," he says. "It's a mechanism to actually do the killing." *Discover* magazine called it one of the top science stories of the year.[16]

Though not looking for a fight with the creationists, Bambach experienced his first such dispute in 1975, when a student group invited Henry Morris back to Virginia Tech. Bambach agreed to be Morris's debate rival, and afterward he decided once is enough. The thrill was gone. "Once you've been around the bases, it doesn't change." He also would give a deposition in the 1981 Arkansas trial, pointing out that members of the young-earth Creation Research Society had to sign a doctrinal statement. This sat wrong with science, Bambach believed, but also with his own more liberal sensibilities as an elder in the Presbyterian Church (USA). Indeed, he was chairman of his synod's campus ministry. "I'm not an evangelistic person, nor am I a traditionally religious person," he says. He rarely uses the word "God" and may be "almost agnostic," but he believes in an ineffable reality that is good, the source of nonmaterial feelings such as beauty. "Whatever the supernatural is, I think it is unknowable, just as I don't think trilobites can figure out what people might think."

What he *can* try to figure out is material nature. It is impossible for science to proceed, he argued, on the creationist premise that God used divine intervention in the past differently from the natural laws in effect now. He intersects with the creationists on only one point: the dogmatic peril of defining science. Once *Science* magazine sent him a clutch of books on the nature of science to review, and Bambach came out as an agnostic. "I don't think you can define science," he says. "Do you want to say something is science or is not science? Or do you want to say something is or is not *good* science? Those are different kinds of things." His solution is to lift from science's shoulders the burden of success. "People can be wrong ultimately but have done good science for the time," he says.

In late 1982, *Science* sent Bambach another slew of books to review, all but one of them by evolutionary scientists debunking creationism. The exception was *Christianity and the Age of the Earth*, by geologist Davis A. Young. The author was an evangelical scholar who, as it turned out, would be a presenter along with Bambach in the 1999 geology forum. In the 1970s, Young had defended the idea of a young earth. In his 1982 book, however, he beseeched believers to accept geological antiquity. He assured them that though the earth could be unimaginably old, "the doctrine of the evolution of man is unscriptural and should be opposed." Early in the five-book review for *Science*, Bambach says belief in science is based on evidence, while belief in religion has the virtue of "faith," a belief held "without substantiating evidence." To conclude his review, Bambach suggested that despite Young's laudatory embrace of an old earth, on evolution he spiraled back into contradiction: "For judging the age of the earth Young requires evidence for his beliefs, for judging evolution he does not."[17]

. . .

Over the mountains from Virginia Tech is the city of Lynchburg, where Liberty University enrolls six thousand resident undergraduate students. As at most colleges, all students must take at least one science course to graduate, and most opt for biology. They also must take the "History of Life" course, a humanities offering, which requires them to read Henry Morris's *Scientific Creationism*.

Much of this teaching load is overseen by Paul Sattler, the chairman of the Department of Biology and Chemistry. He arrived at Liberty in 1985, the year the school gained university status. Like Bambach, Sattler believes he is doing real science; unlike the Virginia Tech geologist, Sattler believes God could have used divine intervention in the past but now sustains natural laws scientists can count on in the present.

Sattler's case in point is the subject of his own research, for which he has received a federal grant. He is studying the population density and life cycle of the salamander *Plethedon hubrichti* in the nearby George Washington and Thomas Jefferson National Forests. "The National Forest Service would like to timber, and they need to know if that will impact the salamanders or not," says Sattler, taking a break in his small office one day. "They can't do anything that's going to endanger the population."[18]

The scientific question is how the creature adapts or springs back from changes in the environment, a process of adaptation that Sattler called microevolution. By training he is a reptiles and amphibians man, having investigated a scientific enigma in the snake and spadefoot toad while earning his master's degree in zoology at Miami University in Ohio and then concentrating on electrophoresis—a technique to study molecules at the genetic level—to get his doctorate at Texas Tech University.

Like every other professor at Liberty, Sattler works under a one-year contract—a policy that, if in force at San Francisco State, would have put biologist Dean Kenyon out on the street. The no-tenure policy assures that faculty members reinforce an evangelical faith in students. Since about 1990, many of the students have come from Southern Baptist families unhappy with what they saw as a secular slide in campus life at the fifty-two colleges and universities—enrolling about 160,000 students in 1990—tied to Southern Baptist associations.[19]

Liberty's largest degree programs are in nursing and education, and though only 1 percent of the students seek a life science degree, nearly everyone takes introductory biology. "We see virtually all of the students," says Sattler. The students use industry-standard college texts, learn about evolution, and are tested on what he calls "the straight evolutionary line." He agrees with Virginia Tech botanist Porter that few of the fields in which students will spend their careers demand knowledge of what evolution, as a historical treatment of biology, says about the structural similarity of organisms. "So in most of our classes evolution doesn't come up," says the diminutive, well-spoken biologist, who sports a thin mustache. "You can study biology without any kind of presupposition of origins. The physiology and the genetic mechanisms are going to be the same regardless. Usually, it comes up only when you are talking about origins on a broad philosophical basis, as in phylogeny"—the study of life-forms over time.

In 1982, the Liberty University biology department underwent its first state accreditation review, a three-hurdle process that began with a visit by a Virginia State Board of Education committee, which after testimony by faculty voted 8 to 1 for preliminary certification. The school wanted its biology graduates credentialed to teach in Virginia public schools, or in thirty-five other states that recognized the imprimatur. Then the chancellor, Jerry Falwell, preached a feisty Sunday sermon on his *Old Time Gospel Hour*, saying the Baptist college wanted "to see hundreds of our graduates go out into the classrooms teaching creationism."[20]

The American Civil Liberties Union protested the state approval, and it stalled for the next two years. Finally, in July 1984, the state board granted accreditation, which required Liberty to divorce the creationism instruction from the biology courses. There was also a concern that the school not be judged to be "pervasively religious," for that would disqualify it from applying for public bonds or receiving state or federal student loans. If the state board was now satisfied, some conservative Christians were angry. "Some well-meaning Christians have accused us of compromising our beliefs for the sake of academic credibility," Falwell wrote in the October 1984 issue of *Fundamentalist Journal*. After acknowledging the "tension" between creationist belief and state biology standards, he stated the new policy: "We do not use the Bible as a chemistry textbook, nor do we use a biology textbook to teach Genesis."[21] The school also announced that the required "History of Life" course, now in the humanities department, would be augmented by a Center for Creation Studies. The center opened the next year as a small museum with offices.

The tensions of that period were gone by 1999, when Falwell declared on CNN's *Crossfire* that Liberty's graduates "go on to be scientists and all the rest." His own daughter, he noted, became a surgeon. "We teach evolution and creation in the classroom, and as far as I know, we've never had an evolutionist graduate from Liberty after having both sets of facts presented to them." In this television theater, Falwell sat opposite Harvard's famous Stephen Jay Gould. Clearly, the preacher had never fixated on the man called "Mr. Evolution," for he called Gould "Jay."

"I am Steve. You called me Jay."

"I'm sorry. OK, I thought it was Jay Gould."

"Oh, he was a rich robber baron."[22]

In Liberty's "History of Life" course, students who have never heard of Gould make a small acquaintance with the Harvard evolutionist. They learn three viewpoints: special creation, gradualistic evolution, and punctuated equilibrium, Gould's argument that history is a story of long nonchange punctuated by sudden changes. "We'll throw theistic evolution in with evolution because the scientific predictions are exactly the same," Sattler says. "Their only difference is on a theistic, philosophical level, not a scientific level."

In his own college days, Sattler did not have any philosophical issues with evolution, for he began his study of biology as a Roman Catholic. When he became a born-again Protestant, it was different: he struggled to reconcile biblical claims with science as he studied for his master's degree. He had his first encounter with the literature of creation science after going on to Texas Tech. "I saw for the first time, 'Hey, the Christian position is not necessarily incompatible with the physical evidence. It's all in how you interpret the physical evidence."

Thus reconciled, Sattler also realized that his scientific role was not to withdraw into a religiously guided institution. From running the gauntlet of Ph.D. studies at a secular research university, he knew that only when creationism had a fruitful research program would university biologists deign a nod of recognition. The long-term problem Sattler has chosen to investigate is the rate at which species can multiply. As a believer, he asked how the creatures of Noah's Ark reached the continents after the Genesis flood. If there were only basic kinds, how did they diversify into the panoply of birds, mammals, and reptiles that face the modern biologist?

The question of the speed of diversification was one Sattler could test independent of his religious beliefs, and biologists already had demonstrated an interest in several cases. One involved six Texas armadillos. Early in the 1900s, their owner had taken them to Florida, where they promptly escaped. "They were put under a basket overnight, and they dug out," Sattler explained. The clan had baby armadillos, and their descendants gradually moved up the Florida peninsula until one branch met its Texas kin in the mid-1970s. A biologist interested in the rates at which new species emerge would ask, Sattler says, "How different are these populations?"

That same question applies to the marine toad *Bufa marinas* in Australia. It was brought there early in the twentieth century, and while its shape has hardly changed, the offspring have spread out, become isolated, and where molecular-level change occurred some became "reproductively isolated"—and thus a new reproductive species. Sattler's interest is in how rapidly this happened, once the marine toad hit Australian shores. "How rapidly can organisms speciate if you've got these wide-open niches?" In a few words, "Do you need millions and billions of years to get the diversity?" Such a long duration matches conventional evolutionary theory. But a rapid speciation fits a recent creation, after which biological diversity quickly covers the earth.

The underlying belief in a six-day recent creation and a global flood that is the starting point for Sattler's research would scandalize evolutionists in the academies. He insists it is a nonproblem as long as he, the scientist, rigorously tests and challenges the hypothesis. Indeed, he admits he could be wrong about the age of the universe. "It's not a matter that affects your salvation, so I consider it an academic issue. I want to know what's true. That's what science is all about. What is the way that it really happened? How does speciation occur now, which is the only thing that we can look at."

On the day I sat in on the "History of Life" class, it was taught by Sattler's colleague in biology, Charles Detwiler, who came to Liberty University in 1991. Detwiler earned his doctorate at Cornell University, studying the biochemical genetics of the fruit fly, so he teaches the "History of Life" section on genetics. "Dr. Sattler and I are not of a mind on every single issue," Detwiler said. "He gives the lecture on the age of the earth because he feels more a sense of closure on that issue than I do."

Though Genesis is holy writ for Detwiler, he is not "emotionally wedded" to any one of the plenitude of creationist models. "Every time I get real intimate with a model," he said, "it starts to disappoint me." His sympathies fall into a category that for generations, from 1830 to 1960, was a dominant view in orthodox, and now evangelical, Protestantism: an ancient earth but a more recent creation of life or, at the least, of man. "Personally, I am happy with a very old earth," said Detwiler. "But I do not like pushing the events of Genesis 1"—the appearance each biblical morning and evening of oceans, plants, animals, and man and woman—"back into that period. The first couple verses of Genesis 1 may refer to a very ancient cosmology, billions of years old. Once we start getting into days that have evenings and mornings, I prefer something more recent."[23]

Detwiler, with his dark, wavy hair and mustache, is as casual and cool as any prof who is teaching freshmen. The bright-faced group before him numbers about thirty, and the lecture opens with Psalm 94, beamed onto the wall by an overhead projector. "Does he who implanted the ear not hear? Does he who formed the eye not see, does he who disciplined nations not punish?" The eye and ear confirmed for the psalmist God's omnipotence, and Detwiler wonders whether Darwin might have thought likewise until he was convinced of evolution. Speaking of Darwin, Detwiler said, "Some nights [he was] awake, wondering what he's done with God."

Because evolution is a "fuzzy" word, Detwiler continues, it is best to begin with a textbook definition: "changes in the genes and the frequency of those genes in a single population through time." Some freshmen might be surprised to hear him say that creationists agree with evolutionists on this kind of gene-frequency "evolution" and on "plasticity"—or changeability—within a species. Therefore, he tells the class, "When we criticize 'evolution' we mean from hydrogen to man. From gas to consciousness—all the way."

Such "big-time" macroevolution clashes with Scripture. Images of the *drosophila* fruit fly are projected onto the wall as an aid to instruction, followed by images of eyes, from simple to complex. "Scripture implies that God is not expecting kinds of creatures to evolutionarily produce other kinds of creatures," Detwiler said. "The implication is that God produces the kinds, and the responsibility of the individuals in each kind is to perpetuate their own kind. In other words, invention is a function of God. Perpetuation, according to some kind, is a function of the organism."

Detwiler goes on: evolutionary biology does not believe in "kinds," but its efforts to push a fruit fly beyond its body plan have failed. Attempts to breed a fly with a higher number of thoracic bristles, or hairlike projections, have worked within limits. But the resulting flies produce fewer offspring, and those they do produce are weak. Darwin, he said, showed that pigeon breeders can produce new variety in the birds but cannot turn them into something else. Organisms seem to have a "cyberwall"—cyber in this instance meaning system—beyond which genetic mixing cannot push their basic forms and functions. "Individual characteristics of organisms do not change significantly because they were created into a perfectly integrated developmental system by God," Detwiler said. "You can't change them without changing the whole system. And that many changes don't happen all at once for anybody but God."

Such is the case with the eye. The illustration projected onto the wall purports to trace the evolution of the light-sensitive spot of a protozoan cell of a mollusk into the camera-like eye of a human. Detwiler's pointer dances on the illustration as he explains why he doubts that a "haphazard sequence of random mutations" could produce such dramatic change. "How do we save up rare, good mutations that move us in this direction while we are waiting for other good ones to arrive?" he asks. "We have to postulate a whole series of selectively positive mutational states in between this one [he points to the mollusk eye] and this one [the human eye]."

The class ends, and Detwiler begins to gather up his materials, confident that he has done two things: taught the students' minds and cared for their souls, not the normal offering of college biology. The university's mission makes it different, he said. He crosses his arms and explains the distinction between faith and academic pursuit. "When I meet my Maker, and he says, 'Look, the earth is only seven thousand years old,' I will end up saying, 'I thought maybe that was so.'" Then again, if his Maker points to the big bang and says it took billions of years to form earth, "I'll say, 'That sounds good to me.'"

. . .

The approach taken at Liberty University will not be found at the nation's secular colleges and universities, which might be said to teach a secular or naturalistic philosophy alongside the sciences. Liberty is one of sixteen hundred independent schools of higher learning in the United States, half of them church-related. The schools with church affiliations are mainly Protestant, some liberal and some conservative in leaning, and their raison d'être is to form Christian characters and intellects to unloose on the world.

The students usually come out as diverse in viewpoint as the schools. One study of evangelical seminary students—many of whom were studying on the campuses of Christian liberal arts colleges—found a line of demarcation on the question of evolution. Slightly more than half the students said the Bible, which they called "infallible," was the word of God but was "not to be taken literally in its statements concerning science." The remainder called the Bible "inerrant." They took it "literally, word for word," and that included words about the Creation in Genesis.[24]

Strict creationists have for years complained that hardly a Christian college in America takes a hard line on evolution, and they are basically correct. During Liberty University's fight for an accredited biology department, school president A. Pierre Guillermin fairly exaggerated when he said, "We are not the only private college that holds to the doctrine of Creation. Mormons, Catholics and other religious groups do as well." Then he asked, "Has anyone questioned the biology education program of Brigham Young or Notre Dame?"[25] The answer is negative, of course, because such courses are taught in as secular a mode as at Virginia Tech. That makes the truly interesting question at church-related schools this: What are the variety of ways students finally reconcile God, Scripture, and nature, and who helps them in this intellectual process?

If the average church-related college or university has no grievance with evolution, some of the most conservative ones don't fight that battle, either. The founder of Regent University in Virginia Beach, Virginia, is religious broadcaster Pat Robertson, who famously backed a legal defense fund for the creation science trials of 1981. He supports the Bible and creation over evolution, yet Regent has no hard-line stance on any particular kind of creationism. "We put evangelism over all these debates on geology," says Vincent Synan, dean of the theological school. "We have a high view of Scriptures, but we don't think it's a book of science. It contains inerrant truth in regard to salvation."[26] The university, made up of graduate schools of theology, government, law, and communications, is far more interested in addressing social policy issues such as federalism, the role of local government, and parental rights—which were Robertson's doorway into the evolutionism-creationism battle. His faith has Pentecostal roots, and that spirit-filled tradition never was bothered by an ancient earth.

Few American theologies are more complex than that of the Church of Jesus Christ of Latter-day Saints, but its flagship Brigham Young University teaches off-the-shelf, industry-standard evolution. That has been the case since 1931, when the church officially said: "Leave biology, archaeology, and anthropology, no one of which has to do with the salvation of the souls of mankind, to scientific research." Still, Mormons have breathed in the general creationism in the American cultural atmosphere. In 1935 only four in ten Brigham Young students said they rejected the theory of human evolution "from lower life forms," but by 1973 the proportion had risen to eight in ten students. What the church requires is only belief "that Adam was the first man of what we would call the human race," says Gordon Hinckley, the church's living prophet. Scientists can speculate on the rest, he says, recalling his own study of anthropology and geology: "Studied all about it. Didn't worry me then. Doesn't worry me now."[27]

American Catholic colleges and universities have worried less than ever about evolution since 1950, when it was recognized by Pope Pius XII's encyclical *Humani Generis*. But while none of the Catholic students who choose these 230 schools, which include 14 "research universities," arrive as Bible literalists, they too have breathed in the general American creationism. For twenty years, science professor Philip Sloan watched students arrive at the University of Notre Dame with the belief that "God works in natural ways" and science will show how. Such an expectation, says Sloan, "leaves them unprepared for a more serious engagement with the deeper challenges." The deeper challenge, he says, is to reconcile Christian faith to the scientific assertion that nowhere in nature is there proof of divine intervention or a special status for man.

To assuage student despair, Sloan teaches that the Christian faith does not stand or fall on an "empirical order to nature," but on three doctrines of Creation. First is that God is outside time and space; in God's eyes eons of evolution are simply an "eternal now." Second is that God created from nothing, or ex nihilo, so in principle, nature never can "exist independently of his continuous creative act"—even though it appears that way, especially to science. The third doctrine elaborates, saying that all empirical laws found by science, be they chaotic or orderly, are found within the limits of nature, and contingent nature is presided over by a perfect God. "I would claim that Darwin's theory does not directly bear on any of these [three] points," says Sloan. In these doctrines, he says, students can find both a scientific haven and, when needed, a weapon against scientific atheism. While it would be "more convenient for theists" if science found a divine order of nature with man at the pinnacle, he says, "I would argue that it is not essential that it do so."[28]

Science professors such as Sloan who teach at church-related schools presumably have fitted God to evolution. Not a few professionals in the natural and physical sciences have done likewise—about four in ten say God "guided" evolution. Far more rare is the natural scientist who, like Detwiler or Sattler, holds a doctorate and also believes in a very recent and direct creation of man. Yet this group, scattered across universities, colleges, and institutions, may amount to three to four thousand individuals, according to one survey of scientists. Four thousand doctorate-holding scientists would be a veritable Roman legion if organized. But where are they? the creationists ask. They seem to be underground. When Dean Kenyon faced his academic freedom case, a few science professors in similar predicaments sent notes. None, however, taught at a major research university.

Kenyon argues that lone dissent is a recipe for trouble. "It's suicide, unless you're able to get a whole lot of people to go with you on this." He suggests that may never happen.[29]

The creationist Duane Gish, who earned a doctorate in biochemistry at the University of California, assisted two Nobel laureates, and worked in top commercial labs, cautions students who follow his footsteps. "What we advise students who want to get a Ph.D. in natural science at a major university is, 'Don't tell 'em you're a creationist,'" he says. "'Just keep your mouth shut. Do your job and get your Ph.D.'" Still, two well-known creationists, Kurt Wise and Paul Nelson, earned doctorates, respectively, at Harvard and the University of Chicago.

What advice, then, would Gish give a young science professor without tenure? "Don't let 'em know you are a creationist, or you won't get tenure. And if you have tenure, you won't get a promotion." This is bald-faced discrimination, says Gish. He posits that the sociological factor of the reward and punishment system of academe is the only thing stopping the evolutionist dam from breaking. "Tomorrow, if you could eliminate every trace [of pro-evolutionist discrimination], thousands of people who doubt Darwinism would come crawling out of the woodwork. Maybe they wouldn't embrace creation, but they'd say, 'Hey look, we've always had serious doubts about evolution.'" Even optimistic antievolutionists say that day will be long in coming to higher education.[30]

College graduates—who make up no more than a third of Americans—once they leave school, join everyone else in the world of informal continuing education by exposure to written material, television, museums, and organized religion. When it comes to evolution and creation, the most obvious temples of learning are the natural history museums and the houses of worship. They are like two great movie theaters, each showing a different creation story, and what is projected there has much to do with American thinking on God and nature.

10. MUSEUMS AND SANCTUARIES

Churches and natural history museums—they are both theaters for a story about life. In the United States, the first outnumber the second by six hundred to one. Yet the five hundred museums tip the scale back when the story is nature itself. Besides sermons on the Creator or hymns on divine majesty in mountains, oceans, and sky, American worship says almost nothing about natural history.

Here is one more example of why the public may see science and religion as entirely separate spheres. What the church curate and the museum curator have in common, however, is a concern about overall public literacy. It is bad news to science museums when four in ten Americans believe humans lived with dinosaurs, and fewer than two in ten understand the terms "molecule" and "DNA." Matters are no better in sanctuaries when only four in ten Americans can cite half of the Ten Commandments, and the Genesis Creation account is picked as the hardest Bible story to understand. "Many of the college educated and religiously involved know little of basic biblical facts," concluded a 1996 survey.[1]

The lack of knowledge has a clear advantage: it allows everyone to gloss over conflicts between the claims of religion and science. Experts in the two fields also welcome this peaceful coexistence, a peace not of ignorance but of "two languages" that are separated—the factual language of the museum and the religious imagination of the sanctuary.

If scientific and religious literacy increase among the public, however, can the disinterested museum and sanctuary endure? This is no small matter for a scientific worldview, or for a religious belief that God intervenes in the material world. Some have already made a "bargain with modernity," acknowledging that all facts and beliefs cannot finally fit together. But for others, the journey between the museum and sanctuary still is to be made, a journey whose poles could be typified by Christ Church, where George Washington worshiped, and the natural history wing of the Smithsonian Institution.

Each year about thirty-three thousand tourists visit the Episcopal parish in Old Town Alexandria, Virginia, to see the Washington family pew. In Washington's day, the nation was founded on a belief in the "Laws of Nature and Nature's God." The historic sanctuary's white enameled interior echoes that mesh of Enlightenment reason and biblical revelation; its lines are classical, its mood austerely Protestant. The looks date the sanctuary perfectly, says church historian

Martin Marty, who sat one day in the Washington pew and talked with me about God and nature.

"Every doctrine of Creation I know in religion is a way of inculcating wonder and awe, the disparity between our smallness and our brevity and the large," says Marty, who would have been the religion expert testifying at the 1981 Arkansas creationism trial if not for a family illness. "Probably the way this comes up more than anything else is in the issue of care for creation. But not on origins." Marty had come to Christ Church to address a national conference on congregational life, lived out in three hundred thousand American sanctuaries, overwhelmingly churches. To make his point, he sang "We gather together" from an old Dutch hymn but quickly added that people in churches actually "gather apart" because of their different viewpoints. When they exit into wider society, they "leave together" in solidarity.[2]

When it comes to beliefs about God and nature, the dynamic in church traditions is similar. To society, the Lutheran Church–Missouri Synod leans creationist by affirming a six-day Creation, man as "the principal creature of God," and a historic Adam and Eve, while the United Methodist Church leans evolutionist, saying, "We affirm the validity of the claims of science in describing the natural world, although we preclude science from making authoritative claims about theological issues."[3] Yet within each tradition, individual members may stand where they please on the evolution-creation topic. Indeed, it is rarely preached on as a definitive doctrine. "I can't remember a sermon in the last thirty or forty years that took that up," says Marty, conversing in the pew. "If you asked a congregation, 'Make a list of twenty things you'd like a sermon on,' I'm not sure that would come up." And the times when the Protestant calendar prescribes reading the Genesis Creation accounts—Easter vigil and New Year's Eve—hardly draw the masses. "They are profound events," Marty says, "but they are a very tiny group."[4]

Another natural history story is being told seven days a week across the Potomac River in Washington, where the Smithsonian's National Museum of Natural History has stood on the grassy Mall since 1910. To start the current century, the museum's IMAX theater began showing *Galapagos*, a 3-D movie on a sixty-by-ninety-foot screen about the fabled islands where Charles Darwin gained insights that "would forever change our view of the natural world," says the narrator. "I always dreamed of coming here, to see what Darwin saw, and to continue the work he began," says marine biologist Carol Baldwin. Her deep-sea vehicle takes her beyond Darwin's reach, deeper than the hammerhead sharks he saw from the HMS *Beagle*'s decks, downward to tiny and strange creatures that have defied extinction. "In the sea as on land, it's eat or be eaten; adapt and survive and pass on your genes to the next generation," she says. She finds several new species of fish, and if the larger story of *Galapagos* is not clear, actor Kenneth Branagh concludes, "Scientists now know that evolution is an ongoing process, and that living things everywhere are constantly changing, adapting, and giving rise to new forms."[5]

When it comes to museums, Americans are most drawn to zoos and aquaria, seeking real experiences such as Baldwin's in the Galapagos. Next in allure are historic sites, such as Christ Church or history museums, and then come art museums. Natural history museums come in last for attendance; but given that "they are the fewest in number," it is no small thing to have twenty-eight million adults a year make at least one visit.[6]

Number of visits to museums is no true gauge of number of minds and attitudes changed. But it is square one in understanding what and how the museums teach the public about natural history. One of the earliest American museums was the red-brick Museum of Comparative Zoology, founded at Harvard University in 1859, the year that *Origin of Species* was published. When the museum opened a new wing—a sleek and separate building—for genetics and molecular biology, Ernst Mayr asked at the 1973 ceremony, "Why do we still have natural history museums?"

He granted that their "image in society" changes from era to era, but the permanent legacy is the mainstreaming of evolutionary biology. The museum as public showcase must continue, he said, but its overriding mission is to make new discoveries; the new Harvard wing would allow research not only on cells and molecules but also on "the whole organism, on the diversity of organisms, and on their evolution." In a world of social needs, Mayr said, the museum might find answers and solutions to be transmitted to the public by "massive education."[7]

The great debate on what American museums would say about nature had, in fact, been both opened and shut at the venerable Harvard site, also called the "Agassiz Museum." When the Swiss naturalist Louis Agassiz erected the complex, he "was spinning visions of a museum in Cambridge to rival the leading institutions of Europe," says his chronicler Mary Winsor. "It is fair to say that Agassiz built his museum as a fortress against evolution." A religious liberal, Agassiz still saw the museum as both sanctuary and collection. It was an era when Richard Owen, the British paleontologist who coined the term "dinosaur," said that the great London museum he directed should be suitable "for housing the works of the Creator." Agassiz looked at nature, even at specimens in the museum, and saw the intelligence of God. He vowed to see God at work in the history of nature, its rocks, plants, fish, and humans, "as long as it cannot be shown that matter or physical force do actually reason"—something still not shown, and still the great frontier of brain and neurological science. Soon after Agassiz's death in 1873, his students and the museum, inherited by his son Alexander, had all become evolutionist.[8]

So had all of American science by 1973, when paleontologist Mac West, his desire to do research strong, left a professorship at Adelphi University on Long Island to enter the science museum field. He would become director of the Carnegie Museum of Natural History in Pittsburgh, with its famed dinosaur collection, and later go independent as a top consultant to museums nationwide.

West began in the museum trenches, however; and he well remembers 1977 and his first exhibit project: to erect an earth history exhibit, *The Third Planet: Earth over Time*, in nine thousand square feet of the Milwaukee Public Museum—all in time for its 1982 centenary.

During the project, the main surprise was a visit by members of the Creation Society of Wisconsin. They urged West and the museum to include a young-earth, flood geology display as well. "I wasn't paying attention," West says about the creationist movement. "They demanded to be heard, and we did listen. It would have been inappropriate to say, 'Well, you're just citizens and we got this bunch of scientists, and we're not even going to let you speak.'"[9] The proposal was evaluated and reported to the board of trustees, but in the end the *Third Planet* went ahead unaltered and remains today a major attraction.

Modern-day museum directors might categorize such an experience as part of a new and general give-and-take with the public, their only audience. "A museum visit is an optional experience," West explained. "So you've got to have a hook to engage people with your subject, whether it be evolution, or astronomy, or medicine." Museums also must have conversations with the public, ever distracted by media entertainment. "Now," says West in 2000, "we are even more interested in what the public has to say than we were in the late seventies." He adds quickly, "But for the most part, this does not alter the integrity of the science that is presented."

The number of natural history and natural science museums—about five hundred—remained stable during West's career, but a new entity called the science and technology center experienced a boom. From twenty-four in 1973, the number of centers grew fivefold by 1997, clocking well over forty million visits (including repeats) that year. Their primary focus is physics, and the allure is all the kinetic and hands-on bells and whistles thereof, but they also include astronomy, cosmology, biology, and medicine. In all, the American temples to science are run variously, by universities, science academies, private owners, municipalities, and state or federal government.[10]

The allure of a natural history museum is bound up in its quest to show the "real thing." It is a quest of necessary compromises, though the public seems forgiving about replicas when it comes to dinosaurs—the unofficial symbol of the natural history museum.[11] The *Jurassic Park* craze of 1993 was only the most recent catalyst. Five years later, Los Angeles County fronted its museum with a bronze sculpture, twenty feet tall, of two battling dinosaurs. Museums can commission a firm such as Dinosaur Exhibitions, LLC, to bring in a *Lost World* exhibit, as San Diego did in 2000, and since the 1980s, robotic latex dinosaurs have joined the inventory of many museums. Just when it seems dinosaurs are old hat, there's always a revival.

For several reasons dinosaurs have become central to telling the story of evolution. The iconic moment came in 1925 in Manhattan. In the days before the Scopes trial, American Museum of Natural History president Henry Fairfield Osborn had invited the Tennessee teacher to pose with him by the dinosaurs for newspaper photos and newsreels. Twenty-two years earlier, the museum had displayed its first great dinosaur, a brontosaurus and to that date the largest fossil skeleton ever mounted. Osborn was a mammal expert, but he "knew that it was mounted skeletons of dinosaurs," says a museum history, "that would attract attention, publicity, and money."[12]

The Field Museum of Natural History in Chicago—which along with New York, Washington, and San Francisco makes up the "big four" natural history museums in the country—traces its dinosaur icons to 1900. That year museum paleontologist Elmer Riggs unearthed the bones of a brachiosaurus, a plant-eating creature that lived 150 million years earlier. The bones were stored away, but not before a model was cast, which had stood for years in the Stanley Field Hall, and at four stories tall the world's largest mounted dinosaur.

The Field Museum marked its centenary in 1993 by reorganizing all its exhibits under two themes: diversity in nature and diversity in human cultures. In those days, botanist Peter R. Crane was vice president for academic affairs, and it fell to him to pull off the centenary and to keep increasing museum attendance, which was at 1.2 million visitors a year in 1996, the year I visited Crane in his office-with-a-view.

"There's a feeling that dinosaurs may have peaked, but that still remains to be seen," the young Englishman told me. That comment came a few years before his return home in 1999 to head the Royal Botanical Gardens, Kew, a three-hundred-acre botanical wonderland on the Thames. Then, in 2000, he returned stateside to speak on "public understanding of evolution" at the American Association for the Advancement of Science, which held its annual meeting in Washington. "There is never-ending fascination with dinosaurs, and that provides the opportunity to introduce concepts of geological time, extinction, and biotic change," Crane told the AAAS session. This conveyance of evolutionary ideas is the challenge, he said. He lauded the Denver Museum of Natural History for its *Prehistoric Journey* exhibit, designed "to tap in the thin edge of the wedge" to broaden the public's thinking about time and environmental change.

To the same end, the Royal Botanical Gardens in 1995 had opened its Evolution House. The enclosed glass house display takes visitors through five zones of evolutionary history, a 3.5-billion-year walk from fossilized early cell life to later tropical and desert settings, rendered with real and artificial plants. The mass appeal of museums and botanical gardens, Crane said, makes them frontline schools of "informal education" in society. "It will require practicing evolutionary biologists to allocate some portion of their time, not just to efforts in the classroom, but for public outreach activity," he said in the lecture. To motivate such work, contributions by scientists in this arena "need to be acknowledged and valued in the system of academic rewards."

In Chicago, the Field Museum's outreach has even gone to O'Hare International Airport, where the lifelike brachiosaurus model has been moved to the

United Airlines concourse. "The towering skeleton," said United, "will help identify Chicago as a world-class destination for dinosaur enthusiasts." That transfer came in 2000, and thereafter the Field Museum had a new dinosaur icon, more ferocious than ever, for in the great hall stood Sue, the largest and most complete *Tyrannosaurus rex* fossil "ever found."

The permanent evolution exhibit at the Field Museum is titled *Life over Time,* and its keystone is the Dinosaur Hall. The hall is the midpoint bridge of a long, U-shaped ambience, a medley of fifteen sections on life's history illustrated by more than a thousand fossils. The challenge is to teach the public with a balance of content and engaging tone of voice, said Crane. Exhibit labels must be accurate yet brief. Their statements must not sound dogmatic. "It's not an easy thing to do in a simple exhibit format," he told me in his museum office in Chicago. "When it's done well, we help people to see there is more than one way to look at a topic."

That is also helped along by talk-back boards, which the Field Museum has used in its evolution panorama and which Crane explained in his Washington talk. "Visitors were invited to note their opinions on paper and post them on the board provided. They can be shared with other visitors as they come through the exhibit."

This is just one of the steam valves for public opinion that the Field and other museums have moved to put in place. Other mechanisms include a requirement that the Field's thirty curators must give public talks, and this in addition to research and exhibit management. Meanwhile, the museum pulls no punches in presenting evolution. One cheerful *Life over Time* pamphlet states, "Human origins are still being explored—our closest ancestors evolved in Africa over three million years ago." The exhibit ends at Homo sapiens but before that presents a bronze bust of Darwin, stationed on a pedestal in an alcove titled "Darwin Got It (Mostly) Right." What he got right was natural selection. What no one knew about in his day was genetics. But by taking a gradualistic view of nature, Darwin had to rule out what, in one modern theory, killed off the dinosaurs and allowed mammals to rule—the importance of catastrophic events in evolutionary history.

The Field Museum stands on public parkland and in the 1990s got a third of its funds from Chicagoans. It was thus beholden to the public, and obviously a public not offended by the great tableau of Darwinian evolution. During my visit to Crane's museum office, he gestured out the window to the Chicago skyline and shores of Lake Michigan. "We depend on that city," he said. "But we don't seem to get pressure on the evolutionary topics."[13]

The pressures have a lot to do with geography, says museum consultant Mac West, and so he understands how the Tulsa Zoo and Living Museum in Oklahoma handles natural history. Darwin's *Origin of Species* had only once used the term "evolve," but the Tulsa museum uses it even less. "They have every other euphemism in there," West says. "They are clearly being careful about the language." Other museums have been opting for evolution synonyms such as "change over time" and "diversity of life," and exhibits wanting to avoid charges of dogmatism say "many scientists believe." In Los Angeles, a new-century exhibit probes the prehistoric oceans, the flying reptiles, and then the evolution of dinosaurs into birds under the marquee "Journey Through Time."

In an age called politically correct, the euphemisms come not only for evolution but also in exhibits on early man and culture so that all suggestions of Anglo-Saxon superiority are gone.[14] Still, Mac West says, credit must be given to a new and informed outlook inside natural history museums. "They are much more inclined to acknowledge that there are other thought systems. So there is more discussion between staff and visitors." No museum has allowed a creationist diorama, but many have carved niches where visitors can freely express opinions. "They save discussion about this cultural side of things, this religious side, for those who want to talk to a docent, or an educator, or a volunteer."

The courts, meanwhile, have backed the right of publicly funded museums to teach evolution in its fullest expression. The major court case was fought over the Smithsonian's 1978 exhibit, *The Emergence of Man,* and one for the following year, *Dynamics of Evolution.* Both exhibits came with great fanfare, but especially the second, which was to run for twenty years. Curators enthused that it was "the first exhibit organized to show how evolution happens." Though using museum specimens, it would focus not merely on fossils but also on concepts of evolution; its subject was "no less than the variety that separates the species and the variations among them . . . and how they got that way."[15]

From across the river in Virginia, a Baptist radio minister sued to dismantle the first exhibit and to thwart the second, or to put a "creation science" exhibit alongside the two. Otherwise, the minister argued, public museum money had established the religion of secular humanism. The trial judge, giving summary judgment to the museum in 1978, said the exhibit had "the solid secular purpose of 'increasing and diffusing knowledge.'" In 1980 the appeals court agreed, adding that just because evolution "may coincide or harmonize with a tenet of Secular Humanism or may be repugnant to creationism," the exhibit did not establish religion. The next year, William Dannemeyer, a Republican House member from California, strolled into the exhibit. "I was surprised how they were so certain in presenting evolution," he says, citing his Lutheran Church–Missouri Synod creationist background. "I'm offended when they teach evolution as fact. It's a theory." Paleontologist West calls this "the common misunderstanding"—a scientific theory is about as certain as science can be. Still, Dannemeyer proposed in House budget hearings that museum funding be curtailed until the evolution exhibit was modified or balanced. But the proposal, perhaps like the dinosaurs, was killed off rapidly.[16]

The Smithsonian is not the only museum to brook a public protest. In San Francisco, the California Academy of Sciences has its Natural History Museum in Golden Gate Park. Its exhibit *Life Through Time: Evidence for Evolution* opened in 1990. Since then, "The Hard Facts Wall" in the exhibit has been targeted by critics of Darwinism, particularly by a science study guide written by evangelicals in the

American Scientific Affiliation. The wall shows branches from one common ancestor to six major taxonomic groups, with real or model fossils glued on each of the six branches. Where the six branches split off, the museum mounts large magnifying glasses, which typically signals visitors to come see a tiny fossil. But "The Hard Facts Wall" has nothing under the glass because no fossils exist. "What are the implications of the absence of the fossils?" the study guide asks. To drive the problem home, the guide illustrates an alternative "hard facts" scheme with six vertical taxonomic lines; there are no branching points because there are no fossils to put there. The two schemes "can lead to different interpretations," the guide says. "The conflicts will be most obvious to students who were strongly influenced by first seeing the museum's way of graphing the data."[17]

In New York, the museum of Osborn has pointed out the shortcomings of its own vaunted horse evolution exhibit. The original exhibit, a set of six horses going from small to large, was heralded as "one of the clearest and most convincing for showing that organisms really have evolved." Now the exhibit is called "A Textbook Case Revisited," and it points to "two versions of horse evolution." The old one, a stately progression, is superimposed by an unstately "branching bush" of ten horses in all sizes. With the discovery of more horse fossils, "the evolutionary story became more complicated," the exhibit states. "This doesn't mean that the original story was entirely wrong. Horses have tended to become bigger, with fewer toes and with larger teeth."[18]

Less obvious to the museum-going public has been the widespread adoption of cladistics: organizing fossils of creatures on branches, for "branch" is what the Greek word *clade* means. In San Francisco, a hands-on video display shows the branching for the animal selected by a visitor. In New York, the branching is on the floor, on which visitors do not walk until after viewing a film narrated by actress Meryl Streep. "The story begins three and a half billion years ago with the first strand of DNA, the basis of all life," she says. "We are now traveling on the tree of life that links all the diverse species that have ever lived on earth. The diagram used to show these links is called a cladogram."

The branchings, animated on film, continue up to the bony fishes, and Streep says, "Next stop is the branching point for tetrapods, animals with four limbs." By now the narrator is asking whether humans, having understood "our family tree," can take responsibility for its preservation—a thought for the actual museum stroll. "Walking along the main black path through the halls is like walking the trunk of the evolutionary tree," explains Streep. "Circular branching points along this path represent the evolution of new features like a backbone, jaws, and limbs. At these branching points you can walk off the main path to explore alcoves containing groups of closely related animals."

All of this is pictorially simple for visitors, and indeed easier to display for the curators. But the cladistic revolution ruffled an entire generation of evolutionist feathers—an earlier generation of museum systematists who felt that true evolutionary relationships wove branches together, sometimes with organisms diverging, other times with their shapes or features converging. It was vastly more complicated—a family tree interpreted by all possible common characteristics across time. But this tree's practitioners believed it truly deciphered ancestries. By 1985, the Hennig Society, named for the East German entomologist Willi Hennig, who in the 1960s invented cladistics, was three years old. "Because cladists care about how many traits various groups of animals share today, not how they got that way, they are agnostic about evolution," reported *Newsweek* in that year, also quoting biologist Steve Farris, president of the Hennig Society: "You don't have to presuppose evolution to do cladistics." Of course, the debating naturalists were all evolutionists nevertheless.[19]

In the early 1980s, the Dinosaur Hall at the British Museum of Natural History was rearranged according to cladistics, raising at least one protest that it was Marxist paleontology, a kind of saltationism in which the branchings occur abruptly like a revolution. While the anticladist protest spoke of the "late lamented dinosaur gallery," the British Museum called the snub "completely mistaken," and a New York City museum curator said of its cladisitics, "We have done this for years and nobody gives a damn."[20]

In picking an approach or design for displays, the New York museum has gotten government grants, and the National Science Foundation supported the California exhibit. In sheer volume of federal funding, however, the Smithsonian is in a different class. Seventy-one percent of its $438 million budget for fiscal 2000 was covered by taxpayers. It was the British scientist James Smithson who paid for the museum's birth in the 1840s. He willed it $500,000, delivered in 103 bags of gold sovereigns, and defined its mission as "the increase and diffusion of knowledge"—just so long as the museum bore Smithson's name. While today the Smithsonian has sixteen units and encompasses history, art, and technology, it began in 1846 in a red sandstone "Castle" under the tutelage of naturalists. At the end of the twentieth century, it clocked more than thirty million visits a year, more than a quarter of them to the gigantic National Museum of Natural History.

Like the Agassiz museum in Cambridge, the Smithsonian was born on the cusp of the Darwinian revolution. Its first secretary was Princeton scientist Joseph Henry, an inventor who made discoveries in electromagnetism, and a Presbyterian with a close friendship to theologian Charles Hodge, a foe of Darwinism. Evolution's claim that nature was ruled by law and chance alone, Hodge said, made it atheism. The Smithsonian's Henry, "after initially doubting the possibility of compatibility between any form of evolution and traditional Christianity, eventually came to a reluctant acceptance of a Christianized form of evolution." When the theologian prayed at his funeral, their disagreement remained.[21]

The natural history museum's modern building opened in 1910 under the second secretary, paleontologist Charles D. Walcott, who was chosen after New York's famed theistic evolutionist Osborn rejected the post. A man with a legendary work ethic, which helped him abide many family tragedies, Walcott had stumbled on a treasure trove of Cambrian fossils in the shale at Burgess Pass in the Canadian Rockies. Amid tens of thousands of specimens, he identified seventy genera and 130 fossilized species, all tiny extravagant creatures such as trilobites and flatworms with soft spines. Walcott dated them to soon after the Cambrian explosion, an epoch when the body plans of complex forms of life seem to have originated, and he classified them as direct forebears of all complex modern phyla, or life-forms. His overly smooth classification was challenged in the 1970s, and his motives were interpreted by Gould's book *Wonderful Life* (1989). Walcott, a Republican and Presbyterian, injected his evolution with beliefs in Providence and progress and ended up "shoehorning" his fossils, said Gould—a political liberal and agnostic—into a perfect scheme from trilobites to man.[22]

The anti-Walcott view prevails now in the nation's museums, including the Smithsonian, but particularly in New York's American Museum of Natural History. There, the second panel of the human evolution exhibit states: "We humans often think of ourselves as the culmination of a steady history of evolutionary improvement. But this idea is wrong, for evolution is neither goal oriented nor merely a matter of species gradually improving their adaptation to their environments." The Smithsonian's permanent exhibit on evolution does not emphasize the word "evolution"; it is comprehensive and conventional, and ends in a modest showcase about the origin of humans. "The science of human evolution is a fast-changing field," said a placard in the 1990s. "Much of the material in this exhibit, which opened in 1974, is now out of date. We're developing a new exhibit based on the latest findings."[23]

For the American public's education on human evolution, the New York museum was the pacesetter. Osborn, who joined the museum in 1891 and served as its president until 1933, had stirred excitement with his theory of human origins in Asia. In the opening decade of the 1900s, the museum unveiled the nation's first human evolution exhibit, a hall "with a series of murals reconstructing the appearance and environment of some of the fossil types known at the time."[24] A "much more detailed and extensive" exhibit replaced it in 1962, but still with no "real thing" in the way of hominid fossils. None of this could foreshadow events of 1984, when the museum opened *Ancestors: Four Million Years of Humanity*. It was the greatest human evolution exhibit of all time, celebrated by a giant yellow banner with a white-bone silhouette of *Homo erectus* and visited by half a million people. Conceived by museum staff in 1979, this "most grandiose idea" was brought to fruition by cultural forces. "We soon began to sense the time was right," the staff said. "Public as well as scientific interest was at a new high due to a spate of well-publicized discoveries [of human fossils] as well as because of the escalation of the 'creationist' assault on evolutionary biology."[25]

Eight countries loaned the forty or so fossils, which had first-class seats in the air, and on the ground were ferried by black limousines into Manhattan. "How can you be anti-evolution when you see so much tangible evidence of our own roots?" says museum curator Eric Delson. The most famous fossils—stars such as Lucy—were absent for reasons of politics or owners. The Leakey family of Kenya came for the exhibit conference but left their fossils in Africa. Mary Leakey's speech lamented that so many hominid bones had been gathered into a single museum room, for a religious "fundamentalist could come in with a bomb and destroy the whole legacy."[26]

No museum can claim the Darwinian legacy, however, more than London's British Museum of Natural History, which today retains its storied Romanesque towers and Victorian halls. Exhibits bear the titles "Origin of Species" and "Our Place in Evolution," and a new science hall, the Darwin Centre, opened to mark the twenty-first century. In the museum's centennial year of 1981, however, a special exhibit on evolution—the one that included a cladistic arrangement of dinosaurs—produced the most open debate on Darwinism since Agassiz, a clash between its curators and *Nature*, the magazine founded in 1869 to champion Darwinism. In the centennial exhibit the curators labeled evolution as "one possible explanation" of life, and a brochure used such phrases as "if the theory of evolution is true." Explaining that the Darwinian answer was based primarily on logic, the literature added that "the concept of evolution by natural selection is not, strictly speaking, scientific."

The editors of *Nature* were furious. "Can it be that the managers of the museum which is the nearest thing to a citadel of Darwinism have lost their nerve, not to mention their good sense?" said their editorial. "Or is it that somebody has calculated that the museum will increase its annual take of visitors by enticing scoffing creationists?" The museum staff replied in words impossible to use in American natural science. "Are we to take it that evolution is a fact," they said, "proven to the limits of scientific rigour? If that is the inference then we must disagree most strongly. We have no absolute proof of the theory of evolution. What we do have is overwhelming circumstantial evidence in favour of it and as yet no better alternative. But the theory of evolution would be abandoned tomorrow if a better theory appeared."[27] No one in science, however, has taken this "better theory" to mean creationism, says West.

The exchange unveiled the diversity that surely exists all across evolutionary biology and museum curators of natural history. Dissent is common enough, but only rarely does it surface in such colorful ways. Meanwhile, evolutionists keep a united front, and nowhere more so than in America, where active creationists are always at the gates. The Martin Marty principle still holds: they gather apart but depart together.

THE SANCTUARIES

Though nearly all Americans believe in God or a universal spirit, only four in ten occupy a pew on any given Sunday. The same number are "unchurched," not having darkened a sanctuary door for six months.[28] Whatever the churchgoers may hear on God and nature is likely to spring from the theological training of the minister.

For most North American Christian clergy, this takes place in the 237 accredited institutions—Protestant, Catholic, nondenominational, and Eastern Orthodox—in the Association of Theological Schools. To assess where faculty and students might stand on questions of religion and science, a 1999 survey asked all the academic deans, "Which view of natural history and human origins predominates at your theological school?"[29] (See Appendix, table 4.) The responses were as follows:

Theistic evolution (God guides unbroken evolution over millions of years)—35% (of all theology schools)

Progressive creation (God specially created new levels of life at points over millions of years)—9%

Young-earth creation (God created the universe and humans only thousands of years ago)—8%

Mix of theistic evolution and progressive creation (millions of years in common)—26%

Mix of progressive creation and young-earth creation (special creation in common)—16%

Unknown—6%

The pattern was not surprising. Those who cited theistic evolution—a stance that would embrace a *Galapagos* film or *Ancestors* exhibit as God's way in nature—were primarily at schools of mainline and liberal standing among Protestants and Catholics, or in theology schools at secular universities. Only the most conservative Bible traditions—Baptist, independent, Seventh-day Adventist, and Lutheran Church–Missouri Synod—exclusively held a young-earth stance, the kind that might have favored a flood geology diorama at the Milwaukee Public Museum.

The progressive creation viewpoint held sway in some Catholic schools and was part of a "mix" in Eastern Orthodox schools and many evangelical seminaries. Believers in progressive creation might walk through the Field Museum's *Life over Time* exhibit and jot these questions on its talk-back boards: At what points did God intervene in nature, performing acts of special creation that seemingly overrode the laws of nature? Was it the beginning of life, the Cambrian explosion, the distinctness of phyla and genera, the ecology of a planet that makes carbon-based life possible, or the fact of human consciousness? If the modern human form did not appear miraculously from the dust, did God change an apelike being into a human by infusion of a soul, or was a "first" human born from the womb of an evolved creature?

One institution in the evolutionist camp was Washington Theological Union, a moderate to liberal Catholic school in Washington, D.C. At a "Science and Religion Workshop" there, Catholic theologian John Haught and ordained Presbyterian scholar James Miller laid out perhaps one of the most perplexing issues: If the Creator is a personal God, why did he take 15 billion years to create humankind, who arrived only recently? "If there is a point to it, why didn't it get right to the point?" says Haught, who teaches at Georgetown University. "When scientists look at this picture, they certainly wonder how you can make religion and theology in any way compatible [with science]." The scientific concept of time definitely can "affect how we think about God," he says. Miller, an officer with the American Association for the Advancement of Science, said that theology schools and pulpits still have to grapple with time. "If I hear another sermon in which God is spoken of as Creator in the past tense, I could scream," he tells the conference. "Who says the universe is finished?"[30]

The survey found that most schools—seven in ten—must grapple with this issue because they teach theology and the Bible against the backdrop of an ancient earth and universe. Time is not the only challenge to theology, for science also posits that no physical law may be interrupted or broken. Belief in God's intervention breaks that rule entirely. Such past interventions in natural history, typically called "special creations" or "miracles," have been seen as revealed in Scripture, since science could not extrapolate back to them. Strictly speaking, a theology school that ascribes to even a single act of intervention, such as a special creation of any kind, has put revelation over the claim of science.

In showing allegiance to science or to revelation, here is how the schools lined up: 35 percent were evolutionist and gave foremost authority to science; only 8 percent were miracle-only creationists, giving top authority to revelation. The majority in between—nearly six in ten schools—lived in apparent contradiction. They wanted both science and revelation to account for natural history. They would join the ironclad and antique nature known by science with a personal God who, known by Scripture and experience, can manipulate nature. These kinds of questions have intensified at theology schools. In 1999, 40 percent of them had at least one course "focused mainly" on science-religion topics. Schools with a "mix" of stances on God and natural history were more likely to have a course.

The theology deans also were asked about theology's "most important or appropriate" mode of contact with science. Eight in ten said theology primarily gives "overall meaning and purpose to life in the material universe." Fifteen percent said theology's main role in science was to "support the biblical account of the human creation and fall." Only 5 percent felt that theology's main contribution was to put "ethical limits on sciences such as biotechnology."

The majority view supports the contention of Haught, the theologian, that religious people can adapt to any scientific claim as long as they retain the most essential element of "purpose" in the universe. Yet the minority view, in which theology impinges on ethics, is what most interests ordinary churchgoers, says Miller, the Protestant thinker. "When you go into most religious communities, the place you want to talk about [God and nature] is not in the theology-science arena, but in the technology and ethics arena," he says. "That is where the rubber meets the road for most people."[31]

Either way, not much preaching on the subject comes from American pulpits. The Protestant liturgical calendar denotes the Genesis Creation text only twice, and in the six-year cycle of Protestant Sunday school lessons, Genesis comes up only a few times as well. The Roman Catholic Church lectionary offers Genesis 1 or 2 on six occasions a year; only one falls on Sunday, the others on feast days. Still, the priest must want to grapple with the topic. "Whether they would choose to preach about the question of human origins, vis-à-vis evolution," says one Catholic liturgist, "might be one choice among many."[32]

Such preaching is far from extinct, however, and often is prompted by current events. In 1996 Pope John Paul II stole the headlines when he affirmed evolution and elaborated that, "if the human body has its origins in pre-existing living matter, the soul was created directly by God." Soon after that, Cardinal John O'Connor mounted the pulpit of Saint Patrick's Cathedral in Manhattan and preached on the Genesis text that God breathed life into earth. "Perhaps that earth was a lower animal," he said. "Is it possible that when the two persons that we speak of as Adam and Eve were created, it was in some other form, and God breathed life into them, breathed a soul into them? That's a scientific question."[33]

Both the pope and the cardinal evaded the contentious question of Adam and Eve. Were they real, symbolic, or obsolete in the face of "polygenism"—the idea of more than one pair of human ancestors—an idea condemned by Pius XII's 1950 encyclical that otherwise affirmed evolution of the body. The doctrinal concern is Original Sin, the entire rationale for a savior. The Genesis 3 story of the Fall may use "figurative" language "but affirms a primeval event, a deed that took place at the beginning of the history of man," the catechism of the Catholic Church states. "Revelation gives us the certainty of faith that the whole of human history is marked by the original fault freely committed by our first parents." The Vatican is wise to avoid specific scientific issues, says biologist Kenneth Miller, who attends mass regularly. "What the church is best qualified to speak about is the origin of the human soul." When the soul appeared in natural history, he says, "is a question that we may never be able to answer."[34]

One occasion for Protestant Bible scholar Luder Whitlock to preach on Genesis was the "Mere Creation" conference, which gathered mostly evangelicals, mostly in science, and mixed in their degrees of doubt about evolution or about special creation. That Sunday in Los Angeles, they sang the Isaac Watts hymn "The Work of Creation" (1719)—"He formed the creatures with his word"—and Whitlock, editor of a Bible commentary series, said the Scriptures are mute on so much science. "The more you look at the data, the more you ascertain we can't know the age of the earth from the biblical text," the conservative exegete preached in the Biola University chapel. Whitlock is president of the Reformed Theological Seminary, bound to a fairly strict inerrancy on the Bible and drawing not a few young-earth creationists to that theological enclave. He nevertheless preached that, in regard to the Bible, "What it does not say clearly, we will not say. We will not take [away] from it, but we will not add to it."[35]

Neither O'Connor nor Whitlock had preached on creationism in public-funded forums, but like so many other issues—refugees, gun control, military defense, school prayer, national health care, affirmative action, the environment, and privatization of government—creationism can divide clergy and laity. "On these [social] issues, even orthodox clergy are usually more liberal than their parishioners," says political scientist John Green. Yet creationism does not polarize clergy and laity quite as much as the others; in fact, it tends to put clergy in conservative traditions a little to the right of churchgoers. "It is a topic of the mind, a religious and theological puzzle," Green says. "So it's a topic that clergy probably care about more than people in the pew."[36]

The clergy who are to the right of the public on creationism typically lead independent churches, which make up from 12 to 18 percent of the nation's total.[37] With independence, these churches do not follow liturgical calendars. Many of their clergy, if seminary educated at all (Bible college is common), would come from graduate theological schools where theology joins science to "support the biblical account."

West of Chicago, the Willow Creek Community Church draws fifteen thousand worshipers a weekend. It is a so-called seeker church, orthodox in theology but casual in demeanor. Yale Law School graduate and former *Chicago Tribune* reporter Lee Strobel was one of the associate ministers who divvied up the weekend preaching. Given his long-term interest in science and belief, Strobel put "Evolution or Creation" in a sermon series. "This is no idle debate," he preached. He surveyed the origin of life, the fossil record, and the idea of finding design in the biological and cosmological world—an idea that Darwin fundamentally rejected. "Friends, Darwin was clearly wrong," he said. He then cautioned against squabbles over details, for Christian belief required only three convictions on Creation: God is Creator, man and woman are in his image, and each is accountable to his or her Maker.[38]

While Strobel fit the topic into occasional sermons, Hugh Ross, a radio astronomer who founded the Reasons to Believe ministry, targets God and nature all the time. A former California Institute of Technology staff scientist, Ross attends a conservative Congregational church, now an independent Bible church, where many science professionals worship. "Science and theology do overlap," he says. "There's nothing in science that is theologically neutral. There's nothing in theology that is scientifically neutral, in my opinion." Ross reconciles an ancient universe with God's successive creative interventions, and he says Bible churches since the 1970s have learned only the "two polar extremes" of atheistic evolution and young-earth creationism. To this group he says that an "error-free" Bible can accept an ancient earth. "We're saying there is a literal Adam and Eve . . . created recently, just thousands of years ago," he says. (His range is from ten thousand to sixty thousand years ago.) "There's more than one literal interpretation of

Genesis 1," says Ross. Or, put differently, "I'm not saying it's literal versus meta-phorical. It's literal versus literal."

As a mathematical physicist, Ross says the Einsteinian revolution and quantum physics allow a believer to understand Psalm 90:4, which says that "with God one day is as a thousand years." In the theory of the "relativistic time dilation effect," a cosmic view of time can be short (as in seven Creation days), but the human view is very long (as in conventional science).[39] These are not easy concepts for people in the pew. But Ross is evangelical in his effort, taking his message to such venues as James Dobson's *Focus on the Family,* Robert Schuller's *Hour of Power,* and the Trinity Broadcasting Network. And for his efforts on behalf of an ancient earth time, Ross has been called a heretic.

The heresy charge is made most overtly by young-earth evangelist Kenneth Ham, who from Florence, Kentucky, leads the Answers in Genesis ministry and publishes literature against "Rossism." In 2001, Answers in Genesis broke ground on forty-seven acres in suburban Cincinnati to build a creationist museum with the nation's largest display of model dinosaurs. "Children and adults alike are absolutely fascinated by these mysterious monsters," says ministry founder Ham, a high school biology teacher in his native Australia. There is really no mystery when it comes to dinosaurs, he typically goes on, for the Bible has the answers. The dinosaur was a land animal created on the sixth day, and it had lived alongside humans—witness Job's account (40:15–24) of a behemoth with a tail like a cedar tree—before and after the global flood, a drowning from which some dinosaurs escaped on Noah's Ark.[40]

Ham worked with the Institute for Creation Research for several years but then beat his own path. His ministry is geared mainly to Bible conferences, pastors' meetings, and Christian youth groups. The most evangelistic of all the creationist groups, Answers in Genesis is trying to develop Creation Clubs at high schools (Genesis-focused Bible groups allowed under the Equal Access Act). It claims to have distributed 150,000 copies of the *Refuting Evolution* booklet outside high schools. For its seminars and "outreach," the group relies on five staff members. In 1999 it claimed to have deployed fifteen speakers who fulfilled three hundred of four hundred speaker requests, reaching ninety-three thousand people.

The Kentucky museum's floor plans show dinosaur models outside and massive displays inside. In condemning "Rossist" teachings, Ham implies that astronomical physics will not win souls the way dinosaurs will. Indeed, "missionary lizards" point to salvation. Once physical death entered the world at Adam's rebellion against heaven, Ham says, every soul needed Christ. "God, who made all things, including the dinosaurs, is also a judge of His creation."[41]

. . .

The museum and the sanctuary tell sweeping stories, but the challenge of literacy for their visitors is in the details. Whether science and religion are separated or overlap in the mind of Americans may depend on detailed conflicts, or precise harmonizations. It is an exercise that has gone on since Washington worshiped in Christ Church.

In 1999 a national poll on evolution found that 68 percent of Americans agreed that "a person can believe in evolution and still believe God created humans and guided their development." This was 20 percent more people than previously thought; Gallup polls had always found 48 percent of Americans believed God guided evolution.[42] The new phrasing of the question guaranteed a high positive response: some may have been speaking for "a person" other than themselves, and in general Americans are unfamiliar with the detailed claims of evolution.

Nevertheless, the survey response represents a massive desire to overlap, intermingle, or relate God and nature. This has been done in three ways by modern biblical believers. The first is to take God out of the details and put him in a transcendent realm. This neo-orthodox approach has been most comfortable with the "two languages," and thinkers such as Paul Tillich have explained how this works by emphasizing that all human perception of God is symbolic. Lutheran theologian Philip Hefner, who studied under Tillich, says that finding God in precise natural events is futile. "I just don't think you can walk down the street and say, 'You see those roses are blooming because God's working there,'" Hefner says. "'Those roses are all dried up because God doesn't want them to grow.'" When *Newsweek* published a poll showing that 84 percent of Americans believe in miracles, Hefner commented, "If a miracle involves God's contravening the laws of nature to redirect the ordinary course of events, I begin to have some problems."[43]

A second approach is to believe that everything that happens in nature is indeed the will of an Almighty God. This classic deity has been handed down by Augustine, Aquinas, and Calvin and is the God of most believers and of the prominent evangelist Louis Palau. To address the question of why an Almighty God may seem not in control of life, allowing suffering and evil, Palau wrote *Where Is God When Bad Things Happen?* The evangelist says that God is indeed on an Almighty throne; he hears every prayer. "It gives me much more peace to know that there is a living God who is loving, who is righteous or just, who is almighty, who makes no mistakes," Palau says. "I do believe God answers prayer. But in answer to prayer he mostly, I think, protects his children, provides through prayer, guides through prayer."[44]

A third view of God, more recent in the science-religion debate, has been called "process theology." It denies that God is all-powerful, which explains why there is freedom and suffering in nature—indeed, why there is evolution itself. "If we are to have any hope of overcoming the long-standing belief that a scientific worldview conflicts with Christian faith," says process theologian David R. Griffin, "we need a form of Christian faith that does not presuppose supernatural interventions." Under a process view, the march of nature as seen in the movie *Galapagos*, with its waste, death, and survival, is nature running free and, for theologians such as Haught, a sign of divine suffering. As in the doctrine that God became man in Christ and died on the cross, Haught sees God as emptying himself of power and humans standing in silence before the awesome sacred. This is a "letting be" by both the divine and the human agents, easily allowing God into any natural history museum.[45]

The influence of process thinking has spread and in conservative evangelical circles has given rise to a view called "the openness of God." Some evangelicals have looked at the Bible and questioned whether God, in all affairs, is a preordaining and all-powerful king and, in practice, a controller of every molecule in the universe. One proponent of this view is William Hasker, an evangelical and editor of *Faith and Philosophy*, the journal of the Society of Christian Philosophers. He takes seriously Bible accounts of divine intervention but also divine inaction or weakness, a state of affairs in which man and nature have real autonomy. Here is a God "who takes risks"—by not preordaining every detail of the future, by living within time, by being open to surprise at what happens next. Hasker would stand amid the *Life over Time* exhibit at the Field Museum and, while calling it somewhat "dogmatic," see the divine presence. The open God allows evolution and, being omnipotent, can still intervene in acts of creation and in response to prayers. "In our view, God is certainly capable of intervening and interacting in miraculous ways," Hasker says. He asserts that process theology provides a far more "personal God" than Tillich's symbolic viewpoint. And the open God may be persuaded to intervene. "There is a great deal of mystery to how and to what extent that happens." Prayers to an open God are not in vain, but believers should be cognizant of God's pattern in biblical accounts and in life. In other words, Hasker says, people should not expect God to "micromanage" nature and human affairs.[46]

Evangelical theology is involved in a turn-of-the-century struggle over what science has done to the interventionist God of the Bible. While one evangelical thinker has lauded openness theologians for their "appeal to the authority of Scripture" and acceptance of the "normative status of the biblical revelation," the more orthodox have condemned openness theology as just plain process philosophy. With openness theology in mind, a Southern Baptist Convention panel in 2000 rebuked the "modern denials of the omniscience, exhaustive foreknowledge, and omnipotence of God."[47]

For believers such as historian Martin Marty, God's intervention is most at issue not in natural history but in the part of the individual that connects with the Creator—the mind or soul. "Frankly, on the evolution thing, the scary part for me would be the brain," he says. "I think that fighting about how old is the world and how big is the universe is small in intensity compared to whether our brain is nothing but chemical processes that evolved. That is a bigger question of soul and spirit and God."

The deans at the theological seminaries were asked about this personal, mind-to-mind contact with God, a "God to whom one may pray in expectation of receiving an answer." Ninety-two percent (see Appendix, table 5) were amenable to this kind of "interventionist God," though with caveats. "This comes close to my belief, but does not touch all points," said one dean of theology. Others said this God was far too simple, while another educator held to a "process theology of God that includes [the] reality of communion," and so, yes, a God who hears prayers.[48]

Scientists, meanwhile, have indeed tried to explain such "communion" with God as the product of purely physical, evolutionary processes. This was the interest of psychologist James Leuba in 1914 and 1933 when he surveyed American scientists on "a God to whom one may pray." He wanted to know: Can the scientist, the purveyor of the natural history museum story, believe in the supernatural? In the 1990s, scientists have given some answers.

11. WHAT NATURAL SCIENTISTS BELIEVE

Nowhere is the quest to separate science and religion more daunting than within the lives of natural scientists themselves. When *Nature* reported in 1997 that four in ten U.S. scientists believed in a personal God, it "stirred a hornet's nest among many scientists, who felt that the 40 percent figure was too high." Says Oxford professor Peter Atkins, "You clearly can be a scientist and have religious beliefs. But I don't think you can be a real scientist in the deepest sense of the word because they are such alien categories of knowledge."[1]

In a religious society such as the United States, institutional science, especially modern biology, has avoided such judgments. The nation's biology teachers declared that science "neither refutes nor supports the existence of a deity or deities," and the National Academy of Sciences explained that "whether God exists or not is a question about which science is neutral." By their very profession, scientists also claim a special detachment. "As a scientist my entire attention is directed to matters accessible to proof," said one natural scientist when queried about deities in 1914. Asked about God in 1998, a National Academy physicist said roughly the same: "I have no interest in this question."[2]

But for all its objectivity, science is a human enterprise. What natural scientists believe, and why, is determined by a variety of factors and forces. Led by the Enlightenment vision, some say that "the religious scientist is an oxymoron" and that science demands agnosticism or atheism. They forecast a "third culture" in which scientists eclipse the literary role of theologians, philosophers, and leaders in the humanities. Yet when ordinary research scientists are surveyed for a predominance of disbelief, "the picture is more complex." One reason for the complexity is how some in science use the term "religion." Einstein said a scientist was the epitome of a religious person, and he spoke of "rapturous amazement"—as if Moses before the burning bush—before the mathematical harmony of natural law. Awe toward the cosmos has often been called spiritual. An outlook known as "religious naturalism" suggests an epiphany of wonder at godless matter. Evolution has also provoked epiphany. Botanist Edgar Anderson, when president of the Society for the Study of Evolution in 1959, had so profound a "religious experience" at the Darwin Centennial convocation in Chicago that he could not deliver his presidential lecture. More recently, E. O. Wilson—who popularized the phrase "epic of evolution"—declared that this story of cosmic and biological evolution will be the new "sacred narrative," since humans need stories of meaning. "We can't live without them," he said.[3]

When scientists adhere to religious outlooks, they "hold no religious views uniquely their own," says theologian Edward LeRoy Long, who studied what American scientists wrote from 1900 to 1950. American scientists, moreover, have attended houses of worship no less than the general public, at least for the social benefits of affiliation. Yet scientists are not just like everybody else, and this is particularly clear when the benchmark is not simply religion but belief in "the existence of a deity or deities" or, more precisely, as Einstein said, "a God who concerns Himself with fates and actions of human beings."[4]

In this arena, scientists believe considerably less than the U.S. public. Where there is belief, it more often veers from traditional ideas of God. Applied scientists (those in chemistry, engineering, medicine) show more belief than natural scientists (those in biology, geology, physics). Mathematicians and physicists believe more than do biologists. In the United States, disbelief among biologists—the heirs of Darwin—is outdone only by that among psychologists, sociologists, and anthropologists who work in the so-called soft natural sciences. Disbelief is higher still in elite science. This realm is occupied by people sometimes called "greater scientists," by members of the National Academy of Sciences, or by Nobel laureates, who were found by one study to show a "remarkable degree of irreligiosity."[5]

In a past "age of faith," the great scientists were all believers in a traditional God, and modern-day creationists have done a service in frequently listing these names. They range from Copernicus and Kepler to Newton and Galileo. These great men of science were led to such a conclusion by reason and emotion but also by an act of the will—the proverbial "leap of faith." Obviously, scientists still take such leaps, whether toward belief or disbelief. Nobel physicist Sheldon Glashow, for example, abides by a "cosmic catechism" that says the universe is governed by knowable laws. "This statement I cannot prove, this statement I cannot justify," he says. "This is my faith." Astronomer Allan Sandage, who late in his career sought the divine through Christianity, opted for God by deciding "to will myself to believe."[6]

The question of belief in natural science, and how that affects the evolution-creation debate, does not hinge significantly on such leaps of faith. Nor does it turn so much on broadening definitions of religion. The key issue remains the existence of God and the ways in which God can be present and active in the world. It was this sort of God that was at issue for two great minds trained in the monotheistic West, two personalities that touch on every debate over science and faith: the Jewish-reared physicist Albert Einstein and the Christian-reared Charles Darwin.

Since 1979, a bronze statue of Einstein has reposed in a bushy grove outside the National Academy of Sciences, hardly a stone's throw from the Potomac River in Washington, D.C. The figure is four times larger than any tourist who visits the

Einstein Memorial. But as he reclines on a ledge, bohemian-like in sweater and sandals, visitors can read on the bronze sheet of paper he holds one of his ultimate beliefs, the equation $E = mc^2$. To Einstein's chagrin, however, his theory of relativity suggested that the universe was either expanding or contracting. In other words, the universe had a beginning. When Princeton physicist John Wheeler dedicated the memorial, the origin of the universe from a singularity was the dominant cosmology, and Einstein was credited with showing "that the universe does not go on from everlasting to everlasting, but began with a bang."

The evolution debate has always raged on the question of what that view of nature says about God. Since then, the scientific theory of the big bang has risen to be one of the other great examples of how a description of nature invariably influences belief. With the big bang, says astronomer Robert Jastrow, "Now we see how astronomical evidence leads to the biblical view of the origin of the world." This "simple and beguiling picture" of a beginning worries atheists such as Sir John Maddox, editor emeritus of *Nature*, because it entices people into Bible belief. He thus campaigned for "down with the big bang" (a cause for which young-earth creationists would cheer, since the bang came billions of years ago).[7]

Einstein also balked at the new quantum theory, with its dice throwing and indeterminate sense of cause and effect, but such a view of matter was a catalyst for many to draw conclusions about theological beliefs. With the death of "strict causality," said physicist Arthur Eddington, "religion first became possible for a reasonable scientific man about 1927." Believers also found in the quantum mysteries "the solution of the agelong problem of the freedom of the will," physicist Percy Bridgman, a science popularizer, wrote in *Harper's* in 1929. But atheism also likes quantum reality, he added, for there "the atheist will find the justification of his contention that chance rules the universe."[8]

Einstein, though an agnostic, has kept God playfully alive in modern science. "God does not play dice with the world," he said. "God is subtle, but he is not malicious." This God, of course, was Einstein's word for the deterministic laws of matter. His disbelief has typically been tied to the power of his reason. Emotional influence was certainly not absent, however. Reared as a religious Jew, he fled annihilation in Nazi Europe because of that heritage. Though he refused to raise his two sons with religious instruction—"something that is contrary to all scientific thinking"—he still spoke of the animating power of "awe" and of a "cosmic religious feeling." He said, in fact, "It is the emotion that constitutes true religiosity." Biographers are quick to remind us that despite these allusions, Einstein abhorred mysticism.[9]

Einstein said he believed, as did the seventeenth-century mathematician Baruch Spinoza, in a "God who reveals Himself in the orderly harmony of what exists, not in a God who concerns Himself with fates and actions of human beings." This so-called deity is, of course, nature itself, as is the God who does not play dice. Einstein's deterministic materialism denied the existence of free will. His antipathy to dogma, priesthoods, and the claim that religion was needed for ethics was as well known as his good-natured tolerance toward people who chose to believe. Most jarring, however, was his 1940 declaration that religious leaders should abandon "the doctrine of a personal God," and indeed a God "interfering in natural events." The rejection was partly moral (for Einstein found fear of a deity repugnant) but mostly rational, since "the more man is imbued with the ordered regularity of all events, the firmer becomes his conviction that there is no room left by the side of this ordered regularity for causes of a different nature." Such a God could not be disproved until all natural events are understood, Einstein said, but he was clearly not holding out hope to traditional believers.[10]

Einstein wrote more on religion than did Charles Darwin, but besides what he said in letters, the Victorian naturalist gave over an entire section of his autobiography to the topic. From being "quite orthodox" on the *Beagle*, Darwin said he came to "disbelieve in Christianity as a divine revelation." He wrote *Origin of Species* as a theist but ended up an agnostic. Darwin's religious doubt was dramatic against a former interest in the Anglican priesthood and love of how theologian William Paley's *Natural Theology* pinpointed God's design in nature. This logic for a "personal God," however, was undermined by how his new theory led him to observe nature. "The old argument of design in nature, as given by Paley, which formerly seemed to me so conclusive, fails, now that the law of natural selection has been discovered," Darwin wrote. Nature runs on "fixed laws" that have no more design than "the course which the wind blows."[11]

Belief that a beneficent and omnipotent God could allow so much creaturely suffering for so long "revolts" human sense, Darwin said. "The very old argument from the existence of suffering against the existence of an intelligent first cause seems to me a strong one." Today, critics of Darwinism have seized on this kind of confession, arguing that evolution is more an answer to evil than to scientific evidence. In his own day, Darwin went on, "the most usual" reason to believe was a "deep inward conviction and feeling." Yet, he reasoned, "I cannot see that such inward convictions and feelings are of any weight as evidence of what really exists," since Hindus and Christians arrive at different convictions. Despite such forthrightness by Darwin in his later years, some writers suggest that he truly wanted to be a believer; many more argue about when his ultimate disbelief dawned, a range of times that stretch from 1832, when he first stepped on the *Beagle* at age twenty-three, to a decade later and even to 1850. "A firm date is impossible to establish," concludes historian Neal C. Gillespie.[12]

Eventually Darwin doubted the uniqueness of the human mind and the human place in the universe. "Why is thought, being a secretion of brain, more wonderful than gravity, a property of matter?" Darwin asked. "It is our arrogance," he said, "our admiration of ourselves." When this self-admiration is organized in a group, he reasoned further, the group invents a God who put humans at the center of the universe. Or, as Darwin jotted in a notebook, "Love of deity [is the] effect of organization, oh you Materialist!" He believed that since the human mind "developed from a mind as low as that possessed by the lowest animals," it was untrustworthy in its theological claims. Yet once these claims are imprinted on the minds of children, he said, it becomes "as difficult for them to throw off their belief in God, as for a monkey to throw off its instinctive fear and hatred of snakes."[13]

The cool reason of Darwin's *Autobiography* may obscure a deeper wellspring of disbelief, says Darwin biographer James R. Moore. "Intellectual consideration weighed heavily with him, but his decisive objection [to God] was moral." Darwin had been overwhelmed by "anger and grief" when, between 1848 and 1851, he lost his father, then lost his ten-year-old daughter, Annie, and "finally renounced his faith," Moore says. "The loss of Annie in 1851 was the point of no return." That "one very severe grief," as Darwin called it, mirrored the experience of the sixteenth-century German reformer Martin Luther. When Luther lost his first daughter, he marveled that a "heart could be so broken." When his thirteen-year-old Magdalene next died, it "overwhelmed him to the point that his faith in God failed." Yet Luther was in the business of wrestling with God ultimately; Darwin had no professional stake in keeping the deity amid human tragedy.[14]

Nine years after Annie's death, Darwin wrote to Harvard biologist Asa Gray that the question of God was simply beyond his ken. "I feel most deeply that the whole subject is too profound for the human intellect," he said. "A dog might as well speculate on the mind of Newton. Let each man hope and believe what he can." For scientists ever after, this could be taken two ways. One is the stance of humility, a kind of benign agnosticism. The other follows Darwin more closely, for he continued to try to explain religious belief in materialist terms. He seemed to hold that human belief in God, under the light of science, was not "too profound" to explain after all.[15]

Darwin's mental struggle has influenced the West. But as an intellectual exercise, it was no more powerful than the arguments of Scottish skeptic David Hume (d. 1776) and Germany's Immanuel Kant (d. 1804) that nature proves nothing about God. Perhaps Darwin was speaking of their influence in 1879 when he noted the rapid spread of "skepticism and rationalism during the latter half of my life." But because *Origin of Species* presented doubts about God as having a scientific basis, Darwin more than anyone had triggered a "crisis of faith" in Western belief. Believing scientists, in particular, were forced "to reassess the relationship between nature and the supernatural," says historian Jon H. Roberts.[16]

Nothing, however, has ever eradicated belief in God, and neither was Darwinism the great scythe that cut down all orthodox faith among Western scientists. The first roster of the National Academy of Sciences, founded by the U.S. Congress in 1863, included the creationist Louis Agassiz and Christian Darwinist Asa Gray. The eighty American naturalists—biologists, geologists, and anthropologists—elected by their peers between 1863 and 1900 also showed "no evidence in either biographical or autobiographical accounts to suggest that a single one of these men severed his religious ties as a direct result of his encounter with Darwinism," reports historian Ronald Numbers. "By and large, the Catholic naturalists in the Academy remained Catholics, the Presbyterians remained Presbyterian, and the agnostics remained agnostic."[17]

Neither did members of the Royal Society in Britain, when surveyed in the 1930s, show a rush to embrace strict materialism. A majority gave credence to a "spiritual domain" that existed along with matter, and seven in ten said that evolution was "compatible with belief in a Creator." Half of the fellows who responded said that a "personal God" had indeed survived the negating effect of science.[18] In such a survey, precise belief is elusive, but the absence of widespread disbelief disappointed materialists. They said this Royal Society sample had gotten soft with age, having peaked professionally in the middle and late 1800s, before positivism and other "acids of modernity" made belief by scientists less acceptable.

The religious beliefs of U.S. scientists received considerable attention through the 1950s, perhaps because of what one theologian called a "spirit of penitence" after America used the atomic bomb, or Christian aspirations evoked by the space program. When theologian Long surveyed writings by scientists in 1952, there was plenty of theological content, and theologian John C. Monsma easily assembled forty essays from believing scientists in 1958, the International Geophysical Year, to argue "that science is wholly consonant with belief in God." Research on these attitudes among scientists peaked around 1970 and then gathered dust on academic shelves. In 1996 biologist Gerald R. Bergman dusted the studies off, and his summary argues that "very few eminent scientists today are devoutly religious, and most do not hold to any conventional theistic religious beliefs." Still, belief had survived, though Bergman says the "relative importance" of the contributing factors is hard to determine.[19]

SPECTRUM OF BELIEF

Early in the twentieth century, Bryn Mawr College psychology professor James Leuba had little doubt regarding what caused the fall of religious adherence among scientists. As a psychologist, he was eager to apply science to the mind as a bundle of material causes. A compatriot in this endeavor was J. McKeen Cattell, a professor of psychology at Columbia University and editor of *Science*, organ of the American Association for the Advancement of Science. To study the makeup of American scientists, in 1903 Cattell asked ten well-known scientists in twelve fields—a jury of 120—to gather names and data for a directory, *American Men of Science* (*AMS*). The 1906 publication posted four thousand names. For the twelve fields, Cattell also gleaned names from the top of lists to designate "greater" scientists (a fourth of the total), also called "starred" scientists for the asterisks that appeared next to their names.[20]

Leuba's particular interest was in how the mind of the scientist dealt with religious belief, so in 1914 he drew on the latest *AMS* (1910) to survey a random sample of one thousand scientists. He asked them, do you believe in (1) "a God in intellectual and affective communication with man [and] to whom one may pray in expectation of receiving an answer" and (2) "personal immortality" (see Appendix, table 1)? With psychological care, Leuba informed his subjects that an imaginary God that produced "the subjective effect of prayer" did not count. Among scientists, he wanted "to separate the believers in a personal God from all others." This was a real "interventionist God," not Spinoza's or Einstein's deity of metaphysics or "the Spirit of rational order and of orderly development" espoused by Nobel physicist Robert Millikan. That God is "not the God of our churches," Leuba said, and "cannot be influenced by supplication, adoration, etc."[21]

When the surveys returned, Leuba announced that scientists (and college students) believed less than the general population. The 1916 report caused scandal, with William Jennings Bryan using it in stump speeches and his antievolution crusades. But Leuba was not shy either. Using the survey from 1914 and another slightly different poll in 1933, he would present an overarching case that the "two cardinal beliefs of official Christianity"—God and immortality—were "apparently destined" to decline under "the diffusion of knowledge." He called for "a revision of public opinion regarding the prevalence and the future of the two cardinal beliefs of official Christianity." The findings drew popular interest. As Leuba explained, "Curiosity as to the beliefs of scientific men is justified, for they enjoy great influence in the modern world, even in matters religious." Between 1914 and 1933, Leuba showed, a "marked increase in unbelief" hit scientists: belief in a personal God dropped from 42 percent to 30 percent.[22]

By the end of Leuba's century, the *AMS* had changed in two ways. First, the directory had expanded its name to *American Men and Women of Science*. It had also (since 1944) stopped picking "starred" scientists, and efforts to revive the designation failed: American science could not agree on how to pick the upper tier. But in the eight decades since Leuba surveyed "biological and physical" scientists in the *AMS*, that group hardly changed in one respect: belief in God. When in 1996 a random sample of *AMWS* biologists, mathematicians, physicists, and astronomers was surveyed on Leuba's questions, four in ten believed in immortality and Leuba's God—about the same as in 1914 (see Appendix, table 2).

This was the finding that "stirred a hornet's nest," since many secularists in American science agree with Leuba that religious belief should perpetually decline. One argument made in Leuba's favor is that his *AMS* sample in 1914 (and in 1933) contained a 40 percent portion of "greater" scientists, while that distinction could not be made in the group surveyed in 1996. In other words, Leuba had surveyed a presumably smarter group of scientists, and this quality that Leuba called "superior knowledge" produced more disbelief. The dispute over comparing the 1914 and 1996 results—both showing about 40 percent belief—has its merits but does not undermine what scientists in the 1990s say they believe, even if 40 percent seems "too high."[23]

Such investigations of religious beliefs of scientists also have produced another common complaint: the way the survey defines God. "Why such a narrow definition?" asked one scientist in 1996. "I believe in God, but I don't believe that one can expect an answer to prayer." A member of the National Academy of Sciences explained, "I consider it quite possible to be a deeply religious person while rejecting belief in a personal God or in personal immortality." Ideally, Leuba would have liked to set apart scientists in three groups, with a "personal God" on one pole, materialist disbelief on another, and scientists who held some kind of "spiritual conception of ultimate reality" in between. But Leuba reports struggling to prepare an effective questionnaire for such a three-part result. Having failed, he stuck to the clearest cleavage of all: belief or nonbelief in a "personal God" who may be worshiped and who hears prayers.[24]

Leuba's three-part spectrum remains valid and was essentially what Edward Long had found in his 1952 review of writing by American scientists. On the disbelieving end, scientists had credos that affirmed only education, democracy, and intellectual wonder. Toward the middle of the spectrum, scientists entertained a metaphysical reality, but it was "fear of anthropomorphizing" God that drove some of them back into humanism, says Long. Scientists without that fear plunged ahead with God as first cause or principle of cosmic growth. The big divide came over monotheism, for it suggested that humanity bore a divine image and a central cosmic role. For some scientists, this God had nothing to do with prayers; it is a kind of deism. Scientists seeking a more "personal" God, however, had to grapple with Bible claims and their reconciliation with scientific facts. To "gloss over the problems between science and faith," Long says, some scientists built mental compartments. Others made the Bible symbolic, not in conflict with natural history, and those who kept the Bible literal rejected or fudged conflicting claims of science.[25]

A sampling of contemporary scientists also gives a sense of the spectrum, beginning with biologist Ursula Goodenough, an advocate of "religious naturalism." It evokes a "fascination" with the natural world, but when Christians include a Creator in that fascination, Goodenough says she must hold back. "But then, on the other side, we converged again," she says. "We differ, that is, in our concept of agency, but not in our joyous astonishment at what has resulted." Harvard astronomer Nathaniel Carleton no more believes in God than does Goodenough, but he gives a pragmatic value to religion and even prayer. To declare nonbelief can produce "an unduly harsh picture," he says in the 1996 survey. "I try frequently to open my mind to an influence of what is good, and the 'subjective and psychological' effects of this can be quite profound, such that I am happy to make contact with the religious tradition by saying that I am praying to God."[26]

A metaphysical theism is proposed by the noted astrophysicist Joel Primack. He has drawn on mystical Judaism to put mind behind the universe, in which he sees a symmetry of magnitude; the human being stationed midway between the smallest particle and largest universe. One day humanity could be spiritually united, he says, by a single, accurate cosmology—compliments of science. The Quakers arose in England as God-talking Christian dissenters and martyrs, but their theism has in many places paled into a kind of metaphysics. In her Quaker testimony, British astronomer Jocelyn Bell-Burnell, a discoverer of pulsars, says: "I don't think God created the world in any physical sense. But that's not to say there isn't a God." The findings of science can indeed change human concepts of the deity, she says. "I allow my science to drive my theology."[27]

Two top American scientists, Charles Townes, a Nobel physicist for the discovery of the principle behind lasers, and astronomer Allan Sandage, touch on a more traditional faith. Throughout Townes's career, he took his family to Baptist, Presbyterian, and Episcopalian churches, settling finally with the "more liberal theology" and ambience of the United Church of Christ. He sees no place for God's "divine action" betwixt the physical laws of the universe, but he gives faith equal billing to science: both take an opening leap, and the leap to faith is often more helpful to life than scientific doubt. Sandage desired to hold a personal, even evangelical, faith late in life, prodded by a puzzle stated by philosopher-mathematician Gottfried Leibnitz: "Why is there something, rather than nothing?" Says Sandage, "I never found the answer in science." But he felt a "divine discomfort" that eludes the mind and ultimately touches on the heart and the will. "The only way I found to answer that is [to] decide to believe, and after the early experience . . . it turned out to be correct."[28]

Quantum chemistry is a relatively new field, and one that Henry "Fritz" Schaefer III had helped to pioneer; his research papers are among the most frequently cited by other scientists. "I became a Christian during my fourth year as a professor at Berkeley. I had gotten the word that I was going to get tenure," he says. "It allowed me to open my eyes a little bit." He has since accepted a chair of science at the University of Georgia and speaks at forums organized by Campus Crusade for Christ. Reared in the Midwest by parents who were scientific skeptics, Schaefer now argues that seeing design in nature is one more buttress to an orthodox and evangelical faith. His mental reasoning placed him opposite Darwin, who said "design" and order could arise without intelligence. Without God, Schaefer believes, "You can't get from interstellar molecules to the simplest self-replicating biochemical system in a hundred million years, or in fact in fifteen billion years, a reasonable guess for the age of the universe. These things just don't work."[29]

PYRAMID OF SCIENCE

The spectrum of belief works well as a shape on which to spread the biographies of particular scientists, but for science as a social institution, the pyramid is more suitable. Building that pyramid has been closely tied to the search for the "greater" scientists. When Cattell founded *American Men of Science*, his quest to figure out what produced the eminent man of science was not new. In England of 1869, Francis Galton's *Hereditary Genius* linked the brightest men by bloodlines, and he calculated that one eminent mind is produced per one hundred thousand population.[30]

Over forty-one years, Cattell's starred list inducted twenty-six hundred greater scientists, ruled eminent by committee. Some have speculated that the number of "greater minds" in a population might be a mathematical proportion that is constant, since Galton's ratio was matched roughly by Cattell's. Then came the rise of American "big science" in the late 1940s. With the number of credentialed scientists mushrooming, statistician Derek J. De Solla Price tried to estimate how much eminence the boom would produce. Based on the number of scientific articles published, he concluded that eminent science minds appear at a far slower rate than ordinary scientists, but he cautioned: "Just as one cannot measure the individual velocities of all molecules in a gas, one cannot actually measure the degrees of eminence of all scientists."[31]

The personification of eminence today is the National Academy of Sciences, which in the 1990s numbered around 1,800 elected members. Back in 1914 and 1933, Leuba's interest in "greater" scientists was to show that eminence correlated with disbelief. His eminent scientists in the 1914 survey, for example, attested to disbelief by 74 percent (52.7 percent disbelief and 20.9 agnostic). To compare the beliefs of top scientists then and now, Leuba's God question was sent to natural scientists in the National Academy in 1998 (see Appendix, table 2). Leuba would have been pleased at the response: disbelief had spiraled in elite science. Ninety-three percent of academy naturalists rejected belief in a personal God (72.7 percent disbelief and 20.2 agnostic).[32]

Such a process of secularization had been Leuba's historic vision for all of society, beginning first and foremost in science. The science profession was to rise like a majestic plateau, set off from society by its collective disbelief. What turned out, however, was a world of science shaped like a pyramid. At the apex, the "greater" scientists embraced the disbelief envisioned by Leuba. The pyramid's middle stood for ranking American scientists who believed in God, but at about half the rate of the population. The base of the pyramid was made up of scientists in applied fields, such as technology, medicine, industry, or science instruction. Their beliefs were closer to those of all Americans, who in vast majorities believe in God and the afterlife.

Three paths lead up the pyramid and into the unbelieving apex. They are the routes of mental acuity (being smart), social connections (friends in the academy), and choice of discipline (especially biology). Leuba had emphasized the first when he extolled "superior knowledge, understanding, and experience." This fits a century of conventional wisdom that learning erodes belief, and it matches the premise of the soft sciences (psychology, sociology, and anthropology) that man is evolving away from a primitive mind that once was home to superstition, hallucination, and fear. As anthropologist Anthony F. C. Wallace puts it: "Belief in supernatural powers is doomed to die out, all over the world, as the result of the increasing adequacy and diffusion of scientific knowledge."[33]

There is growing dissent on this front, largely because surveys since the 1950s have not been turning up disbelief among all educated Americans. In 1969 a

Carnegie Commission survey of sixty thousand U.S. college professors found that from 55 to 60 percent of physical and natural scientists were "religious." Some years later, a review of standard testing said, "There are no great differences in intelligence between the religious and non-religious." Others disagree, and a classic rebuttal comes from two Oxford dons, beginning with Atkins's assertion that "a real scientist in the deepest sense of the word" could not possibly believe. Also, Dawkins points to the "compartmentalizing," or self-deceit, required in a believing scientist's brain. "Anybody who really does live with the contradiction in his own head is being intellectually dishonest," he says. He sees the contradiction every Sunday when scientists of faith leave the laboratory and go to church. "They really do shut the lab door and then switch into a different intellectual gear."[34]

Scientists who are religious tend to "go for the most literal kinds of religion," says the Astronomer Royal of Britain, Sir Martin Rees, a nonbeliever. "That may be because they are the kind of people who crave certainty." The very uncertainty of science, he goes on, "makes me suspicious of anyone who claims to have more than a very incomplete metaphorical understanding of anything deep and important." When Ernst Mayr polled his National Academy fellows during a regular Cambridge lunch, they all disbelieved. "I found that there were two sources," Mayr says. One was the aversion to supernatural explanations. "But others would say, 'I just couldn't believe that there could be a God with all this evil in the world.' Of course, most atheists combine the two. This combination makes it impossible to believe in God."[35]

As a psychologist, Leuba operated on the theory that scientists who worked closely with living matter could more easily imagine that nothing supernatural was at work. Today this is called the "scholarly distance" hypothesis: scientists distant from investigating human phenomena can more easily believe in divine influence on human affairs. In his surveys, Leuba has noted that many scientists who worked in organic science "limited God's action to the psychic world," the world of human consciousness. Still, he said most believing scientists accepted divine intervention in physical processes as well as the mind.[36]

Back in the 1960s, sociologist Rodney Stark had actively researched the same "secularization hypothesis" that appealed to Leuba. Stark found some confirmation, such as that in 1963 "a major religious phenomenon associated with being a graduate student is loss of faith." Yet the "loss" did not appear all the time, or even usually, so he began to doubt. Overall, he noted, religion has survived the twentieth century "despite a tremendous increase in average educational levels, revolutionary growth in technology, and explosive increase in both the stock of scientific knowledge and the fraction of the population engaged in scientific research."[37]

He also disputed the "scholarly distance" hypothesis. He attributes it to a biased view in human sciences—from August Comte in sociology to Sigmund Freud in psychiatry—that religion arose from a "primitive mind." Stark wonders why, if the distance hypothesis is true, more physical scientists are not atheists.

He does not accept the response that somehow they are too distant. "Biologists and physicists," he protested, "routinely address religiously charged questions about human evolution and the origins of the universe." In contrast, he says, psychology, sociology, and anthropology are far less scientific, and "their semi-religious reliance on non-testable claims puts them in direct competition with traditional religions." Two hundred years of scientific marketing, meanwhile, has made higher education inhospitable to religious belief. "In research universities the religious people keep their mouths shut," Stark concludes. "And the irreligious people discriminate. There's a reward system to being irreligious in the upper echelons."[38]

Such a reward system may be at work in the second path up the pyramid of institutional science in America. Intellectual merit could be secondary to the necessary politics of scaling an organizational ladder. For Leuba, the top scientists had "traits making for success in the professions concerned." One of those skills was in conforming and making alliances socially, which were important for being starred by Cattell, says Stephen S. Visher's study, *Scientists Starred*. "If a man's research was not rather widely known, or if he was personally known to only a few, he did not receive enough votes to win a star," Visher says. Universities also viewed the starred list as a tool for institutional advancement. Says one 1939 study of starred psychologists, "The various universities employing starred scientists . . . not only attempt to retain and attract men already starred, but also to have local men not yet starred win this high honor."[39]

Visher also touched on what members of the National Academy of Sciences in the 1940s pointed to as the number one requirement for election: "proper connections." Next was "possession" of at least two influential friends in the right academy section; after that, an eligible age of "about fifty." Merit must also be a basis for selecting such an elite group, and current inductees have spoken of the impact of their scientific work as a major criterion. Yet science-as-social-club is invariably woven into these judgments. "Scientific excellence is defined by whom we elect to membership and invite to our scientific colloquia," academy president Bruce Alberts says. Election from within makes sure a science academy is "relatively insulated from the twists and turns of political systems."[40]

No one is quite sure whether this insulation also works against scientists who are outspokenly religious. "Getting into the National Academy is a terrifically political thing," says Michael Ruse. "A born-again Christian could be elected, but I would think that there might be tension." The issue arose in an earlier American debate on social classes in Protestantism, framed by the question: "Is there something about Episcopalianism or Presbyterianism that makes for eminence or achievement, or is there something about eminence or achievement that suggests a conformity to the most respectable (sic!) traditions or class stratification?" That question had been asked by a Protestant editor in 1933. Asked differently for science today, it goes: Do National Academy people disbelieve because of their higher intellects, or does the academy welcome only disbelievers? "It is a bit of both," Ruse says. Those in the younger sciences such as biology and sociology might feel more animus toward religion, he says. "That's because they are more insecure about their status, and partly because so many of these people would themselves be anti-Christian."[41]

Working in biology, in fact, seems to be the third well-trod path to science's apex of disbelief. Biologists believed less than physicists and mathematicians in the Leuba surveys of 1914, 1933, and 1996, and the survey of the National Academy found that only about 5 percent of biologists believed (a third of the rate for mathematicians). This confirms a general measurement that belief increases in the direction of physical science, engineering, and space science. Of scientists listed in *AMS* in 1959, "The chemical engineers attended church in a larger proportion [and] were much more likely to avow a belief in life after death." Next in belief come physicists. That resonates with British astronomer Anthony Hewish, a believer and Nobel winner. "I always felt, looking around me in Cambridge, that the physicists were more inclined to be believers than biologists," he says.[42]

Disbelief among biologists has been linked in various ways to the "distance hypothesis" as well. When compared with physicists, biologists think they can comprehend organic nature, says biologist Lewis Wolpert. "The modern biologist really thinks that if we go down to the level of DNA we understand things," he says. "If you are a physicist, in a world of quantum mechanics and the big bang, it is so bizarre and ludicrous that the concept of understanding almost disappears." Leuba elaborated a similar point, saying physicists and mathematicians believe more because they are "the scientists who know least about living matter, society, and the mind." Biologists, however, "have come to recognize fixed orderliness in organic and psychic life," Leuba said.[43]

Biologists seem to know all of these implications, especially in the face of the evolution-creation debate. It was the biology teachers (and not physics teachers) who declared neutrality on religion, and the National Academy's release of *Teaching Evolution and the Nature of Science* was the occasion for Alberts to say, "There are many outstanding members of this academy who are very religious people, people who believe in evolution, many of them biologists." Alberts was correct, given a wide range of meanings for religion. Yet after the 1998 survey of academy biologists, which found 95 percent disbelief, *Scientific American* commented, "The irony is remarkable: a group of specialists who are nearly all nonbelievers— and who believe that science compels such a conclusion—told the public that 'science is neutral' on the God question."[44]

In light of the pyramid of science described here, full-time career biology teachers in public schools might be less believing than chemistry or physics professors. Teaching, meanwhile, tends to fall into the applied fields at the base of the pyramid of science—where the difference in religious belief between science professionals and citizens is not as great. Moving up the pyramid, ranking scientists in biology would be more inclined to disbelief; this sector might include textbook writers or curriculum directors. At the apex, the level of science rarely comes in contact with the public. Research is done in top facilities, and National Academy advice goes to government panels or federal science consortia.

Some have argued all along that what scientists believe has no bearing on religion, and after the 1960s the religious musings of scientists may have lost all cachet. "It proves nothing to point out in the pulpit, as some preachers do, that God exists because a famous scientist says so," Edward LeRoy Long said back in the 1950s, when scientists enjoyed social prestige. With the technology of war, however, came a stark exposure to the human side of science, and now Long says that people became even more aware as they saw scientists disagree on policy, much like politicians, and recognized that knowledge has limits. "All of these have changed our tendency to appeal to science, or to scientists, as a way to endorse conclusions in other areas," Long says. Or as scientist Matt Cartmill wrote in 1998, scientists—"being only human"—persuade themselves that their science requires agnosticism or disbelief. "It's an honorable belief, but it isn't a research finding," Cartmill says.[45]

Some biologists are calling for honesty in one other area of this debate. "The legend grew up that biology established itself against the beliefs of Christians and the church," says Arthur Peacocke, an Anglican priest and Oxford biochemist.[46] The truth was, he says, that the Victorian biologists were cautious and skeptical about Darwin's new theory, whereas the clergy fawned over it almost immediately. Yet the story comes down backward, he says, thanks to Thomas H. Huxley and his early mastery of one particular tool for his cause—the great debate.

12. THE GREAT DEBATE

Biology professor Kenneth Miller arrived at Brown University in Providence, Rhode Island, in 1981, the year his own legend was born. Fifteen years later he works in his university lab, tanned and fit in polo shirt and denim shorts. Tall and lightly mustachioed, Miller is an unlikely heir to the sideburned, overdressed, and extroverted publicist and debater Thomas H. Huxley.

According to Victorian legend, Huxley had roundly defeated Anglican bishop Samuel Wilberforce in a public debate on evolution in 1860. The Miller legend began in 1981, when he was credited as being the first to blunt the 1970s juggernaut of creationist debaters, whose winning record on campuses and in auditoriums had its own legendary mystique.

Ever since the Huxley-Wilberforce duel, the winning or losing of evolution debates has relied as much on strategy, skill, audience, and personality as on facts, logic, or proof. Debates have shown that science and religion are difficult to separate in such public forums, especially when the debate is over the ultimate origin of things. Miller had the strategy, skill, and persona and was called "the best evolutionist debater to surface to date" by Henry Morris, the creation science leader. "Evolutionists claim that his debate with me in April 1981 was the turning point in the current evolution/creation warfare."[1]

The young Roman Catholic biologist—age thirty-two to Morris's sixty-three years—had prepared for the 1981 showdown, unlike other scientists. He had also "paraded as rapidly as possible a long series of evolutionary claims and anti-creationist charges, far too many for the creationist opponent to have time to answer," Morris complained. "He is handsome in appearance, charismatic in manner, and very glib in speech." Says Miller, "I'm also a natural ham." Son of a working-class Catholic family, he was a high school politician and student body president. "I was elected the governor of New Jersey boys state." From 1981 to 1983, Miller, an up-and-coming biology professor and textbook writer, would star in a total of five debates. He would make a few more cameos in the next decade, especially on television and the Internet, and some called his book *Finding Darwin's God* (1999) a joining of his skills in class lecturing, textbook writing, and debating. "He would do well in politics," says Michael Behe, whose ideas are criticized in Miller's *Finding*.[2]

Evolutionary biology has produced both publicists and recluses. Though Darwin pitched the *Origin* to a popular audience, he shunned the limelight, preferring to write letters, while Huxley took the debate public. "He is the best talker whom I have known," Darwin gushed. "He never writes or says anything flat. From his conversation no one would suppose that he could cut up his opponents in so trenchant a manner as he can do and does do." How much these verbal clashes have educated the public or altered policy is a matter of debate itself. But as public spectacle in modern America, there have been obvious dividends for creationism. "With the exception of the legal battles to outlaw evolution or to get 'scientific creationism' into the public schools, nothing brought more attention to creationists than their debates with prominent evolutionists," writes historian Ronald Numbers.[3]

Noteworthy debates were few when, in 1941, American biologist W. M. Smallwood looked back and found only three worth mentioning. Writing in the *Quarterly Review of Biology*, he began with Paris in 1830, when "two intellectual giants"—Georges Cuvier and Etienne Geoffroy Saint-Hilaire—debated twice before the Académie de Sciences. Cuvier rejected evolution, much debated in French science, and contested that organisms make sense only as an integration of functions, indeed as the Creator had set them in place. Warm to evolution, Geoffroy argued for his "principle of connections," the idea that similar body patterns existed across many or all organisms.

"The Cuvier-Geoffroy debate, like many celebrated scientific disagreements, was a highly personal affair," says historian Toby Appel. A long, prickly friendship between the two men ended in bitter reproach. "Although Cuvier exhibited a better command of the details of anatomy during the 1830 debate, and while his arguments were presented in a more logical fashion, Geoffroy and his followers were by no means routed," Appel says. Indeed, Geoffroy's view bolstered the study of homology, or common shapes, and French evolutionism, which Darwin in turn drew upon long before he published *Origin of Species*. And so, says Appel, the debate was "an important chapter in the history of evolutionary biology."[4] Smallwood, moreover, says that if Cuvier won the debate before the academy, "The man on the street was with [Geoffroy] Saint-Hilaire."[5]

Though the most memorialized debate in history, the Huxley-Wilberforce clash did nothing to jar science. It was, Smallwood says, "the least important as an analysis of the real problems involved." Each year the British Association met for a week in a different city and drew a curious citizenry. At Oxford in 1860, "The public flocked in, thousands of top-hatted gents and ladies in their new tent-like, crinoline dresses," writes historian Adrian Desmond. This was the Saturday public session, dedicated to the topic of Darwinism and society, and spectators from clergy to raucous students filled the university's Gothic Revival museum.

Upon his turn to speak, Bishop Wilberforce targeted the *Origin*. He derided its weak philosophy, evoked Egyptian mummy paintings to disprove that animals changed, and drew a line between man and beast. Two hours into the program, the bishop—recalling Huxley's reference to a gorilla to embarrass a Darwin critic—turned to ask him if apes were on his grandfather's or grandmother's side. "In the heat of the moment no one could remember his precise words," Desmond reports. Huxley recalls saying that if the choice was between a miserable ape and a man of means and influence who "introduced ridicule into a grave scientific discussion, I unhesitatingly affirm my preference for the ape." Huxley's autobiography calls it the winning blow, but Desmond concludes, "Perceptions of the event differed so wildly that to talk of a 'victor' is ridiculous."[6]

Looking back, Oxford biochemist Arthur Peacocke, an Anglican priest like Wilberforce, can appreciate Huxley's goal, if not the legend. "From 1860 to 1880, the event was never mentioned in any document, journal, newspaper at all," Peacocke said in 1998. "It was a nonevent." Yet after 1880 it became a titanic symbol handed down by the autobiographical Huxley: the triumph of the scientist over the priest. "Huxley was terribly keen, and I can't blame him for this, to establish science as a profession not under the thumb of the church," Peacocke said.[7]

The Oxford legend has obscured a debate that took place an ocean away in Boston, also in 1860. The debate pitted antievolutionist Louis Agassiz, head of Harvard's Museum of Comparative Zoology, against geologist William Barton Rogers, soon to found MIT. It was the first debate to seriously dissect Darwinian claims. Staged at the Boston Society for Natural History, it ran in four sessions through the spring of 1860. "Agassiz was a poor debater," Smallwood says. Once before, agitated by Asa Gray, the botanist and Christian evolutionist at Harvard, Agassiz "became sufficiently angry to challenge Gray to a duel." While the Boston debate marked the official divergence of two American views, it no more touched public opinion than did the Saturday episode at Oxford. Smallwood looked for coverage of the debates in the Boston newspapers, but they "failed to show that it was even mentioned."[8]

With the help of radio, the debates in America became mass public events in the 1920s, the decade of the first antievolution crusades. New Yorkers flocked to Carnegie Hall in 1924 for a debate broadcast on radio, pitting pro-evolution Unitarian minister Charles Potter against Baptist minister John Straton, both of New York City. A three-judge panel from the New York State Supreme Court gave Straton the victory on technical competence. A month before the 1925 Scopes trial, science popularizer Maynard Shipley debated two Seventh-day Adventist editors around the West Coast. For two nights at the Native Sons' Hall in San Francisco, a large crowd "filled the auditorium long before the meeting hour." The first debate was on evolution, the second on teaching it in public schools. Before a panel of California jurists, Shipley lost the first one for sniping at William Jennings Bryan and won the second for espousing freedom—that is, to teach evolution.[9]

The number of evolutionists drawn into this exchange increased in the 1970s, when the creation science movement organized a full-time speaking circuit. There were preludes, however. In 1961 Morris's *Genesis Flood* stirred sharp exchanges within evangelical circles and publications, and then in 1966 came news of the "great debate" in Arkansas, planned after the state supreme court struck down the state's unenforced, antievolution law of 1928. It was an election year, and seven days after the ruling the state declared it would be appealed.

Former Arkansas speech professor H. Brent Davis, who had become field director of the Anti-Fraud Committee of Texas—whose mission was "exposing faith healers"—arranged with Harding College professor of Christian doctrine James D. Bales for a two-debate session at the city auditorium in Little Rock. Davis said the debates would give "the world what the Scopes trial should have given it"—a full intellectual airing—and he promised the appearance of pro-evolution scientists who were "the best in the nation that are available." For the evolution side, the Nobel geneticist Hermann J. Muller, then professor emeritus at Indiana University, recruited two young evolutionists, Carl Sagan and Richard Lewontin. In the days before the debate, Davis circulated a statement on evolution, signed by 177 scientists, that Muller had written during the Arkansas court ruling. Muller also recruited for the debate his friend Ernan McMullin, a Catholic priest and chairman of the philosophy department at Notre Dame. "Muller was looking for a believing Christian to fend off the lions," McMullin recalls.

He and Sagan, then at the Harvard College Observatory, had quickly disassociated themselves from Davis, who carried a gun and was agitating for a third debate on "Resolved: That the Bible is the word of God." On their arrival in Little Rock, where an atmosphere of contention mixed with hundred-degree weather, they were greeted by an *Arkansas Gazette* reporter who commented, "I think you two guys are crazy to come here." So onstage, before sixteen hundred people, Sagan began by reading a disclaimer that said evolution and the theology of revelation are not connected. "We would never appear in any context in which supporters of evolution were to be cast in the role of critics of the Bible," he read. "Our debates have been given a carnival-like atmosphere by the public statements of Mr. Davis, who is not and never was empowered to speak on our behalf." They came rather than cancel, Sagan added, for the "opportunity of repudiating both extremes."

In the quiet before the debate—which turned out to be more like separate monologues—the moderator exhorted, "There are to be no alleluias." The topic was "Resolved: That Genesis provides the most probable explanation of the origin and nature of the universe," so Sagan spoke for twenty minutes on cosmic evolution. With slides he showed the scale of the universe, from a mother and child to a satellite view of Earth to millions of distant galaxies—a scale the Genesis writers did not comprehend. McMullin calls it Sagan's "dry run" for the *Cosmos* series he later produced. Sagan at first had wanted to debunk Genesis as prescientific, McMullin says, but the priest counseled restraint. "In those days, Carl was not well known," says the priest, who had told the audience, "Genesis is not in its intention an attempt to give the sequence of events of the origin of the universe." He recalls later, "I'm sure the debate didn't change any minds. They never do."

Lewontin arrived the next day for a second night of debate— "Resolved: That the theory of evolution has been scientifically established"—faithfully there because Muller had "put the screws on us to go down." He argued that change on the earth is indisputable, and that "if our opponents do not deny that the world is changing, . . . they must think that there is a continual intelligent intervention because the original creation was not good enough."[10]

It was a vivid trip to the Deep South for Lewontin, where he and Sagan—"two New York atheist Jews"—faced the hundreds of fundamentalists on separate nights. "Despite our absolutely compelling arguments," Lewontin wrote later, "the audience unaccountably voted for the opposition." He and Sagan drew different conclusions from their foray into the Bible Belt. Sagan believed that education would change the rural folks from superstition to knowledge. Marxist that he is, however, Lewontin believed the rural folk had turned to religion to compensate for feeling a loss of control over their fates. "When I was a student, we didn't learn about evolution," Lewontin recalls. "Evolution was jammed into school by a group of university scientists, who wrote new curriculum for the schools. The school accepted it all. But what the southern and southwestern populace recognized is that the Northeast and ultimately Europe were the fonts of a kind of culture which was being imposed on them, and they didn't like it. My heart goes out to them."[11]

Lewontin's social analysis had sympathy for William Jennings Bryan's left-wing populism in the 1890s. The sympathy was not wanted by creationist leader Henry Morris, but like Bryan he, too, believed in democratic science. So he took creation science to the heartland in the fall of 1972, his first public debate. Held at the University of Missouri in Kansas City, it was sponsored by Campus Crusade for Christ. Morris faced off with a geologist, Richard Gentile. "That one turned out so well that people all over the country began setting up these debates," Morris recalls. "I never have liked to debate, and I still don't." But it worked like nothing else. "We did find that by having a debate we could attract ten times the number of students to come as you would in a lecture."[12]

Without doubt, the lead debater of the creationist cause was Morris's chief partner, the biochemist Duane Gish. Between 1973 and 1999, Gish would participate in 247 debates, well over half of them at U.S. colleges and universities. When Gish was a Ph.D. chemist at the Virus Laboratory at the University of California at Berkeley from 1956 to 1960, he had read a little booklet called "Evolution: Science Falsely So-Called," and "that kind of got me excited."[13] His Baptist pastor in nearby Walnut Creek invited him to speak at a study series on Genesis. Word got around, and Western Baptist Bible College asked him to speak to the faculty. He moved to the research staff at the Upjohn Company in Michigan, where a Sunday school newsletter on his lectures spawned invitations to speak in Ohio and Indiana as well. He was also going to meetings of the American Scientific Affiliation and in time became one of the ten young-earth creationists who left to form the Creation Research Society in 1963. He left industry in 1971 for full-time work with the Institute for Creation Research.

The lecture that Gish gave in 1972 to the National Association of Biology Teachers meeting, which received a required fifteen-minute rebuttal, was nevertheless published in *American Biology Teacher* and stands as a rough model for his debate logic in the future. Pointing to the general theory of evolution—which he called "molecules-to-man"—Gish argued that when a theory like evolution "can, indeed, explain anything," then it is really not testable, as science requires. He went on: evolution is questioned within science; its theoretical validity is strongest when paired in a "two-model," evolution-creation, approach; it fails to explain the Cambrian "explosion" of new body plans and "discontinuity" between vertebrate classes. Evolutionists, Gish said, had resorted to new theories about jumps and abrupt appearances to explain the fossil record but had ruled out creation merely on the grounds of "authoritarian materialism." Throughout, he cited evolutionists to make his case.[14]

What was frequently added to this scheme was an argument from the second law of thermodynamics. The law says that closed systems degenerate. That means life could not create itself, Gish would argue. If time permitted, he would throw in the unlikelihood of dead chemicals organizing into self-replicating life. Nearly twenty years later, with 20/20 hindsight, Gish says of his format, "I begin by saying, 'This debate is not about anybody's theology and not about the age of the earth.'" Creationists disagree on the age question, he explains, so the only issue with the evolutionists is evolution. "Now, the evolutionist almost always spends time on the age of the earth, because he wants to convey the idea that, 'Oh, if the earth is old, that proves evolution.'" The evolutionists also know it is the best way to label creationists with Bible literalism. Indeed, it was the young-earth creationists' rule of avoiding debates on "time frame" that allowed them to "debate other issues," says science professor Walter Bradley, and "meet with a fair measure of success."[15]

In a team format, Gish and Morris had typically demanded one hour for each side, and then rebuttals. "Usually nobody gets up and leaves until you get to the question-and-answer period," Gish says. Creationists usually won as well, but there were reasons. "You see, in the early years they came unprepared. They thought, 'Oh, we are going to debate a preacher or an empty-headed religious nut.'" A chronicler of the first decade of Morris and Gish debates, Marvin Lubenow, says creationists like Morris and Gish could stay on script. They had mastered the use of slides and humor, while the evolutionists were neophytes. And it was great theater: "Creationism benefitted from this addiction to fads on the part of modern man," writes Lubenow. "Creation—although it is old—is just being rediscovered. Evolution has been around a long time." He estimated that in the first decade of debates—from 1972 to 1982—Morris and Gish had reached nearly 130,000 people. The total went up to 5.6 million when radio, television, and newspaper coverage was included.[16]

The redundancy eventually made the debate content and strategy easy to quantify; that was first done by Washington state biology professor David H. Milne, who debated Gish in the spring of 1979 at Evergreen State College in Olympia. "It is not easy to make evolution seem plausible to a skeptical lay audience in a one-hour presentation," Milne recalled. "My approach was successful." He wrote it down in the now-famous 1981 article "How to Debate with Creationists—and 'Win.'" He says that creationists hinged their debates on two points, the second law of thermodynamics and the "scarcity of transitional fossils." What is more, he says, "creationist debaters do not rely upon tangible evidence in favor of their position." They only attacked evolution. To blunt their approach, Milne showed easy-to-understand evolutionary fossils and put the burden on creationism to explain them. He also used a "barrage" of one-sentence examples to overwhelm any creationist response and cited odd creationist quotes. The evolutionist debater should not dismiss religion or "lose your cool," he says.[17]

Kenneth Miller, the Brown University cell biologist, was first to apply the Milne program. "In the spring a group of students in my introductory course said, 'There is this guy coming to campus and he says evolution is wrong and he's challenged anyone on the Brown faculty to debate him.' It was Henry Morris." Miller, who had been an assistant professor at Harvard for six years, went to Brown to teach upper-level cell biology. At first Miller declined, but when the Christian students suggested that the creationist might be right, he threw in his hat. "They got me audiotapes of two debates, one with Henry Morris and one with Duane Gish." The Gish debate took place in April 1980 on the Princeton campus of his opponent, anthropologist Ashley Montagu. "It was a stunner to listen to the debate," Miller says. "Montagu was destroyed by Gish."

In these tapes, Morris and Gish had defended a young earth by citing two scientific measures that implied youth: the Earth's weak magnetic field and a thin dust layer on the Moon. "I was flabbergasted," Miller says. "How do you answer that?" To prepare a rebuttal, Miller asked geologists about the measures and read Morris's book. When the day came, the debate had to be moved to the hockey rink to accommodate twenty-five hundred people, keeping them there for four hours. "By his own account, I just flattened Morris," says Miller.

Later that year Miller faced off with Morris at a high school in Florida, where a creation science bill had been introduced. "It was packed. And once again I very easily had my way with Morris." On a roll now, Miller teamed up with Milne to take on Morris and Gish before a largely evolutionist audience. "And that was a blast," he says. Chronicler Lubenow says Miller "often gave the creationist argument and then attempted to refute it before Morris or Gish had an opportunity to present it initially. Psychologically, it was quite effective. Miller was also an insulting debater, but he had a way of mixing it with charm and warmth, as well as with a few compliments." Despite Miller's headway, Eugenie Scott of the NCSE knew the talent was limited. "Avoid debates," she advised in 1996. "If your local campus Christian fellowship asks you to 'defend evolution,' please decline. Public debates rarely change minds; creationists stage them mainly in hope of drawing large sympathetic crowds." She added that they have the art down pat, while a teacher or professor would find it hard to simplify a course curriculum into a one-hour coup de grâce.[18]

The opponents in the creationism debates were not always evolutionists. The young-earth creationists had also galvanized opposition from evangelical theistic evolutionists and old-earth creationists. Since the days when the young-earth advocates had left the American Scientific Affiliation, the theistic evolutionists were not interested in their argument, though some campus debaters were professors with Christian convictions—like Kenneth Miller, a Catholic. The old-earth apologists, however, felt the debate was worthwhile. In 1977, for example, Fuller Theological Seminary in Pasadena, California, organized a debate between Gish and Jerry Albert, a Christian theistic evolutionist.

Such an interbeliever debate was on display in 1992, when the young-earth Gish clashed with old-earth Hugh Ross of the Reasons to Believe ministry on James Dobson's *Focus on the Family* radio show. Ross had been on this Christian show the year before, Dobson said; it had provoked a thousand supportive letters, a few charges of heresy, and "some very emotional reactions," such as a station threatening to drop Dobson. On the age question, "Some people feel like they absolutely know," Dobson said to open the Ross-Gish debate. "I'm not one of them." With Dobson's Bible-believing audience, Gish and Ross contended most over which scientific interpretation honored Genesis. Belief in the big bang, Gish argued, put Ross in company with all the atheist astronomers. Near the end, Ross said that Gish's Institute for Creation Research was highly uncharitable, calling him "an apostate who has no heart for evangelism." Dobson seemed to show an old-earth leaning. "How could there be a twenty-four-hour-day before the earth was revolving around the sun?" he asked Gish.

"Well, I just believe," Gish said, "that God had the ability" to rotate the earth. "Hugh does not believe in a global flood," said Gish.

"I believe in a universal flood," Ross said, using "universal" in a figurative sense.

"Did the floodwaters cover the earth?" Gish challenged.

Said Ross, "All of mankind was destroyed, and all the animals associated with him."

Gish rejected that explanation as insufficient, saying, "Ah, no, no, no."

Years later Ross, while giving credit to young-earth debaters for taking on the materialists, says it had become a rut. "In one sense, they are comfortable debating one another," he says. "Both sides have a vested interest that nobody else with a different viewpoint gets into the debate. The media also has a vested interest in making sure nobody else gets into the debate, because it takes away the sensational nature of what they are trying to report."[19]

. . .

Effective terminology has always been important for the debate, especially for the side that wants to talk about a principle without evoking God. The key example is the semantics of intelligent design, which emerged and matured in a window between 1984 and 1992.

The book *The Mystery of Life's Origin* (1984), cowritten by historian and chemist Charles Thaxton, raised the curtain on intelligent design by concluding that five options answered the origin of life dilemma, and one was that an "intelligent Creator informed inert matter." Eight years later, the argument was put center stage in a first public debate, "Darwinism: Scientific Inference or Philosophical Preference," held at Southern Methodist University in Dallas. The critics of Darwinism made no reference to creation and trafficked in terms such as "naturalism," "rules of reason," "teleology in nature," and "inference of design." One other impetus for the symposium was Phillip Johnson's *Darwin on Trial*. On his speaking circuit, Johnson deployed a new semantic tactic: he avoided the protean term "evolution." Instead, he debated against the "blind watchmaker thesis," which claimed that pure chance—or "chance of the gaps" instead of God of the gaps—gave rise to everything in nature that looked designed.[20]

While the Dallas debate may be considered the public launch of the intelligent design movement, it had been shaping up in publications since the 1970s. Texas publisher Jon Buell, who organized the Dallas event as president of his Foundation for Thought and Ethics, was the movement's Gutenberg. After a decade of campus ministry, he concluded that Christians needed first-rate intellectual products to match the worldviews of the big universities. So in 1972, at age thirty-two, he helped to found Probe, which produced booklets by Christians with science degrees and other academic credentials. These, he felt, could persuade far more students than "beating your gums to death" in campus talks. "The naturalists had realized the importance of origins for a worldview," Buell says. So he turned to three believers with doctorates in science—Thaxton, Walter Bradley, and Roger Olsen—to write a booklet on the subject. Their text soon became too advanced for the Probe series, and the project moved with Buell when, in 1980, he created the foundation to publish high school textbooks. The three authors completed *Mystery of Life's Origin* in 1983, and after inquiries with 176 secular publishers, it was released by the Philosophical Library in 1984. "We were determined the book would not be published by a Christian publisher, and therefore ignored," Buell recalls. "It was the first book favorable to creation by a reputable secular publisher in over five decades."[21]

Thaxton was a prime mover from the start. With a Ph.D. in biochemistry, the Texas native traveled to Switzerland to live in Francis Schaeffer's evangelical community of L'Abri. Filled with a sense of mission, he then completed two postdoctorates, one in science history at Harvard, another in biochemistry at Brandeis. He realized "you can't throw stones at science from the outside," but you can criticize theories from within, especially if the correct terms are used. "If you could avoid using terms like 'certainty' or using 'absolute proof' or 'knowledge with finality,' and you discuss your presentations in terms of 'probabilistic' themes, then you can say we 'would make an inference to this' rather than 'we prove this,'" Thaxton explains. He took other cues from the famous article "Life Transcending Physics and Chemistry" by chemist-turned-philosopher Michael Polanyi, a Hungarian émigré in England, and drew on what he had learned about information theory and advanced engineering at Harvard.[22]

Finally, on the problem of how information got into the DNA molecule, he applied standard uniformitarian thinking—"that the present is the key to the past"—thus, the information, with its marks of intelligence, must have been operative at the origin of life. When Thaxton and his coauthors wrote *Mystery*, he says, "We asked ourselves, 'What is being missed through the mind-set of those doing the [origins of life] work?' It was the question of information. It was almost singularly not discussed, or it was minimally represented." One leader in the research, Leslie Orgel, admitted to DNA being an information-loaded molecule, describing its quality as "specified complexity"—as if a person wrote down a specific onetime instruction for a given situation.[23] It was a term the intelligent design movement would build upon.

The year of *Mystery*, Thaxton organized a conference on atheism and theism. The British atheist Anthony Flew spoke, as did Dean Kenyon, a biochemist, who addressed the topic "Going Beyond the Naturalistic Mindset: Origin of Life Studies." Thaxton assembled another forum in 1988 in Tacoma, Washington, titled "Sources of Information Content in DNA." This event drew a new circle of interested parties, from Michael Denton, the Australian biochemist, to Notre Dame philosopher Alvin Plantinga. The question was becoming: What umbrella term for the nascent movement? "The word 'design' had been avoided because in biology it was forcefully resisted," says Thaxton. But few alternatives proved better.

Going back in history, biologists had spoken before of "creative intelligence" at work in living things. Louis Agassiz spoke of "rational intelligence." Theologians writing about science in the 1930s evoked divine intelligence, and nonbelieving British astronomer Fred Hoyle, discontented with Darwinism, wrote *The Intelligent Universe* in 1984 to suggest that outer space contained a quality of mind. Then, in December 1988, Thaxton sped off to lecture at Princeton. Needing overhead visuals, he grabbed a July news article he had clipped with the headline "Space Face." The *Viking I* spacecraft had photographed a sphinxlike face on Mars, and one scientist mused seriously about deciphering "intelligent design" in nature. Thaxton used the term in his lecture, and it was well enough received. "As the saying goes, the rest is history," he says.[24]

From Buell's point of view, *Of Pandas and People*, a high school biology text with a publication deadline of 1989, had to quickly put a name on its use of design theory. "Finally, the day came when we were going to have to decide," he says. So "intelligent design" went into print, and Buell looks back, saying, "I don't think anybody alive today coined the term." The term coined a movement, however, and a new approach to public debate. Says Thaxton, "I had already worked for fifteen years—and other people had before me—developing the ideas but, boom, it just reached a certain point and it was just like tinder meeting a spark."

With philosopher of science Stephen Meyer "kind of like a Johnny Appleseed," Thaxton says, others were drawn to the movement and branched into other theoretical problems. One new recruit was the young mathematician William Dembski, who had met Thaxton around 1991. Before that, Dembski had been contacted by Meyer and his colleague Paul Nelson, then a biology doctoral student at the University of Chicago. Meyer and Nelson had met at the Tacoma conference in 1988. They had taken on a project for the Pascal Center, a religion and science unit at Redeemer College in Canada, and in 1991 were looking for a mathematician. In separate incidents, they read pieces by Dembski in *Nous*, a philosophy journal, and *Perspectives in Science and Christian Faith*, the journal of the American Scientific Affiliation. They liked his ideas and credentials: a Ph.D. in math from the University of Chicago and a postdoctoral stint at MIT. "I was in Chicago working on a Ph.D. in philosophy and did the design work on the side," Dembski recalls. "I was really pretty isolated." He joined Meyer and Nelson on the Pascal project but first was recruited to be a presenter at the 1992 symposium in Dallas.[25]

Meyer turns to the year 1996 to typify the new debate scene. Atheist zoologist Dawkins, with his *Climbing Mount Improbable*, and intelligent design biochemist Behe, with his *Darwin's Black Box*, treated the mainstream science book market to opposing views. "It is more in the evolutionists' interest to deal with the young-earth creationists," Meyer says. "It establishes a false dichotomy. They say it's either Elmer Gantry or it's us." He pointed to the airing on PBS of "Inside the Law" in 1996, in which Dembski was paired with Kevin Padian, University of California at Berkeley paleontologist, on what might be taught in schools. "That's not a mismatch anymore," says Meyer. The next year, four members of the intelligent design group faced off with four evolutionists on a PBS *Firing Line* special.[26]

"These debates seldom settle anything, but they do often show viewers where people are coming from," Padian says of the PBS event. He says the intelligent design group was off base: "They're attempting to isolate evolution from the rest of science. It's futile." As a backer of the intelligent design side, Meyer says the PBS debate further undermined "the Scopes trial stereotype." Debate participant Behe says that while there was no time for details, the face-off "might stir up interest for people to look at things on their own." Scott drew this conclusion after being in the debate: "There are educated people out there"—namely, PBS viewers—"who are still clueless on the high level of activity on antievolution." So *Firing Line* was a wake-up call. "Antievolution is not only young earth."[27]

The main anchor of the movement was the Center for the Renewal of Science and Culture, run by Meyer at the Discovery Institute. In 1999 it operated on an annual budget of roughly $750,000, with forty-five affiliated fellows doing research on evolution and design topics. The other organizational entrepreneur of the intelligent design movement was Dembski, who added a Princeton divinity degree to his mathematical background. He drew on computers, information theory, probability, and complexity theory to argue that intelligence could be an alternative scientific model to Darwinism. "You just can't shift into a vacuum. You need a new conceptual grid." Other alternative concepts, from chaos theory to early information theory, generated an initial hoopla, he says, but no revolution in science thinking. "Specified complexity does have the potential to shake up our worldview."[28]

To introduce that revolution to academia, Dembski founded the Polanyi Center at Baptist-affiliated Baylor University in 1999. Faculty opposition would shut down the center after a year, but not before Dembski pulled off a major conference, "Nature of Nature," which featured two Nobelists—Christian de Duve and Steven Weinberg—other naturalists, theists, and design theorists. One session was a face-off of evolutionists. Robert Wright argued that evolution is progressive and provides human meaning, while his opponent, *Skeptic* publisher Michael Shermer, championed nature as aimless and meaningless. But finally, the center itself became the topic of debate, caught in the maelstrom of Texas Baptist politics and academia's deep suspicion of "creationism." Still, it had produced one first-rate exchange between theists and naturalists. Dembski says: "The important thing is to get the conversation going. It's not to set up a debate and stack the deck so that one side wins."[29]

. . .

Since the contemporary debates got started in the 1920s, how have they changed? The social settings have evolved, of course, but much has stayed the same in debates between atheists and believers, evolutionists and creationists, and modernist and orthodox Christians, as well as between evangelicals who interpret Genesis differently. Nine brief summaries provide perspective.

Straton versus Potter (1924). At Carnegie Hall. "Resolved, that the Earth and Man Came by Evolution." The Reverend Charles Potter of West Side Unitarian Church said evolution is acknowledged by science, change is evident to all, early humans had simple beliefs, and a personal God can work in natural law. The Reverend John Straton of Calvary Baptist Church in New York said evolution is a philosophical dogma that opposes God and the Bible, misconstrues man, relies on chance, and, scientists admit, has no answer for the origins of life.[30]

Shipley versus Nichol (1925). At Native Sons' Hall in San Francisco. "Resolved, that the earth and all life upon it are the result of evolution." Maynard Shipley, founder of the Science League of America, said the process of evolution is "fact," and only the "cause" of the process is debated, that living beings had orderly evolution, that Bryan's protests are absurd, that apes are only cousins of man, and that evolution predicts facts of nature. Francis D. Nichol, associate editor of the Adventist *Signs of the Times*, said evolutionists must "prove" the origin of earth, life, and species; he cited authorities saying that Darwinian "theory" is more psychology than science and is not backed by anatomy, embryology, or fossils.[31]

Riley versus Rimmer (1929). The Northwestern Bible Conference in Minnesota. "Resolved, that the creative days of Genesis were aeons, not solar days." William Bell Riley, a fundamentalist leader, said "day" in the Bible has broad meanings, progressive creation is more probable, it is demanded by geology and ancient cosmologies, and is favored by conservative scholarship. Harry Rimmer, a Presbyterian evangelist who believed in a six-day Creation on an ancient earth, said a Bible "day" means one rotation of earth, God does not need eons to create, the fossil record came from Noah's flood, evolutionary geology is secular, Moses wrote literally and later talked of a Creation week, and if the days were ages, how could the earth's biota survive on the third day if the sunshine came on the fourth?[32]

Bradley versus Morris (1982). Chicago meeting of the Council on Biblical Inerrancy. Professor Walter L. Bradley, a material scientist at Texas A&M University, said that "create" in the Bible can mean miracle or process, the Genesis "day" allows for the ancient time of modern geology and astronomy, and thus progressive creation is most credible. Henry Morris, president of the Institute for Creation Research, said the burden is on those who are reinterpreting "day," that progressive creation makes a "God of the gaps," it has no answer for death and suffering, rejects the catastrophism of a global flood, and is a slippery slope to religious liberalism.[33]

Gish versus Saladin (1988). At Auburn University. Kenneth L. Saladin of Georgia College's biology department said that by science's criterion creationism is supernatural, historic science need not "see" events, Bible belief leads to young earth and search for Noah's Ark, and creationists misquote experts. Duane Gish, a biochemist and vice president of the Institute for Creation Research, said the debate is not over the age of the earth, that a self-powered evolution and rise of complex DNA is impossible under the second law of thermodynamics, that the "theories" of evolution and creation are equally valid, and that intermediate fossils are absent; then he cited the Piltdown man hoax among the fossils of human evolution.[34]

Johnson versus Provine (1994). Stanford University. Phillip Johnson, University of California law professor, said mindless evolution denies theism, evolution museums deceive by lacking empirical evidence, natural selection cannot equal animal breeding by humans, and Darwinism is philosophical naturalism. William Provine, science professor at Cornell University, said Johnson presumes God for meaning, that after Darwin's first insight about common descent he posited natural selection to replace a Designer, that biological change is not limited and has intermediate forms, that extinctions question a Creator, and that Darwinism does indeed lead to atheistic humanism.[35]

Behe versus Dennett (1997). Notre Dame University. Lehigh University biochemist Michael Behe said the hardest part of biology is to explain differences, especially irreducibly complex novelties, that design is a detectable quantity, and that natural selection does not explain what seems designed in nature. Daniel Dennett, a Tufts University philosophy professor, showed computer-generated evolution as an example of nature, said natural explanations inevitably will be found, that design cannot be absolutely ruled out, but that critics of natural selection have the burden of proof and that ignorance of how evolution works is not a good argument against it.[36]

Weinberg versus Polkinghorne (1999). Washington, D.C. "Cosmic Questions" conference of the AAAS, asking, "Is the Universe Designed?" Weinberg, a particle physicist, said no because you first need an idea of a designer, that fine-tuning arguments are unconvincing, human suffering refutes a benign design and purpose, and religion insults human dignity. John Polkinghorne, an Anglican minister and Royal Society physicist, said accepting either design or no design is a metaphysical stance, that nature's intelligible order, human consciousness, and human comprehension of mathematics all suggest design, and that evil is no simple matter because of freedom in nature and the human will.[37]

Gould versus Falwell (1999). CNN *Crossfire*. "Creationism versus Evolution: Which Should Be Taught in Our Schools?" Harvard paleontologist Stephen Jay Gould said polls showing two-thirds support for both views indicate the public's fairness, but learning is not by majority vote, good teaching is not dogmatic, a theory is an extremely strong category in science, creationism is a literal Bible religion out of place in science class, and science has no conflict with religious ethics. Jerry Falwell, a Baptist minister and chancellor of Liberty University, stood by the Bible; said that Americans are smart, are not intimidated by Harvard elitists, and should get their way in tax-supported schools; and that academic freedom covers creationism, which is as scientific as evolution.[38]

In his 1981 article, biologist Milne said the debate route was worth the trouble. Teachers can learn to teach better, and the cause of evolution is advanced. They will, moreover, gain "a sense of continuity with biologists of other generations that cannot be acquired any other way."[39] He was speaking, of course, of standing in the shoes of Huxley at Oxford and biology teacher Scopes at his Tennessee trial in 1925.

Not all modern-day evolutionists agree. Gould had done television spots when news was breaking, but his appearance on *Crossfire* was rare. He had discouraged Miller from diving in but acknowledges that some evolutionists have the knack for debate. "But the point is, it's politically stupid," Gould says. "This is a political struggle. It's not an intellectual struggle. I'm talking about the right-wing creationists. Debate is absurd. It's dishonorable in science to think that an issue will be solved by rhetorical skill. Young-earth creationists want a sign of respectability they cannot have, because they are wrong." Johnson has taken such comments and Gould's refusal to debate with him as a sign of the fragility of evolutionist arguments. He would not disagree that it has tremendous political symbolism. "If they agree to meet me in a public forum at a university, which means it is an academic event, on equal terms, the game's over," Johnson says. "We don't even have to hold the debate. The point has been made. See, they all know that." He has debated on several occasions with William Provine, but that's different, he says. "Now, Provine sort of doesn't care. But they do."[40]

Not long after Johnson's *Darwin on Trial* came out, the vice chancellor of the University of California, John Heilbron, invited Francisco Ayala to debate him. But Ayala declined. "I am quite willing to debate a scientist who has beliefs different from mine, but not somebody who is not a scientist," says Ayala. He says he found Johnson's book "not scholarly, not serious, so I said no." He does go to public forums, however, where others can pummel him with criticism if they please. "I don't want to be engaged in debates where it's all a matter of rhetoric, and where it's your word against mine. You have your twenty minutes to say something, I have my twenty minutes. What do I do, call you a liar, say that you are wrong?" Michael Ruse finds that in religious America the evolutionist is always at a disadvantage, but he is not without hope. "When I'm in these debates, I'm not going to convert Phil Johnson or Duane Gish. But I'd like at least one undergrad to go home, and say, 'Gee, you know that Ruse, he had a couple of good lines. I wonder if he's got a point?'"[41]

The debates have definitely ignited an interest in some Americans to pursue the topic further, and even to take sides as activists. To a considerable degree the debates also preach to the converted. Their audience reach is limited, except when they can hit the PBS airwaves, as with the *Firing Line* special. The mass media may even overshadow the debates as the place where Americans get their ideas about evolution and creation.

13. MEDIA-EYE VIEW

The *Baltimore Sun*, a venerable East Coast daily, had a unique stake in the Kansas State Board of Education vote over the status of evolution in science standards in 1999. Decades earlier, *Baltimore Evening Sun* correspondent H. L. Mencken had gone down in history for his colorful coverage of the 1925 Scopes trial, which he had "skewered," the modern-day *Sun* told its readers during the Kansas controversy. "In the spirit of the old saw that those who forget history are condemned to repeat it, we bring you excerpts from what Mencken described as 'The Tennessee Circus,'" offered the *Sun*.[1]

The history of U.S. news coverage of the evolution-creation debate suggests that the Scopes trial has been nearly impossible to forget. Its symbols and themes have dominated the press's handling of the topic. The blame that Mencken cast on southern Bible thumpers for the "circus" endures, even though historians now have shown that the event was staged as much by cosmopolitan news media and New York City interest groups. "The press has made this story," said one correspondent on the scene in 1925. "Its spotlight has been turned upon Dayton as if by a common agreement among all editors everywhere." The task of reporting facts with fairness and accuracy was sorely tested, *Editor and Publisher* reported. "Some of the reporters are writing controversial matter, arguing the case, asserting that civilization is on trial. The average news writer is trying to stick to the facts as revealed in court, but it is a slippery, tricky job at best."[2]

The press as a modern institution—with its mixed bag of objectivity, interpretive reporting, and pack journalism—in many ways came of age while covering the "trial of the century" in that hot Dayton summer. The trial drew more than two hundred reporters. It was the first for which live transcontinental radio was used—for eight days straight. Newsreels were shipped to theaters overnight.

The days of trial coverage, one church historian has argued, were enough to force an entire population of Protestant fundamentalists culturally underground for the next forty years. Interpreters of the trial have looked back selectively at the news clippings and recast its meaning in books and movies such as *Inherit the Wind*. Thus dramatized, the evolution question has become what one *New Republic* editor calls a veritable "IQ test" for Americans. With the IQ test in hand in early 1980, *Boston Globe* columnist Ellen Goodman took as "absurd" stories of a creationist resurgence. "It was as if some people had exhumed poor John T. Scopes and put him back on trial."[3]

In its modern incarnation, the evolution-creation debate is a story that reporters can love and hate. Having covered the progress of a 1996 antievolution bill in the Tennessee Senate, one *Memphis Commercial Appeal* reporter said, "The [Jay] Leno quotient was very high. . . . It was just a fun story, who could resist?" Newsman Lou Boccardi, while president of the Associated Press, ranked the topic with the most explosive social issues of the day. "Reporting on creationism or gay issues or abortion or gun control are areas where the mandate for fairness is especially strong . . . and difficult to execute," he said. Quoting the extremes is part of balanced reporting: "I don't think it makes you unfair if you interview somebody who's on one end of a spectrum on a controversial public issue. It probably strengthens the story."[4]

The news media's discomfort with this topic arises from being caught amid its readers' and viewers' conflicting interests. Nearly all scientists think creationism should be given "no" or "low" priority in the news, while most of them say evolution has "moderate" or "high" news value. Most of the newspaper-reading public, meanwhile, believes that fairness dictates that both creation and evolution be taught in public schools. General news editors and science writers are caught somewhere in between: they nearly all agree with scientists that creationism merits no serious coverage, and yet only 3 percent of them give scientific research on evolution a "high" priority as news.[5]

Many topics get low priority in newspapers, newsmagazines, and network news. The media quest has been described variously as trying to "reflect" society "without fear or favor" and meanwhile attempting to educate the public while staying in business. It has had a mixed record on the evolution-creation topic since the Scopes trial set the tone for most coverage thereafter.

．　．　．

The "Darwin industry," a worldwide group of scholars, publicists, and publishing houses that make a living on Darwinian studies, has a small subset that could be called the "Scopes industry." Journalism professor Edward Caudill of the University of Tennessee has made some of its most interesting contributions.

He sifted through indexes and samples of the *New York Times* and the monthly *American Journal of Science* between 1860, the year after *Origin of Species* was published in England, and 1925. "Darwin (or evolution) tended to be a hot or cold topic, depending on the existence of events that made the theory newsworthy," he says. In those years, the number of annual stories on Darwin or evolution ranged from none to 460. Coverage spiked with the 1876 visit of Thomas H. Huxley, Darwin's death in 1882, his birth centenary in 1909, and then—shooting up like a rocket—the first antievolution law in 1922. "The most dramatic change in the nature of coverage was a shift from portraying evolution as a radical idea in 1860 to presenting it as a valid scientific theory in 1925," Caudill says. "Articles in the *New*

York Times were in complete opposition to the skepticism expressed sixty-five years earlier." The *Journal*'s coverage was more regular but less explosive on political ferment than the general-interest *Times*.[6]

For the Scopes trial, Caudill then looked at widespread news coverage in July 1925, a period described as "sensationalist" in the annals of journalism. The factual question of the trial was simple: Did Scopes teach evolution in class, flouting the state law? But it became a trial of ideas and public opinion. Given that outcome, the news media could have discussed "the validity of ways of viewing material evidence and drawing conclusions about its meaning," Caudill says. The trial evidence was not so material as the vying claims of religion and science. Bryan held out the revealed knowledge in the Bible, while Darrow attacked those "data" and fell back on the authority of science.

"The press had for the most part abandoned idea-based reportage for fact-based coverage of events, people, and places," Caudill says. "Hence, Bryan's philosophical argument was doomed before the trial began." The news media's preference for verifiable fact rather than even rational ideas had become "the institutional bias of the press," says Caudill. Rather than the news media dissecting two kinds of knowledge, the *New York Times* "devoted a lot of attention to the idea of empirical knowledge." In the *Chicago Tribune*, "Bryan was condemned for not accepting empirical thinking."[7]

This shift toward empirical fact and news as an event had taken place well before Scopes, Caudill says, but the trial set up a certain template for stories on religion and science. In the context of courts and politics that would mature, but first came the savvy publicity machine of the Darwin Centennial at the University of Chicago in 1959. The event gained unrivaled headlines for evolution. Putting an emphasis on celebrity, the university gave Julian Huxley a brief professorship and drew Sir Charles Galton Darwin, grandson of the naturalist, if not succeeding with the royal prince. The hard-sell press agent promised that "it's going to be a scientific and intellectual world series," and later an *Encyclopaedia Britannica* documentary set this tone: "Newspaper headlines, stories in magazines and television interviews have drawn attention to the celebration; excitement is in the air this week in November." While the five panel discussions at the heart of the celebration came off "surprisingly flat," says historian Vassiliki Betty Smocovitis, twenty-seven reporters thronged to the university hallways for *The Darwin Exhibit* or reported on an original Darwinian musical, *Time Will Tell*, whose Victorian overtones had to suffice, since the hoped-for staging of *Inherit the Wind* had fallen through.[8]

The stories carried headlines such as "Scientists Toast Darwin's Ghost" or "Evolution a Fact Darwin Fete Told." But it was Huxley's "evolutionary vision" speech (which drew "unfavorable" coverage for its atheism) and a session on science and theology that were, the publicists recorded, the "outstanding spot news events." In all, the centennial generated 196 printed news items from fifty-seven publications. The publicity staff put it this way: the 4,147 inches of news copy amounted to "more than enough to fill a twenty-three-page newspaper without ads." Historian Smocovitis summarizes: while the "popular press capitalized on the occasion with trivial and sometimes inflated reports, it also frequently transmitted the scientific consensus." Evolution was reported as "a recognizable fact," she says, and under the theme of knowing the past to control the future, "Evolutionary progress and social progress were thus inextricably linked for American popular audiences."[9]

No political issues were wafted upward by the centennial, but in a few more years that would reverse amid clashes over school boards, textbooks, and state laws. The political and legal reporting practiced at the Scopes trial now became established procedure, for the evolutionists were making a final push to abolish the nation's three remaining antievolution laws in Tennessee, Arkansas, and Mississippi. In 1965 the Arkansas Education Association asked biology teacher Susan Epperson to file a suit challenging that state's law. Biology teachers had ordered the *Modern Biology* textbook, which covered human evolution, and she sued to gain her freedom of speech on the topic. The trial judge ruled in her favor. As the state began an appeal that wound its way to the U.S. Supreme Court, Tennessee legislators saw the handwriting on the wall and overturned their state's statute. Mississippi lawmakers did the same to their law after the Court ruled in favor of Epperson in 1968.

The court said antievolution laws had a religious purpose because they banned one kind of scientific topic "for the sole reason that it seemed to conflict with a particular religious doctrine."[10] This was an unconstitutional establishment of religion, the court said. By commenting that public opinion favored the teaching of evolution, the court also gave a cue to the national press. To the media, *Epperson* was "a curious but welcome victory over the dead hand of a bygone era," says historian Edward Larson. *Time* said it "seemed to come from another era, a benighted past," and *Newsweek* marveled that it took forty-three years since Scopes before the court "struck down one of the monkey laws." *Life* magazine ran a satire titled "Arkansas Against the Onrush of the 20th Century."[11] Combined with the decade's Supreme Court rulings against prayer and Bible reading in public schools, the evolution ruling signaled to the news media that in public schools, religion in any form was out, but evolution was in.

Not until the 1980s were some of these assumptions shaken. The decade before, creationists had experimented with a new strategy: they claimed that teaching the "religion of evolution" violated religious freedom, or that Genesis should get equal mention to balance the viewpoints. Fundamentalists and secular humanists were already doing battle on these topics, but when presidential candidate Jimmy Carter acknowledged being "born again," many of these religious issues finally became fair game in newsrooms. So NBC anchorman John Chancellor came clean in 1976 with, "By the way, we've checked this out. Being 'born again' is not a bizarre, mountaintop experience. It's something common to millions of Americans—particularly if you're Baptist." Other "unknowing political reporters," says news veteran John Seigenthaler, "wondered whether the former governor of Georgia was some sort of religious nut."[12]

President Carter was not a creationist, for the Christian right soon accused him of "secular humanism." Yet the anecdotes suggest how unprepared the national media were for the creation science movement when it began hitting the courts again. That began in 1979, when Kelly Segraves, an activist creationist parent in San Diego, petitioned the courts to overturn the California science curriculum for violating its own "antidogmatism" policy. Getting no satisfaction, he sued the California Department of Education for violating his thirteen-year-old son's religious liberty. In opening the March 1981 trial, son Kasey "solemnly testified that his teacher insisted that he was descended from an ape." The *Washington Post* called it a "modern version of the 1925 Scopes monkey trial" and quoted the older Segraves as saying, "I am told this could be the trial of the century." Began the front-page story: "After 56 years of fitful slumber among the old, passed-over issues, it is back: evolution vs. the creation. The literal Bible vs. the accumulated judgments of scientists."[13]

Network cameras flocked to the superior court in Sacramento expecting to see up to twenty expert witnesses, including Carl Sagan, but news interest dropped precipitously at the end of the second day. Plaintiff Segraves narrowed the complaint from wanting evolution diminished or creationism augmented to merely asking that the term "hypothesis" be added to the California science framework. Until 1974, in fact, the framework had allowed classroom affirmation of the creationist view, and since 1972 it had carried an "antidogmatism" policy on evolution. Roughly a month after the Segraveses' day in court, the judge was reported to have told the parties on June 12 that the ruling allowed "each side to feel that it had won."[14] While the Segraves boy had lost no rights, he said, the antidogmatism policy was not being enforced. The remedy: circulate the policy to all school districts.

The California trial lent itself easily to Hollywood's *Inherit the Wind* characterizations, but news coverage also disclosed some new developments. Many of the new creationists had advanced degrees in science and teaching experience in public schools. "In the past," offered *Newsweek*, "the strongest advocates have been fire-and-brimstone preachers. Now a new group of spokesmen has rallied to the cause—born-again scientists who use concrete evidence rather than threats of damnation to convince doubters." The trial had been previewed as a clash in which attorneys "are ready to discuss science and religion—to debate whether religious freedom requires that public schools must teach, or at least not contradict, the biblical story of creation, taken literally."[15]

Rough-hand histories of the Scopes trial appeared. The "new creationist movement" behind the lawsuit was described as "a direct descendant of the successful [movement] of Scopes's day"—which was only partly true. Many youngearth creationists looked back on Bryan as a traitor for his belief in an old earth and animal evolution. The new creation science movement behind the California litigation was ardently opposed to the big bang theory and upheld Noah's flood.

The Scopes trial, moreover, raised no constitutional issues about religion or free speech. California's titanic court battle of 1981 finally whimpered over what the judge called the "semantics" of fact and theory. Segraves's lawyer Richard Turner explained the withdrawal as a fender-bender strategy in which a victim asks for "a million bucks" without expecting it. "They end up settling for a few thousand. That is what has happened here." Still, the ruling was "the closest thing to a court-room victory for creationists since Scopes."[16]

News coverage worthy of the Scopes legacy finally came in December 1981, when a "balanced treatment" law that had been passed in Arkansas that summer was tested in a federal courthouse. Though the Little Rock courtroom was tight, seventy-five news organizations wanted credentials to get inside. Looking at their trial reportage later, a content study ruled that it gave positive coverage to the embattled law, mainly because of its appeal for "balanced treatment." While conventional media theory puts the courts and press elite in one cabal, this time Judge William Overton's ruling against creationists did not hurt their credibility in the news. An occasional headline such as "Creationist Tells of Belief in UFO's, Satan, Occult" resuscitated stereotypes. Still, news reports "worked indirectly to legitimate the populist discourse of creationism" by shunting aside "expert testimony." The news echoed a "public idiom" of fairness in matters of local autonomy. And to the chagrin of elites in science, each side was quoted as equivalent. To the contrary, elites held that "science education is an enterprise best organized by experts, and notions of pluralism and fairness are irrelevant to the discovery and transmission of scientific knowledge."[17]

Science writer Marcel La Follette, later an educator at MIT, took another look at the Arkansas trial coverage with an eye to how science fared. Sadly, she says, though several scientists had testified and the judge had ruled according to their testimony, newspapers and newsmagazines omitted discussion of the science issues. "Scientists were quoted only rarely," La Follette says. "Accounts stressed the difficulties that court and counsel encountered in grappling with highly technical testimony; yet none of the science testimony was described." She says, "John Scopes, in fact, was mentioned far more than any of the Arkansas witnesses." La Follette added that scientists, though on the side of the court in the end, suffered from being portrayed as "dogmatic" elitists, and some of that was their own fault. The science news media, La Follette says, sometimes came off as snide toward creationists, which "may contribute to public sympathy" for the creationists' cause.[18]

Other studies said the rhetoric of the trial showed the classic power struggle of one "profession" against another. A leading creationist, biology professor Wayne Frair, says creationists got a "lot of mileage" from the trial, despite humiliating ACLU tactics of questioning about Satan and UFOs. "Creation was discovered because of the trial," Frair says. "There was hardly any publication that dealt with public issues that did not discuss it." Involved creationists were "busy for the next five years dealing with things related to this particular trial."[19]

The Arkansas trial came at a time of transition in some theoretical studies of mass media. Media pundit Walter Lippman had gotten the analysis rolling around Scopes's time when he said that news may be "facts," but media-generated "pictures in your head" turn them into reality. By the 1960s, mass-communications studies had gathered around the idea that the news media were "agenda setting." News people were gatekeepers of those mental pictures Lippman had discussed. It was not total power, but it was power, said one team of media analysts: "If the media cannot tell people what to think, the media can demonstrably tell them what to think about."[20]

This idea has not been toppled, but mass-communications experts have augmented it by saying "frames" imposed on the facts created a new media reality. "Events do not speak for themselves but must be woven into some larger story line or frame; they take on their meaning from the frame in which they are embedded," says William Gamson. "A frame is a central organizing idea, suggesting what is at issue." The new assumption was that authorities, activists, and other newsmakers could control media by promoting certain frames, what the press calls "spinning" the news. The strategy came to involve polls, focus groups, "media events," and "leaks," and every savvy politician had a "handler" to research which frames won public majorities—and which tarnished an opponent's appeal. There were limits to the spin, especially for less powerful players, for journalists still are the gatekeepers, Gamson says, "deciding which frame sponsors will be granted standing and selectively what to quote or emphasize."[21]

The national media and their scholarly probers also have come to acknowledge that certain kinds of stories will always get coverage. Herbert Gans's book *Deciding What's News* (1979) presents a list of story-line categories that dominate newsrooms and speak to the public. The groupings are broad and include altruistic democracy, responsible capitalism, social order, and national leadership. Deviations from these tracks stay beyond the pale of news interest. For the Arkansas trial, La Follette matched up Gans's values of "moderatism," that moderates are heroes, and the news of "moral order," the attempt of a law to bolster moral instruction for children.[22]

Three years after the judge's January 1982 verdict in the Arkansas trial, a Louisiana federal court refused to reverse a lower court rejection of a similar "balanced treatment" law in that state. While some creationist leaders opposed appeal of the Arkansas trial, fearing a media backlash, others believed the Louisiana ruling lacked some of Arkansas's problems, so they went ahead. The Supreme Court ruled narrowly in 1987 on whether "balanced treatment" had a secular purpose and decided the opposite. The *Washington Post* and *New York Times* described the 7-to-2 ruling as a "major defeat" or "major blow" for "fundamentalists," linking the story to a political season in which religious broadcaster Pat Robertson was running for the Republican presidential nomination. The *Post* devoted a tenth of its story to Justice Antonin Scalia's combative dissent, which constituted a third of the *Times* report. Scalia, saying academic freedom was a valid secular purpose for the law, chided the bench for its "instinctive reaction that any governmentally imposed requirements bearing upon the teaching of evolution must be a manifestation of Christian fundamentalist repression."[23]

The next year, Robertson lost in the GOP primaries, making 1988 the zenith for top-down efforts by religious conservatives, which had begun during the Reagan administration. The new strategy became bottom-up, or what was being called "board fundamentalism." School boards were viable targets, and religious conservatives learned the value of winning in low-turnout elections. Opponents labeled the successful challengers in such races "stealth candidates" to suggest secret agendas that, dictator-like, they would impose once in power. The "stealth" frame was an effective spin in the media, though the facts usually were ambiguous.

Such was the case in the Vista Unified School District, north of San Diego, where in November 1992 two conservative Christians won seats, joining a third to form a majority bloc on the five-member school board. The election became one of the biggest creation-evolution stories of the decade, helped along by the fact that one elected school board member was the bookkeeper at the Institute for Creation Research. The flap lasted two years, was featured in a PBS documentary, prompted a failed recall election, and evoked the ghost of Scopes: "The setting will be Vista, Calif., rather than Dayton. Other than this, one might conclude that little has changed in nearly 70 years," a columnist wrote in the *San Diego Union-Tribune*.[24]

The creationism issue first arose at a January 1993 board meeting when educators who were keen on evolutionary biology demanded that the three conservatives disclose their agenda. The simple denial of a plan to mandate creationism produced a national news story. Said the Associated Press, "Hundreds of parents showed up for a debate on teaching creationism vs. evolution before a school board dominated by conservative Christians. Evolution won."[25] The two-year saga involved much more than that issue, because abstinence-based sex education, immigration policy, and use of a nonsectarian prayer to open school board meetings were also at issue. A creationist agenda did unfold, however, when the science curriculum came up. A recommendation to adopt *Of Pandas and People* for biology was rejected in May by a panel of science teachers. That followed an April decision to change a sentence in the science curriculum for kindergarten through eighth grade to say, "Living things are diverse, interdependent and changing" rather than "evolving."

Then, in August, the majority passed a resolution saying that, based on the state's antidogmatism policy, no theory was to be put forth dogmatically, no student should be forced to express belief in a theory covered in the curriculum, weaknesses in evolutionary theory would be presented, and "divine creation, ultimate purposes or ultimate causes" would be discussed "at appropriate times" in history, social studies, or English classes. Said the *Los Angeles Times*, "The embattled but resilient Christian right majority . . . formally opened the door to the teaching of creationism in the city's public schools."[26] The Vista initiative complied with Supreme Court strictures, but protest was great, and the ACLU was poised to sue. Topics such as creationism, sex education, and prayer at board meetings had become so divisive that a recall effort was mounted, and though it failed, four of the five board members were replaced in the regular November 1994 election.

During the two-year episode, the *San Diego Star-Union* and the *Press-Enterprise* of Riverside, California, carried the most regular coverage. Wire services and newspapers such as the *New York Times*, the *Washington Post*, and the *Los Angeles Times* produced roughly thirty stories or editorials, which also went out on their news services. Growing knowledge of the Vista case spurred several other stories about how religious conservatives attacked moral issues on the nation's school boards.

Well afterward, a local reporter analyzed the two-year media frenzy. "They came in search of the religious right. And they got a lot wrong," Randy Dotinga wrote in the *Columbia Journalism Review*. Unlike in Arkansas, the media had sided with judicial elites, emphasizing a story of religion and coercion and reporting the August 1993 resolution as forced creationism in the classroom. Dotinga cited how the Associated Press said teachers were told to teach Genesis "as an alternative theory to evolution," the *New York Times* goofed by saying the school board's prayer was mandated in classrooms, and the *Los Angeles Times* "expanded the myth" of coercion with reports that evolution became "just another theory," while creation was required. "Neither assertion was true," wrote Dotinga. Local papers were better on the facts, and the *Washington Post* "correctly reported that the board's new science policy was largely symbolic," he said. "Fundamentalist Christians in Vista say biased reporters contributed to their defeat in the November election, when they lost control of the board. That's debatable. But there is no doubt that chances for a debate on the real issues were diminished."[27]

Not too long after the Vista flap ended, the Alabama State School Board took a step with perhaps even greater national implications—but it received almost no coverage. In a 6-to-1 vote in November 1995, the board approved use of a disclaimer in all state biology textbooks. The disclaimer said evolution was a "controversial theory" and not a fact, and it listed problems in the theory. Its length and careful wording have made it the Cadillac of disclaimers, and anticreationists say its adoption was "the first statewide success of the 'evidence-against-evolution' approach." The hearing on the disclaimer made television news because the governor stopped by to endorse it. "TV cameras showed Governor Fob James shambling across the floor imitating an ape," reported *Science*. "Many in the audience wore badges saying, 'Don't monkey with my children.'" The AP dispatch was the only national report, and its third paragraph quoted a Baptist preacher as saying, "Evolution comes from the devil." The disclaimer's statement that evolution was a "controversial theory" was reported, but none of the specific problems it alerted students to consider was mentioned. The school board vote was uncontested, low-key, and not dramatic enough to activate the national press.[28]

Protestant creationism has plenty of company within the constraining frames put on issues that concern both religion and science. When the papal commission on Galileo Galilei said in 1984 that "church officials had erred in condemning" the Italian astronomer, it was embarrassingly evident that the church took 350 years to make this admission. The delay has been explained by the protectiveness and snail's pace of Vatican affairs. And revisionist history at the time of the 1984 statement was stressing how hard Galileo had made it for his friend Pope Urban VIII during the time of the Inquisition, and how, back in 1741, the church exonerated Galileo by granting an imprimatur to publish his writings. Only a church mea culpa was left undone. This public act was "an unprecedented breakthrough in the history of the Church," says one papal biographer. The news media treated it with a mixture of glee and sympathy for the Vatican, emphasizing the church's inclination to suppress science but also noting that Galileo was an aggressive publicist and self-promoter who, in some ways, had it coming.[29]

The papal statement on November 24, 1996, that evolution was "more than a hypothesis" received far more news coverage. It flowed onto many front pages, perhaps because the Tennessee legislative debate on evolution that year had primed the media pump. The *New York Times* story, run by many papers, opened: "Nearly a century and a half after Darwin's 'Origin of Species,' Pope John Paul II put the teaching authority of the church firmly behind the view that the human body may not have been the immediate creation of God, but is the product of a gradual process of evolution." The *Washington Post* informed U.S. readers that the statement was a "significant step beyond" the papal acceptance of evolution in 1950.

Months later Gould argued that popes since *Origin of Species* had been basically amicable to evolution, so the 1996 papal statement was no big deal, despite the media flurry. Richard John Neuhaus, a Catholic priest and editor of *First Things*, also took the front-page treatment as a little overdone. "The Holy Father wasn't saying anything substantially different from what the church has said for a long time," he says. He believes the 1996 headlines had two goals: to dust off the embarrassing Galileo affair and to needle creationists. "The papal statement was exploited to serve those story lines." The *Baltimore Sun* editors combined dispatches on the story but in the opening paragraph noted that Rome's action was "likely to raise howls from the religious right."[30]

The papacy frequently gets major coverage, but for riveting stories on evolution, Rome is trumped by Scopes's state, Tennessee. In 1973, Tennessee legislated that evolution must be taught as a theory and that Genesis must be given "equal" consideration. A federal court struck down the law in 1975, and the movement to demand "equal time" for creationism was stillborn as activists in other states dropped the initiative. Then, in early 1996, state senator Tommy Burks, a Democrat and pig-farmer statesman, and state representative Zane Whitson, a Republican who needed the religious vote in the fall, introduced in both houses a bill to penalize teachers who presented evolution "as a fact." The legislative drama focused on the state senate and stretched over eight weeks, ending with the bill's 20-to-13 defeat in a March 28 vote. Tennessee already was getting ink for a bill to post the Ten Commandments in public schools, but front-page coverage of the evolution issue brought a "media frenzy" upon the Nashville legislature.[31]

Two years later, former reporter Cynthia McCune pored over the in-state newspaper coverage and found that one side had gotten three times more press backing than the other. "The bill's opponents did a better job than did the bill's supporters of framing the debate in terms of their own worldviews," the McCune study concluded. The episode was about politics, religion, the state's image, public education, and science "in roughly that order," the study says. McCune's study distinguished twenty-five themes that framed stories, headlines, and cartoons. Most frames were for or against the bill; a small number were neutral. News coverage peaked in week four, but for the entire period, the dominant frames were, in order of frequency, "history/Scopes," "monkey bill," and "state's image." "Monkey" stood for backwardness, and "image" for the state's embarrassment. What is more, "the negative connotations of the Scopes/history frame provided a foundation for several news frames that were unfavorable to the bill."[32]

The three next most prominent frames on the bill also were negative: right-wing politics, unconstitutionality, and intimidation of teachers. Two frames in the bill's favor ranked below these: that evolution was "not a fact" and concern for the "moral values" of students. The bill's foes had argued strategically that the state law would trample on "local control"; and that frame was next in frequency, followed by the neutral theme of the conflict of science and religion.

The science aspect received little coverage. Scientists emphasized biology's role in medicine and genetics and said the law might prohibit the study of genetics and selective breeding at agricultural colleges. Others risked explaining the meaning of theory, fact, and hypothesis. "That's when I started rolling my eyes and saying, 'Oh, when will this be over,'" recalled a seasoned reporter for the *Commercial Appeal*. The McCune study showed how frames are a way to quantify media impact. But it also argued that today, newsmakers will boldly try to manipulate the frames—and the Tennessee news coverage was just such a case. Besides the frame dynamics, however, McCune pointed to other factors that worked against the creationist side: either a lack of grassroots support for the bill, a bumbling strategy by sponsors, or an excellent strategy and more emotional commitment by opponents. "Liberal bias in the press" also needed to be considered.[33]

The rap for expounding liberal ideology, pinned on the national media in 1980, has been softened by the diluting effect of "alternative media" on cable television, radio, and the Internet. The height of the "liberal press" debate evoked a range of justifications, especially that the press is reformist, so change gets stronger play than tradition. *New York Times* culture critic John Corry traced the liberal slant to a "dominant culture" in the news business. It grows out of an "intellectual and artistic culture" on the left and is so primary that a reporter's beliefs don't matter. That culture's membership had either little affiliation with religion or much aversion to its conservative forms, says *Time* political reporter Laurence I. Barrett. "Those of us who do attend religious services go to mainstream institutions," he says. He cited a study of religious practice in professions: "Among the 'media elite,' the study found zero practitioners professing to be fundamentalist, born-again, or evangelical." Harking back to 1925, historian Gary Wills argues that the *Baltimore Evening Sun*'s Mencken was not just secular or liberal. He drenched his trial dispatches in social Darwinism and the "will to power" of German philosopher Friedrich Nietzsche, holding out Dayton as proof that superior people must assert themselves over inferiors. It was a "Superman trial" more than a monkey trial.[34]

News collection for general readers can be seen as a social leveler, treating equally "any idea, however noble or humble, intelligent or inane, provincial or universal."[35] In Arkansas the great leveler arguably put creationists on the same plateau as evolutionists. It turned out differently, though, in the news coverage of the Kansas State Board of Education vote in 1999. The *Kansas City Star* told of state senator Rich Becker's trip to England, where an *Evening Standard* headline goaded a Londoner to say, "Boy, Kansas really is the laughingstock of the world now." The tone of most national and state reporting may have been mocking, as noted in a study by the Center for the Study of Religion in Public Life, but a call for retribution also appeared. The editor of *Scientific American* urged all college and university admissions boards to "send a clear message" that Kansas high school graduates would be scrutinized. *Time* gave Gould a page on which to lambaste the board vote as part of the "long struggle by religious Fundamentalists" driven by their "narrowly partisan religious motivations."[36]

Kansas had become "the butt of jokes from coast to coast," said columnist Jim McLean of Topeka's *Capital-Journal*, who then asked what was the big deal. "Why all the furor over a decision that some insist does little but leave the question of what to teach on macroevolution up to local school boards?" He suspects the furor streams from a clash of worldviews about humans. Public television "Science Guy" Bill Nye contacted the Associated Press to offer that Kansas had taken a "nutty" course of action. "Bill Nye is one mad science guy," said the wire service. Said he: "To reject this fundamental, beautiful thing about the world around us is harebrained."[37]

The general public reaction to Kansas and to Scopes shared two similarities. First was the populist-media split. The media had crowned Scopes a hero, but the public voted for more laws against the teaching of evolution. In Kansas, a large majority agreed with the "Science Guy" on schools teaching evolution, but in national polls, a far larger majority wanted creationism as a belief or alternative view covered somewhere in school, either in science or in another course. In both Kansas and Scopes, moreover, the "meaning" of the news swamped the facts of the news. "It has been kind of a confusing thing," CNN's Melissa Bruener reported from Topeka. "Nationwide, people are hearing that Kansas students will no longer be learning about evolution and that's not entirely true." The board removed questions about large-scale evolution from state assessment tests. "What that means is that local districts can still teach evolution in their science classes."[38]

Historian Mark Silk, a former newsman, says a few different news frames were competing for the Kansas story. The drama of "intolerance" had defined news coverage of the Scopes trial, he said in his book *Unsecular Media*: the Scopes story was dramatized as "the imposition of narrow religious dogmatism on the population at large." When I asked Silk about Kansas, he said the media had attempted to replay the 1925 drama. "There was an effort to make Kansas a Scopes story, but 'Kansas bans evolution' does violence to the actual rule that was passed. It was a local option, and that came across as tolerant. Rhetorically, local option is a strong position in America. People don't respond to arguments from authority." Other media research has confirmed this. When the public watches a science controversy in the news, the side that asserts an "ideology of expertise" loses a lot of sympathy.[39]

From another vantage, the Kansas story was never truly a "national story" because it never made prime-time news, said Tim Graham when he was research director at the conservative Media Research Center, whose Alexandria, Virginia, office monitors networks and combs print coverage to look for trends. "Kansas became a story because it confirmed stereotypes of religious know-nothingness," Graham said. "You could argue that Kansas was news because it challenged Darwinism, but that's not an acceptable angle."[40] The Kansas board's vote, in other words, could have been reported as a successful response to court rulings against mandated creationism and against dogmatism: the decision making was moved to school districts in harmony with the law.

The story also could have been "Kansas School Board Strengthens Evolution." The new standards increased the study of evolution over what the 1995 standards had mandated. Yet the increase did not match the expectations of the *National Science Education Standards*, and so the story defaulted to "Kansas Bans Evolution." That story was "easy work," says biologist Jonathan Wells, a critic of Darwinism. He saw better stories in questions such as "How many millions of your tax dollars will be spent this year by Darwinists trying to find evidence for a theory they claim is already proven beyond a reasonable doubt?"[41]

In the dust of the Kansas controversy, the *New York Times* explored the antievolution landscape and came up with a young-earth evangelism ministry in Kentucky known as Answers in Genesis. Its leader was Kenneth Ham, an Australian and former biology teacher, whom the *Times* called "Captain Creation." The otherwise evenhanded story hinted strongly that creationist evangelism can be lucrative indeed. At the time the piece ran, members of the intelligent design movement were holding a conference, and titular leader Phillip Johnson complained that the *Times* story was chosen, as usual, to play up the Bible versus science stereotype. "They want the public to believe that the only argument is one between open-minded, fact-gathering scientists on the one side, and dogmatic, science-disregarding biblical literalists on the other side," he told the gathering. He almost spoke too soon, however, for in early 2001 the *Times* gave the intelligent design movement a Sunday front-page story, which Johnson portrayed as "the most valuable intellectual property in the world."[42]

Meanwhile, it would be wrong to say the media are never hard on evolution—or never ambivalent. When top scientists announced at the White House in June 2000 the completion of the human genome sequence, "the extensive press coverage the next day had very little reference to evolution," biologist Joseph McInerney recalls. As to being negative, the news frame is usually about infighting evolutionists or, occasionally, that creationists with academic credentials might be reputable scientists. For example, science writer Robert Wright, the author of the recent evolution book *Nonzero: The Logic of Human Destiny*, contributed an attack on Gould to the *New Yorker* in 1999. Titled "The Accidental Creationist," the article said Gould confused the public about evolution, and this gave aid and comfort to folks such as those in the intelligent design movement. Wright said Gould's trademark idea—that evolution is utterly aimless and human life is a onetime lucky fluke—drained all meaning from the science and had given that handle, or weapon, to the creationists. Wright's solution was to clarify that evolution has "direction" toward complexity, making the arrival of humans not only inevitable in nature but meaningful. "This indictment of Gould will no doubt surprise his large reading public," said Wright. His comment suggested that, true or not, the public did not have a clue about infighting among evolutionists.[43]

Arguing evolutionists have appeared on the front page of the *Boston Globe*, where a local Cambridge flavor came with the conflict between Gould, Dennett, and psychologist Steven Pinker at MIT. Headlined the *Globe*: "Survival of the Theorists: Professors Battle over Darwin's Concept of Evolution." Perhaps more famous is a *Newsweek* article from 1985 that became a tool for creationist propaganda. The magazine had reported on "heated" clashes in evolutionary research. One scientist, the article said, was tempted to bolt for "a field with more intellectual honesty: the used car business." The article was waved at creationist debates, and one evolutionist shot it down by excoriating "journalistic opinion" on matters of science: "I've been written about hundreds of times by journalists," said the evolutionist, "and I know how inaccurate they are."[44]

Finally, a story in *U.S. News & World Report* in 1997 was prominent enough for the evolutionist lobby to mount a counteroffensive to sway public opinion against it. The story featured a software program, Terra, "used by geophysicists around the world" to study heat convection. Its designer, a scientist at Los Alamos National Laboratory, also used it to argue for a young earth and a global flood. The article was swept up as a lobbying tool by flood geology creationists, so the National Center for Science Education rapidly produced a counterreport. Center interviewers talked to the same people, garnering quotes that said the software did not measure age.[45]

The Scopes-trial template has endured because most coverage of the evolution-creation debate is about public disputes in courts or on school boards. As Professor Caudill argues, actions, not ideas, have become the staple of news coverage of this science-and-religion debate. The best magnet for coverage, therefore, is an event, and if it is a conference, some ideas can be slipped in as well. Two assemblies funded by the Templeton Foundation in the 1990s illustrated that the science-religion debate could ring media bells. The "Science and the Spiritual Quest" conference met in Berkeley and drew 68 reporters, assigned to either science or religion beats. The event generated 112 print, radio, and television stories. Next, the "Cosmic Questions" forum met at the Smithsonian Institution in Washington, D.C. It attracted 110 reporters, mostly on science beats, and generated 164 reports.[46]

These events cost money and required sponsorship. They also had the imprimatur of official science and worked on the presumption that Darwinian evolution was an undisputed fact. Conferences by antievolutionists clearly do not get the same coverage. No conference, however, can ever be as ripe for the front pages as a politically charged school board vote. The Kansas event, as one columnist exaggerated, generated "thousands of headlines." Media attention was intense and advocates galvanized, but in the end most of the public was not paying attention. Soon after the uproar, half of Americans told a national poll that they had never heard the word "creationism," though most had heard the term "evolution." People for the American Way commissioned the national poll, which, like a *Kansas City Star* survey in Kansas, found disapproval from most Americans (60 percent) on the Kansas action when it was described as "to delete evolution from their new science standards."[47]

The school, rather than the academic salon, is the storm center of evolution-creation issues for obvious reasons. It is the place where young people are trained and where emotions get churned over questions about human nature and morality. Did human nature evolve from apes, or was it designed and given a rule book by God? Does human nature make us free, or make us accountable? If schools and school boards cannot handle the topic effectively, scientists and theologians may be just as conflicted. Take the British evolutionist David Lack of Oxford. He helped pioneer the New Synthesis and wrote *Darwin's Finches*, the book that made the birds famous in America's science classrooms. But after a Christian conversion later in his career, Lack could not paper over an "unresolved conflict." Though he might have wished otherwise, he concluded that "Darwinism seems irreconcilable with Christian, and perhaps with any, moral standards."[48]

Being a philosophical problem, the "unresolved conflict" between evolution and morals would never appear on the front page of a newspaper or headline the evening news. Still, it may be the most sensitive nerve of all in the evolution-creation debate.

14. THE GOOD SOCIETY

Before the Louisiana Senate voted in 1981 to approve the teaching of creation alongside evolution in public schools, a woman's plea rang through the statehouse chamber. "I think if you teach children they are evolved from apes, then they will start acting like apes," she said in hearings. "If we teach them possibly that they were created by an Almighty God, then they will believe they are a creature of God, and start acting like one of God's children."[1]

Her testimony was put before the Supreme Court to show that the legislation had a religious intent. The court agreed and struck down the law as unconstitutional.

Human evolution is the most passionate aspect of the evolution-creation debate. While modern school boards curtail teaching it to "avoid limiting young people in their search for the meaning of human existence," evolutionists are offended by charges they are corrupting society. They have been solicitous, however. The national science standards, for example, omit the subject, and when it is explained elsewhere, there is a qualification: "Humans did not evolve from modern apes, but humans and modern apes shared a common ancestor, a species that no longer exists." For creationists this distinction on apes is without a difference. Or as William Jennings Bryan had written, "It is all animal, animal, animal with never a thought of God or religion."[2]

When the Louisiana woman spoke about what the child is and how the child behaves, she was repeating the timeworn categories of scientists and philosophers. The debate over "what human nature is" and "how humans should live," is complex but by now a few basic approaches have been fixed upon. First is the decision on whether to define human nature with a top-down or bottom-up approach, the first a typical religious stance and the second naturalistic. Two texts make this distinction clear:

> What are human beings that you are mindful of them, mortals that you care for them? Yet you have made them a little lower than God, and crowned them with glory and honor.—Psalm 8:3–5

> Fully human sensibilities emerged only after the appearance of anatomically modern people. Nonetheless, were we to meet certain of our ancestors in the flesh we might well see enough of ourselves in them to want to de-

scribe them as human. There is thus not a definitive use of this elusive term.—American Museum of Natural History.

A second decision on human nature is this: Will it be described as monistic, in which all things human arise from biology, or dualistic, a state in which something human operates outside of nature, as if split off, "uncoupled," "free floating," or "emergent"? This floating reality may be identified as human culture, human consciousness, or, in theological arguments, the soul itself. The third and final choice is to decide for a pessimistic or optimistic outlook on human nature. Is human nature inclined toward a war of "all against all," or is it sociable?

These themes—top-down, bottom-up, monistic, dualistic, pessimistic, and optimistic—are one set of ways to analyze approaches to human nature. Society does it all the time, as with the killings at Columbine High School in 1999. Soon afterward, political scientist Francis Fukuyama argued in the *Atlantic* that the biological sciences had opened up new vistas to control social deviance because "we human beings are by nature designed to create moral rules and social order." American Bible Society president Eugene Habecker recoiled from such reliance on biology for a moral lesson. "We believe Scripture tells us otherwise," he wrote. "It is God, our heavenly father, who has set boundaries and gives us 'golden rules' for living in community."[3]

Darwin knew well this clash between bottom-up and top-down views. A student of the Bible, he knew the proposition that God designed human nature with a purpose. Human purpose was evident in what theologians call general revelation or natural law, and clearer still in the special revelation of the Ten Commandments, the Sermon on the Mount, and Scripture. This top-down view of human nature was not simple, however, for that nature was presumed to have been good at Creation but was now alienated from God. Now, God the designer had to become God the redeemer. The West was thus built on this dynamic of a good nature and "sin" in combat, with the intervention of God as the final liberator.

Reluctantly, Darwin decided that such theology explained nothing. Wearing bottom-up spectacles, he saw human nature as the child of Mother Nature, and she was "clumsy, wasteful, blundering, low, and horribly cruel." When Darwin joined this evolved human with his own Victorian utilitarianism, his "emphasis lay to a very considerable degree upon selfish motivation," says historian Loren Eiseley. In *Descent of Man*, Darwin accounted for human morality as a byproduct of the survival instinct. Survival worked best in groups. Group interaction taught the benefits of cooperating, and human conscience sprang forth. "Man prompted by his conscience," Darwin said, "will through habit acquire such perfect self-command, that his desires and passions will at last yield instantly and without struggle to his social sympathies and instincts." Such sympathy will extend to tribe and nation until "there is only an artificial barrier to prevent his sympathies extending to men of all nations and races."[4]

Looking back on Darwin, some commentators have seen in him both a monistic view and an optimistic view of human nature. His friend Thomas H. Huxley would choose the opposite, later in life declaring that evolution puts all of nature under "a gladiatorial theory of existence." The only path to moral salvation was to resist nature, which included human nature. "The ethical progress of society depends not on imitating the cosmic process, still less in running away from it," said Huxley, "but in combating it."[5] These two vantages on nature, one seemingly morose, the other almost cheerful, infuse the debate even today. The Darwinians did not invent such outlooks, of course. Without a theory of evolution, England's Thomas Hobbes painted human nature just as darkly; France's Jean-Jacques Rousseau limned it as innocent and noble. Even earlier, Christianity had melded the two, the angel and beast, into one anthropology.

Unique to the nineteenth century, however, was its idea of human progress. And what Darwin, a man of his times, had added was a mechanism in nature—namely, struggle—by which only the fit survived. Here was social Darwinism, and with its aura of being a law of history, human evolution was taken as a force of either darkness or light in the United States. Bryan had crusaded against the Darwinian trampling of the weak by the mighty, and Christian evolutionist Henry Fairfield Osborn rebutted that with a gloss of goodness: "The moral principle inherent in evolution is that nothing can be gained in this world without effort; the ethical principle inherent in evolution is that the best only has the right to survive."[6]

Not a few notable historians have doubted that Darwin's theory revolutionized any social morals that were not already being shaken by the modern era. Historian Gertrude Himmelfarb says that evolutionism and agnosticism had enamored Victorians well before Darwin's *Origin of Species*, which thus only gave a "scientific" justification for current feelings. She notes further that even England's ardent, agnostic evolutionists nevertheless harped on duty, morals, and churchgoing. Revisionist historian Peter Bowler, in turn, explains that the modern West underwent a "non-Darwinian revolution," meaning its dark twentieth-century impulses of war, racism, and nihilism arose from "developmentalism"—a mysterious, organic evolutionism of another kind—not Darwin's stately idea of natural selection. "The chaos of political, moral, and artistic philosophies that seems to characterize modern thought stems from this loss of faith in the power of reason to uncover the orderly foundations of natural development," Bowler writes. "It was the developmental, not the Darwinian, view of evolution that played the most vital role in the origin of the most disturbing facets of modern thought."[7]

When this debate took place in 1920s America, all the theories of evolution intermingled. The "acids of modernity" were not Darwinian in particular but a naturalism that undercut supernatural authority as a basis for morals. Historians have argued that amorality did make headway, but only among social elites, for whom Victorian sexual norms had been broken by new scientific ideas. This had been called scientific progress by some anthropologists, but that optimism about liberating inhibitions was chastened by the Second World War. The science of human nature became characterized far more by ambiguity than by certainty.[8]

Such was the ambiguity of paleontologist George Gaylord Simpson's famous lectures, *The Meaning of Evolution* (1949). Simpson had attended a Unitarian church, but in his rejection of all things supernatural, he argued for an ethic that was no longer bound to religion, introspection, authority, or social convention. Society must find "a naturalistic ethical system," said Simpson. His bottom-up approach was clear, but in grappling with monism and dualism, he ended up with an uncertain patchwork. Monistically, Simpson argued that "what is ethically right is related in some way to what is materially true." Dualistically, however, he split human ethics off from nature: "Evolution has no purpose; man must supply this for himself." From where, then, does man get this purpose? He gets it in the new evolution of culture, Simpson answered, not the old evolution of biology. "The best human ethical standard must be relative and particular to man," he prescribed, "and is to be sought rather in the new evolution, peculiar to man, than in the old, universal to all organisms."[9]

The Simpson ambiguity—humans are part of nature and yet separate from it—is the conundrum that endures for evolutionist theories today. Some in science and religion, however, prefer utter clarity.

. . .

Leaving the Museum of Creation and Earth History, visitors pass down a hallway of black-and-white photos of famous evolutionists. Here is the unambiguous moral lesson, the Christian museum claims, of what evolution has wrought. There are Hitler and Lenin and Freud, and comments from great American capitalists like John Rockefeller, "a ruthless developer," and Andrew Carnegie, a "cruel and heartless" robber baron who said, "I got rid of theology and the supernatural but I had found the truth of evolution."[10]

North of the museum is the University of California at San Diego. It is poised between the cold Pacific currents and the hot desert breezes, a stance between extremes that science philosopher Philip Kitcher says he has tried to navigate on the human nature question. "I think Simpson saw something when he said, 'We need a connection between evolution and ethics,'" says Kitcher, who taught in the sunny Pacific climes until 1999. "We need to make morality part of the natural world, to de-supernaturalize it, if you like. But doing that is more than just making a simple connection with biology. It's also a matter of making a connection with the history of human culture."[11]

Kitcher has approached this complexity by marking off the extremes. His *Abusing Science* (1982) criticized creationism and its unadorned moral scheme. Then, in 1985, he wrote *Vaulting Ambition* to upbraid the other extreme of genetic determinism, which said morals are hardwired in biology. "I thought, 'This is an interesting phenomenon,'" Kitcher told me in our interview, recalling his time at the University of Vermont in the 1970s, when creationist literature was in circulation. "Their ideas were really catching intelligent academics off guard." He wrote *Abusing Science* to supply school administrators and teachers with arguments against creationist claims such as "The evil fruit of the evolutionary philosophy is evidence enough of its evil roots." For a rejoinder, he turned to Huxley, who "appreciated the complexity of moral thinking." Kitcher said creationist moral logic was dubiously simple, such as "We learn of our kinship with other animals, so we turn savage and promiscuous, tear down our social institutions, and abandon our ordinary attitudes to personal relations." The point was, Kitcher concluded, any belief system can be misused for evil. "Both the Bible and evolutionary theory can be misread and their principles abused," he wrote. "Hitler's anti-Semitism is no more a fruit of evolutionary theory than it is of Christianity."[12]

By siding with Huxley's dualism, Kitcher already had taken up arms against what he called another simplistic extreme: the monism in genetic determinism, also called "sociobiology." He attacked it in his *Vaulting Ambition*, with E. O. Wilson in its crosshairs. In 1975 Wilson had published *Sociobiology*, with a final chapter saying humans could be understood in the same way as gene-driven insect societies. Or, as Wilson wrote later, "The biological principles which now appear to be working so reasonably well for animals in general can be extended profitably to social science." Though Wilson has since hinted at a dualistic position, speaking of a "gene-culture evolution," he staunchly puts biology first. Human habits, tastes, and beliefs are "epiphenomena," or secondary results, of genes. For understanding human nature and culture, he says, "There is intrinsically only one class of explanation." And that is the biological one.[13]

"I was somewhat ungenerous in my criticism of E. O. Wilson," Kitcher told me two decades after the publication of *Sociobiology*. "And yet I was right to see too crude a connection between evolution and ethics. If there's a connection there, it is going to be something that's very hard to work out, and it's going to encompass and take into account the historical development of moral thinking. It's not going to be, 'We're programmed to want this, therefore we ought to do it.'" Kitcher, as a philosopher of science and society, is as much concerned about the social power of scientific theory as he is about "the truth" it purports to find. So, in *Vaulting Ambition*, he spoke of "two faces" to sociobiology. One face (with pursed lips) was skeptical and prudent in animal research. But the other face (a face behind a megaphone) announced "with great excitement" exactly why humans behave the way they do.[14]

Two highly publicized examples of something like sociobiology have been MIT psychologist Steven Pinker's assertion that women have been driven by genes to kill their children, and the argument by two Arizona biologists that rape is compelled by male biology, not by a so-called psychological need for power, a favorite feminist argument. Mothers as habitual baby-killers and men as robotic rapists—all of it a matter of genes rather than free will—are only a sample of the possible social stereotypes that could arise from the overly sweeping claims that genes control human behavior. And it is this slippery slope, this vaulting ambition of science, Kitcher has suggested, that might do violence to the open and liberal society in which he believes.[15]

Such a liberal society has been the aim of evolutionists since 1933, the year of the first Humanist Manifesto. The manifesto (updated in 1973 as Humanist Manifesto II) has been the bogeyman of creationists ever since, though typically it is signed by only a few and dismissed as too credo-like by other secularists and naturalists. Its third version, Humanist Manifesto 2000, continues to show how scientific naturalism goes from evolution to human ethics. "Moral tendencies are deeply rooted in human nature and have evolved throughout human history," the manifesto states. "Humanist ethics thus does not require agreement about theological or religious premises—we may never reach that—but it relates ethical choices ultimately to shared human interests, wants, needs, and values. We judge them by their consequences for human happiness and social justice."[16] Among the highest human values are choice and autonomy; thus the manifesto supports abortion, euthanasia, and homosexual families, but it also endorses international entities to redistribute wealth. Endorsement signatures were topped by E. O. Wilson and Richard Dawkins.

Unlike Wilson, Dawkins is the Huxleyan dualist par excellence. Having proposed the "selfish gene" in 1976 as the very engine of evolution, Dawkins then sounded a Huxleyan call to arms to "rebel against the tyranny of the selfish replicators." During his Atlanta visit, Dawkins says, "I'm in favor of the struggle of Darwinism on how life got here." But, agreeing with Huxley, humans "need to struggle against 'nature, red in tooth and claw,'" Dawkins says, quoting the words of Tennyson. "Thank goodness," he says, this struggle has produced civilization as a "cushioning" against brute nature. "As to where morals came from, that's a genuinely difficult question. . . . We do need to try to understand why we respond to a moral sense. It's not always the same moral sense in different cultures. But it seems to have grown out of something biological, though it's no longer in any simple sense biological."[17]

During the humanist conference in Atlanta, Dawkins peeks in on the session "Does the Social Life of Primates Offer Insights into Human Behavior?" Doubtless it does, but his name is more closely tied to the logic of genetics. This logic swept the evolutionary field after 1964, when Oxford biologist William Hamilton, based on the study of ants, said that they sacrificed for offspring who carried the greater number of genes. This "kin selection" was a feigned altruism; ants that helped other ants did so solely to promote their own gene survival. The math was alluring, as portrayed in the fabled comment of one biologist, "I'll gladly give my life for two brothers or eight cousins." Kin selection, says biologist Robert Trivers, was "the most important advance in evolutionary theory since the work of Charles Darwin and Gregor Mendel," who discovered heredity. Trivers then turned to the apparent altruism in close social groups, saying it was a self-interested survival tactic, a détente between organisms that recognized one another's predictable behaviors. This was called "reciprocal altruism."[18]

With this new focus on the survival of the smallest units, namely, the individual and genes, Darwin's group selection theory was overthrown. The centrality of individual survival was argued so masterfully by State University of New York biologist George Williams, who wrote *Adaptation and Natural Selection* (1966), that "group selection became one of the most rejected aspects of evolutionary theory," says biologist David Sloan Wilson. In this world of greedy genes and self-serving organisms, where creatures fake "altruism" in brutal calculation, the pessimistic view of nature found a home. "I account for morality," wrote Williams, "as an accidental capability produced, in its boundless stupidity, by a biological process that is normally opposed to the expression of that capability." And this from E. O. Wilson: ethics "is an illusion fobbed off on us by our genes to get us to cooperate." Science writer Robert Wright echoed that comment: "What is in our genes' best interest is what seems 'right.' Moral guidance is a euphemism." Finally, from biologist Michael Ghiselin: "Scratch an 'altruist,' and watch a hypocrite bleed."[19]

While Darwin had mused about an evolved sympathy "extending to men of all nations," this did not work in the logic of gene survival. Trivers went on to explain so-called universal love, the traditional province of religion, as a self-deception that benefits the deceived person's own survival. There is no easy answer. In Darwinism, Dawkins says, "What's harder to make work is any system of universal altruism, a system where you are good to everybody." Yet he is confident that one day this human inclination will be accounted for. "It's no more of a challenge than anything else in our culture, like music or mathematics. These are not things which, on the face of it, assist your genetic survival." And yet they are here.

Evolutionists trace the modern human brain, with its capacity for ethical thinking, to the Pleistocene epoch, a span of time beginning 1.6 million years ago and ending only 10,000 years ago. It was a time when the upright, bipedal, hand-using, and big-brained human emerged in an African patchwork of woods and grasslands after the Ice Age had ended. Back then, Dawkins says, the brain was "originally selected to be selfish or very guardedly altruistic on the Pleistocene plains of Africa." Now, the same thinking organism functions in the space age, in a patchwork of technology. "Our Darwinian explanations have got to be filtered through that," says Dawkins.

That filter is human culture, which has evolved more rapidly than human biology: books are read faster than babies are born. What is culture made of? Dawkins has called its building blocks "memes," a substance separate from nature which, like packets of ideas, circulate in society much as genes are passed down in reproduction. These memes, which have the power to possess human minds and behavior, have remained mysterious in terms of what they are made of; Jeffrey Schloss, a theist in biology, has called it the "new demonology"—the study of how memes possess unwitting souls.[20]

Other naturalists like Dawkins have similarly spoken of an evolutionary substance, like a secretion from a gland, that creates a hierarchy of nonphysical things, or "emergent properties," important for human life and well-being. None of this "substance" has yet been captured in a test tube or in a mathematical equation, irking some theologians who say scientists are poaching on their turf. What irks others, however, is the vision of nature as a coliseum of gladiators.

. . .

Frans de Waal, tall, with curly blond hair, works with primates at the Yerkes Primate Center in Atlanta. He disdains the Huxleyan dualism, its pessimism, and its war against nature. De Waal stands before a Boston conference on natural sciences and altruism, openly criticizing the theories of utter biological selfishness. "Huxley was a very combative person who saw combat everywhere," he says. "What you observe in nature is quite different."[21]

De Waal had summarized his work in *Good Natured*, a book that makes the case that some branches of primates have an innate morality akin to that of Homo sapiens. "I started with aggressive behavior," he says of his research. But he found the opposite, such as selective food sharing. "Reconciliation is now one of the best studied areas of primate behavior." And among the hominids—humans, the apes, and the gibbons—acts of consolation, or sympathy for the bereaved or dejected, take place. De Waal pulls no punches on the human implications. "Given the universality of moral systems, the tendency to develop and enforce them must be an integral part of human nature."[22] In some ways, de Waal's empirical conclusion has joined a far older philosophical argument: it is the older protest that Darwin's "law" of struggle in nature is not a law at all but only a piecemeal observation having little to do with how life diversifies and develops.

The dark overhang of utter struggle and selfish genes finally had gotten to be too much for some in natural science, de Waal among them. Some biologists, believing that organisms were the real stuff of natural science, began to bridle at the emphasis on "gene replicators"—genetic theories that the body of an organism was irrelevant. So, in rebellion against the nature-as-wicked approach, a sizable number of naturalists have shifted toward nature-as-beneficent. The shift includes a reemphasis on organisms and a leaning toward monism; there was no need to fight nature (dualism), for biological life is sociable and its neurological systems empathetic (monism). Reflecting this same optimistic, good-natured mood, other biologists revived Darwin's theory of group selection. It is championed in particular by biologist David Sloan Wilson and philosopher Elliot Sober. Sober had been a dualist and had coined the term "free-floating." But now they are keen on the multiple sources of good-naturedness. "Darwin's pluralism as a way to view nature is making a comeback," Sober says.[23]

Some Darwinians of the old school, such as Ernst Mayr, were part of the pluralism before genetics and the "selfish gene" seemed to take over the field of human nature. Mayr believes Darwin basically had it right in group selection.

Adhering to the conclusion of the British naturalist C. H. Waddington, he holds that humans are "born with probably a certain innate capacity to acquire ethical beliefs, but without specific beliefs in particular." In humans as with animals, he says, the specific behavior is imprinted by a parent onto offspring. If animals have fixed instincts, humans have "an open behavior program." This allows different choices and different ethical systems, the best of which, Mayr says, are tolerant, liberal, and egalitarian. By comparison, he says, theories such as kin selection use more "gimmicks" than are necessary.[24]

Others in this good-natured trend of thinking base their conclusions on observing humans, without trying to reduce explanations down to the genetic level. Social scientist James Q. Wilson, for example, had studied human criminal behavior. "What most needed explanation, it seemed to me, was not why some people are criminals but why most people are not," he says. Relying on the natural sciences, much as the political scientist Francis Fukuyama has done, Wilson ascribes to humans an innate "moral sense." Anchored in biology, this moral sense is made up of sympathy, fairness, self-control, and feelings of duty. The sense is best nurtured in the two-parent family. And what of universal love? "Humans are biologically disposed to care for their young, but, unlike some other species, what people find attractive in their own young are traits that, to some degree, are also present in other people (and even other species)," writes Wilson.[25]

Modern intellectuals have strayed from this natural, positive, and universal inclination, Wilson says, by adopting warped ideologies. These have included Marxism and Freudianism, but also the moral relativism of modern cultural anthropology and philosophy. In particular, Wilson aims this criticism at Ruth Benedict's *Patterns of Culture*, in which any cultural pattern is as good as any other, a stance identical to that of pragmatist philosopher Richard Rorty. "What counts as a decent human behavior," Rorty says, "is relative to historical circumstance, a matter of transient consensus about what attitudes are normal and what practices are just or unjust."[26]

From the field of sociobiology, too, there had meanwhile emerged a more optimistic view of the gene-centered science. By 1990, under sociobiology's new name of evolutionary psychology, the controlling power of chemically coded genes was "softened"—a softening accomplished by filtering genetic determinism through developmental psychology, or the process of emotional and mental development in children and family generations. In developmental psychology, ideas and beliefs do make a difference. And here, sometimes, is found a positive dualism in which ideas held by human minds are considered so real and tangible that they may override biological instincts, "that is, the immaterial overpowers the material," an idea that contradicts the very premise of modern science.[27]

In the field of sociobiology, a favorite tool of analysis has been "game theory." Invented to analyze how nations in the cold war might determine the gains and losses associated with particular moves against an opponent, it now has been applied to explain how self-serving biological organisms calculate the best chance of survival. Sometimes that calculation may lead to cooperation or compromise, and thus the appearance of so-called altruistic behavior.

Robert Wright, whose evocative science writing once emphasized the darkness of human motivation, has turned toward the light. In the game of Darwinian survival among primitive humans, he now argues, life began to produce fewer cases of win-lose and more incidents of win-win. These he calls "non-zero" outcomes. "Over the long run, non-zero-sum solutions produce more positive sums than negative sums, more mutual benefits than mutual loss or exploitation," he writes. "As a result, people become embedded in larger and richer webs of interdependence."[28] His rebuke of life as utter evolutionary accident and his claim that biology unfolds with purpose hopeful to humanity are a clear departure from the ambiguity of Simpson: "Evolution has no purpose; man must supply this for himself."

. . .

Theology, of course, has been a primary provider of human purpose, often infusing an ethos of meaning into scientific pursuit. Yet nothing is so implicitly top-down as the theological: God shaped human nature in the divine "image." Classic theology puts an eternal soul inside a perishable body. What is more, the theological aspiration of love is alien to the selfish allusions and ersatz "altruism" of genetic theories. By all measures, the theological stance seems bound to collide with the bottom-up, monistic, and self-serving fact of biological evolution.

Into this breach, however, has come theology without dualism. Theologian Nancey Murphy, a member of the "science and religion" dialogue, is pushing this option to its maximum, dramatizing what is possible in her study of neuroscience and the soul. "It seems that the brain does most, if not all, of what was once attributed to the mind or soul," she told the 1997 "Epic of Evolution" conference.[29] "So we are heading for a major conflict between science and religion, right?" In fact, she says, "Wrong."

Wrong because a new monistic view of God-given human nature can open doors to science: "There is no metaphysically distinct part of us that is immune from scientific investigation." What theology offers is not a supernatural parallel to matter but the overall purpose and meaning to human existence, a meaning that eludes science in its empirical focus on cause and effect. Murphy, who was reared Catholic and now works in the Protestant tradition, is part of a new Christian school that says, "Monism, yes," but not the "nothing but" genetic reductionism so common to philosophical materialists.

"Our core theme is a monistic, or holistic, view of humans," she writes with her colleagues in the book *Whatever Happened to the Soul?* "In order to avoid confusion with reductionist or materialistic forms of monism, we have chosen the label 'nonreductionist physicalism' to represent our common perspective."[30] With neuroscience explaining more and more details about human thought and emotion, she argues that the theory of a "spirit" apart from the body is no longer necessary. The Hebrew Bible, she goes on, did not see the person as having a soul or an afterlife; they were concepts added on later by Greek thinkers, New Testament writers, and finally theologians of great influence such as Augustine and Aquinas.

For these new Christian physicalists, the human spirit is "an emergent property." It is a whole that is greater than the sum of the parts, and a quality unique to humans in their desire for relationship with others and with God. They argue that an "emergent property" cannot be reduced to a chain of events beginning with atoms and molecules, so it is a kind of "top-down" property within which human free will, self-consciousness, and even the influence of God may be found.

With science on their side, the Christian physicalists exude confidence in the face of dualism, and Murphy is no exception. "It's clear that the burden of proof has shifted to the dualists to explain why we need to postulate an additional entity, the mind, when accounts in terms of brain activity are becoming increasingly more powerful," she told the 1997 conference. "Anyone who has adopted this [dualism] strategy to reconcile the scientific and theological accounts of human origins ought to be very nervous about the developments in neuroscience." Yet Murphy is still a Christian theologian, teaching at Fuller Theological Seminary, founded by the neoevangelicals and now the largest theology school in America. So she must also emphasize that the new holistic, physicalist approach "is not at all incompatible with the picture we see of the nature of human persons in the Bible, both Hebrew and Christian." If there seems to be a conflict, it is because of dualist Greek Christians: "We have been misled by the way early Christians translated the Hebrew text."

The immediate rejoinder to this new Christian monism is the classic dualism, which is coming from a growing school of Christian philosophers. Even Murphy, for example, concedes that neuroscience is so complex that neither monism nor dualism can be proved absolutely—and Nobel-winning brain experts have come down on all three sides: the mind is dualistic, is mechanistic, and is an "emergent property."[31] The Christian dualists try to take the high ground by honing in on the mystery of personal subjectivity, the proverbial "I" of human identity. They want to show the failings of scientific materialism on this front and instead argue that a nonmaterial person, or soul, can account for the self-consciousness of human nature.

The Christian philosopher J. P. Moreland, who once studied chemistry, has challenged Christian physicalism's eagerness to surrender the soul to science. "The classical understanding of the human person as consisting of body and soul has come under attack for some time," write Moreland and Scott Rae, noting that rejection of a soul erodes morality as well. "We hold that metaphysics and morality are intimately connected and that our dualistic view of the body and soul provides the most compelling account of human personhood and its moral dimension." Not having science to rely on, however, the Christian dualists must build on the obvious gaps where materialists have not yet explained the person by cells or nervous systems. The dualists offer a philosophical argument, such as that the person is a "substance" uncontained by biology, or that the person is made up of the interacting forces of reason, will, and emotion—what Moreland calls "the doctrine of internal relations."[32]

Faced with the insurmountable complexity of the mind-brain problem, scientists have begun to speak of the person as an "emergent property," a sum that is greater than the parts, and thus untamed by mechanistic science. Yet Moreland and Rae argue that "emergence" is still physicalism, just as when Dawkins speaks of "memes" that transmit human culture, or when Huxley said humans can muster something extra to fight their biological nature. The Christian physicalists may claim that "emergent properties" are a top-down influence on the human, but Moreland argues that they are more like an "outside-in" influence—an influence of the physical environment, not the real person. The top-down of the soul, he said, is the only way to get "genuinely new kinds of properties contributing to lower-level effects" of the physical human being.[33]

In the Christian context of this debate, there also has to be room for the theological claim for the self-giving love, or "agape," of the New Testament. Especially problematic for biological theories such as kin selection is the love shown toward a stranger or for people who return no benefits. While a Christian ethic does not require total imitation of Christ's death on the cross to show love, it leans in that direction, prompting evolutionists to call it biological insanity. Between these extremes of self-immolation and survival of the fittest, religious thinkers and psychologists are trying to find the proper balance of human altruism and human biology.

On one occasion, the annual science and theology session of the American Academy of Religion, which has taken place since 1986, focused on the altruism topic. One regular participant is Karl Peters, who teaches at a Unitarian Universalist college and has drawn on world religions to balance altruism with nature. His 1999 paper argued that an alternative to martyrdom is the Asian religious virtue of self-beneficial harmony, of giving and getting. He said it could make theological inroads into science, whereas total self-denial may not. For example, Peters argued, the Darwinian idea of group selection (cooperative survival) and reciprocal altruism (survival by trust) are not in conflict with Asian religious mandates to harmonize with others.[34]

And yet because people still do lean toward altruism, if not martyrdom, the question remains of whether it can be gotten at by experiments: Is such behavior biological calculation or mental sincerity? That has been the problem worked on by experimental psychologist Dan Batson. Authentic altruism is based on a motive in the person, a motive Batson has tried to measure in blind scientific experiments. What he has found is a natural sympathy in people that can lead to conscious acts beyond "the self." Self-serving biology is not everything, Batson has concluded. "Empathy-induced altruism is part of human nature."[35]

A similar mystery arises in the study of "righteous Gentiles," people who hid Jews at the risk of Nazi punishment. These studies further complicate the idea of a saint. The act of self-risk does not always spring from virtue, skill, or faith, according to political scientist Kristen Monroe. She found that these true altruists shared many general traits but no one cardinal virtue. "Altruists look much like other human beings in their education, religious and socioeconomic characteristics," she says. "All the altruists I interviewed saw themselves as individuals strongly linked to others through a shared humanity." Where that link comes from, she says, is elusive. She did not find it in kin selection, religious commitment, or the "rational choice" often associated with economic self-interest. Finally, she says, the source was an ability to identify with others, which led to an "altruistic perspective."[36]

Ethicist Donald Browning of the University of Chicago Divinity School has asked the same questions, but in the context of Protestant belief, which has much to say about goodness and nature. Reformation theology begins with a fallen world in which only supernatural grace can produce acts of love. It is a love, moreover, that must be "self-sacrificial activity." And in Protestant tradition, according to Browning, the actor must see the good deed as "made possible by God's grace and with no thought for the good returned to oneself."

Such a theology poses obstacles to joining human altruism with nature, or even faith with science, Browning has argued. So he has turned to the ideas of Thomas Aquinas, who built upon Aristotle's observations of nature. "The two worlds of biological functionality and divine transcendence coexisted in Aquinas's pre-Darwinist thought," Browning says. A Christian ethic that harmonizes with nature, he says, cannot always evoke agape. A workable ethic would shift more "toward the synthesis of Aristotelian friendship and New Testament love found in a variety of Roman Catholic formulations." In this synthesis, "love [is] mutuality or equal regard, with self-sacrifice serving as a transitional ethic to restore love as equal regard."

Browning's bridge to science is daunting, however. He wonders if this Christian idea of love might work with evolutionary psychology, which is a softer materialism. Meanwhile, the hardened law of genetic survival would leave no room for theological forces. Browning recognizes the conflict, and he cautions against compromise. "Christian love as mutuality or equal regard can never be reduced to the logic of [biological] reciprocity," he says. What is more, a Christian ethic must believe that "all good (and all specific goods) comes from God, and that the ultimate meaning and direction of all finite loves is the overarching love and enjoyment of God."[37]

These efforts to bridge religious values with biological evolution will finally force theology, with its tools of philosophy, to choose between two paths. The first is to embrace scientific monism, but at the risk of having all "spiritual" attributes of human nature explained away by genes, the brain, the nervous system, or an "emergent property." On this path, theology will struggle to explain behavior at a time when Darwinism and genetics are boldly giving their "answers" in ever more clever and mathematical ways. Theology's second path is to fight the materialist hegemony over human nature. It must make a persuasive case that consciousness, subjectivity, and the "I" are transcendent of chemistry, electromagnetism, and what the brain scientists now call a "computer of meat."

. . .

When the Louisiana woman put forth her stark options for human nature—evolved ape or child of God—she was drawing on some everyday theological dualism. Her real concern was pragmatic, however. Which view of human nature makes children behave?

The Founding Fathers were equally pragmatic. They had human nature in mind when designing a government to rule over its inclinations. They joined a top-down and bottom-up stance, a melded outlook prevalent in the pre-Darwinian era of the Enlightenment. It was an approach that drew on both the dark and the light estimations of human nature. From the classic liberalism of the eighteenth-century English and Scottish thinkers—Adam Smith and David Hume among the better known—the Founders framed people as a mixture of self-interest and "moral sentiments" that foster sympathy toward others, a point Darwin picked up in his *Descent of Man*.[38] From the broad swath of Calvinism, moreover, they borrowed a darker Christian anthropology. They never defined this human nature but rather pointed to its sources and its implications in the Declaration of Independence. Human nature arose from the "Laws of Nature and Nature's God," and these made it "self-evident" that "men are created equal" and "are endowed by the Creator with certain unalienable Rights."

The words of the U.S. Constitution, in contrast, are godless. Neither do the *Federalist Papers*, which chronicle the debate over the drafting of the Constitution, record a "full philosophical view of human nature," says political scientist William Galston. The clash between good-natured utopians and sin-suspecting realists, however, was not lacking. The anti-Federalists had Calvinist leanings and warned against centralizing national power. In the end, the founding documents show only a "rough-and-ready" view of human nature, says Galston. "The Founders argued for a balanced view; we have a capacity for virtue but we also need a framework to restrain our natures." What is more, George Washington and Thomas Jefferson clearly were utilitarian in their views of religion, and how its flourishing would brace a more virtuous public.[39]

There is continuity between that founding experience and the evolution-creation disputes of today. In his classic work *Scientific Creationism*, Henry Morris is as utilitarian as Washington or Jefferson. He presents to teachers the moral benefits of teaching the recent six-day Creation as described in Genesis. "There is no greater stimulus to responsible behavior and earnest effort, as well as honesty and consideration for others, than the awareness that there well may be a personal Creator to whom one must give account," he writes. Indeed, more than four in ten

Americans adopt a utilitarian approach toward creationism in public schools, agreeing that "teaching creationism will help instill children with a more moral base." Yet there is more at stake than just morals, for creationism in general can quickly become a matter of precise religious doctrines. In *Scientific Creationism*, Morris goes on to say that the ultimate goal of presenting this "scientific theory" in the classroom is so that "the student can be led into a comprehensive, coherent, and satisfying world view centered in his personal Creator and Savior, Jesus Christ."[40]

This blend of moral utility and religious evangelism in public schools gets mixed reviews from others besides the secularists. In many theological assessments, religion is strictly about salvation, or getting "right with God." Religion is compromised, in this view, when used for utilitarian effect, such as conforming human behavior. Indeed, an orthodox theology would assert that only God's supernatural work can reconcile man and make his conduct "good." Not only evolutionists, therefore, have applauded the defeat of the Louisiana law. Many creationists believe that the sole purpose of teaching the Genesis account of Creation is to reap salvation. They would be dubious about using it to help students "start acting like God's children." Nearly all Christians, meanwhile, agree that government schools should not teach anyone's particular version of salvation.[41]

The naturalists, who do not believe in a supernatural reality, vary in their appreciation of religion, either as a product of biological evolution or as a blunt fact of life. When I met Kitcher in San Diego, his Jaguar bumper was decorated with the "Darwin fish"—the Christian ichthus given legs and the name "DARWIN." He was sent the symbols by people who had read his anticreationist book. "And of course, I'm very happy to put them on the car," he said. He does not like fundamentalist or doctrinaire religion but has nothing against more liberal and tolerant religiosity. "The history of morality, which is intertwined with the history of religion, is an attempt to find more and more delicate and satisfying ways of entering into stable relationships," he said.[42]

Mayr also believes that the best human ethics to come from evolution are liberal. But in a time of moral breakdown, the imprinting of mores such as self-control is of great value. And he points to strict religious groups such as Mormons, Adventists, and Mennonites as examples of such successful imprinting of morals, from which society could well learn. "Quite a few readers will smile at such seemingly old-fashioned advice," Mayr writes. "Is that the best science can come up with, they might say." Mayr is not out to debunk religion in favor of some naturalistic Enlightenment for all of society. Whatever morals work, he seems to say, are acceptable, even if many scientists say those morals have their origins in prescientific ignorance and superstition. "If they believe these things, and it is harmless, why shouldn't they believe in superstitious stuff?" Mayr says. "As long as it isn't interfering with society."[43]

There is a view in science, however, that calls for the expulsion of superstition. The Victorian evolutionist Herbert Spencer, a contemporary of Darwin, argued that evolution had taken humans from a state of atheism to one of spirit worship, polytheism, and then monotheism. The final stage, he said, was the rise of science, whose positivism—that all is material—would finally displace the superstition of monotheism.

Sociobiologist E. O. Wilson would not disagree, but his prescription for reaching a secularized world is mild, not ideologically aggressive. He argues that something like religion—a sacred epic or myth—will always be with humanity. It just so happened, he has mused, that people evolved to believe in gods and not to believe in biology. Biological evolution, in fact, has ended up creating some kinds of religions and made other kinds impossible, he says. "There is a biologically based human nature, and it is relevant to ethics and religion. The evidence shows that because of its influence, people can be readily educated to only a narrow range of ethical precepts. They flourish within certain belief systems, and wither under others. We need to know exactly why." He forecasts a future in which "religious transcendentalism" is overshadowed finally by "scientific empiricism."[44]

Dawkins is more aggressive in his call for secular enlightenment. He allows his daughter to read the Bible, but only as literary training, and otherwise has called religious instruction "mental child abuse." Of the memes, or ideas, that evolve through human culture, religion is the one most like a virus. That does not argue for carefree immorality, Dawkins explains, for there is also the meme of moral philosophy. Like science, he says, moral philosophy can be a human enterprise that relies on rationality. "And there are strands of moral philosophy which I do think are respectable."[45]

The humanists, in their manifestos, also call for an end to religious fundamentalisms and superstition, which should bow before the triumph of scientific knowledge and reason. Humanist Manifesto 2000, in fact, has called on humankind to move beyond the Simpson ambiguity—the seeming contradiction that people are in nature but also separate from nature. The humanists hold out scientific knowledge and education as the way to unite humans and nature, a state of monism that lacks ambiguity. In calling for such a monism, the humanists run smack into the so-called naturalistic fallacy. It is not good to harbor this fallacy, they say; what is a *fact* should not necessarily be a *value*. What is an *is* should not necessarily be an *ought*. To get around this, the humanist vision argues thus: "The growth of scientific knowledge will enable humans to make wiser choices. In this way there is no impenetrable wall between fact and value, is and ought."[46]

In other words, by means of greater knowledge, what is materially true in nature can also become morally true for human ethics and behavior—an idealistic, total solution to George Gaylord Simpson's quandary.

. . .

This quest for human nature and morals takes place against a social backdrop and is always colored by people's assessment of whether society has gotten better or worse. In modern America, that assessment is as contested as in other eras. While former education secretary William J. Bennett has created a map of leading cultural indicators that has shown sharp moral decline in terms of violence, sexual promiscuity, family relations, and learning in America, others have pointed to improvements in tolerance and social justice.[47]

Interpretations of human behavior also vary, with political scientist Monroe concluding, for example, that few people are real altruists. Truly evil people are the exception, too. Social commentator Charles Murray, one of a group of thinkers who look expectantly to biological science for answers to human behavior, argues that human behavior averaged on a bell curve clusters mostly between "fairly selfish" and "fairly altruistic." Despite this averaging effect, when crime soars or a Columbine High School killing erupts, society will despair of its effort to teach virtue. The despair spurs again the debate on whether the causes and remedies are in nurture (education) or nature (genetic makeup). "While scientists can measure the observed shape of these behaviors, they have been stymied by the nature/nurture problem," says Murray.[48]

Murray takes an empirical stance and places hope in what science will inevitably learn about how genes determine personal and social outcomes, and how this can help everyone. This is an optimism shared by other empiricists such as Fukuyama and James Q. Wilson. Accordingly, religious belief is taken by them to be not as fundamental as biology, though religion serves a valuable role as a cognitive guide for people and, for those who need it, a power to transform individual lives. So says Wilson, the criminologist. "Religion, I think, can neither be the source of morality nor provide the support for morality," he says. "It cannot be the source because it is obvious that there are highly moral people who are not religious and fanatical extremists who are deeply religious." He goes on, "Nor is it the case that religion provides the main sanction for a moral code." In defiance of creationist Henry Morris's claim that fear of God promotes good behavior, Wilson says that morality based on fear of punishment or hope of receiving a reward is not true morality but calculated benefit. Thus, he prefers to see basic morality in the biology of human nature. On top of that, he says, "Religion's chief contribution to morality is to enable people to transform their lives."[49]

Political scientist Guenter Lewy is also an empiricist, and an agnostic, too. He would accept that religion can have a transforming role in individuals but is persuaded also that religion is a primary source of morals for all society. His research, documented in *Why America Needs Religion*, found a strong correlation between morality and religious instruction, public religion, and private religious practice. Religion is a significant inhibitor, he argues, of several negative behaviors: juvenile delinquency, adult crime, prejudice, single parenthood, teen pregnancy, and divorce. "There exists a significant relationship between religiousness and the observation of certain moral and social norms," Lewy says.

For the individual, degrees of intensity determine morality. Thus, Lewy says, "The minority of Christians who take their religion seriously are different from nominal members of Christian churches." Generally, however, individual adherence does not necessarily distinguish one person's behavior from another's. What gauges the difference are two other factors: a person's church attendance and participation in religion, and the presence of community-wide religion. Humanist ethics, laid down in manifestos, may or may not guide and satisfy some intellectuals, Lewy says, "but for the majority of people morality cannot command conviction without being linked to some transcendent reasons for goodness." Lewy questions the strongest claim embodied in Dostoyevsky's warning that without God everything is permitted. "But a weaker version of Dostoyevsky's thesis may nevertheless be true: a society that tries to cut itself off from the religious roots of its moral heritage is doomed to moral decline."[50]

Evolutionists disagree with this picture of doom, even as they argue that Darwinism genuinely undermines the top-down morality of religion. "Every society has a myth of creation, such as Genesis, and that story invariably carries a moral message," says the British evolutionist John Maynard Smith. "Darwin replaced those myths with a story that was utterly amoral." The humanists are aware of what creationists say about their scientific naturalism and relativistic ethics. "Indeed," says their manifesto, "humanists are often blamed for the alleged moral breakdown of society."[51]

The same blame has fallen on evolutionists, who argue back that such a cause-and-effect connection is erroneous. "Give us a break," says Eugenie Scott. "You can't blame us for the deterioration of society when close to half of the population doesn't accept evolution. Even a lower percentage had it taught to them properly. And it's very difficult to argue that it's people who accept evolution who are out raping, pillaging, and stealing."[52] Scott speaks as a naturalist and an agnostic, but thanks to great numbers of theistic evolutionists, there is a substantial group who say God and morality can be imbued even in humans who have risen from biological predecessors such as apes.

At this point, one fairly strong rejoinder comes from creationists. They look at the God of theistic evolution and question whether that Creator, so distant and usually constrained from intervening in the evolving natural world, can inspire awe in humans. Is this a God who can command moral assent? This was a point made in 1925, when the populist Bryan debated evolutionist Osborn in the *New York Times*.

Bryan would have endorsed the Louisiana woman's assertion that belief about God may well determine whether young people will be "acting like apes." Said Bryan, "The evolution that is harmful—distinctly so—is the evolution that destroys man's family tree as taught by the Bible and makes him a descendent of the lower forms of life." He said that human evolution was at best a "hypothesis," a mere "guess" of some scientists, and thus not worth risking national morality on. In his candor, however, he also said: "If it could be shown that man, instead of being made in the image of God, is a development of beasts, we would have to accept it, regardless of its effect, for truth is truth and must prevail."[53] He was counting on guesses to be long-lived.

In science, the notion that the theory of organic evolution is a guess is scoffed at. Yet in the search for human nature it is often no better than that—a lot of guesses. The brain's one hundred billion neurons match the number of stars in the Milky Way, and the number of connections active in the brain's functioning verge on the number of stars in the entire known universe. Yet science declared the 1990s "The Decade of the Brain," mainly in hopes of medical advance, but still as if pinning down consciousness is just another moon landing. Some in neuroscience say they will figure it out someday.

With the scientific and theological viewpoints on the table, and no ruling authority at its head, Americans make a choice in how they define human nature. For now, most citizens adopt a top-down view of who they are; a minority turns strictly to natural science. On the question of human morality, there may be only one empirical laboratory. History is this laboratory, and it alone seems able to test whether a society, based on its beliefs and behaviors, rises toward the heavens or slouches toward the jungle.

15. SEARCH FOR THE UNDERDOG

The summer of 2000 marked the seventy-fifth anniversary of the John Scopes trial. It was a calendar date not to be missed by evolutionist or creationist partisans.

At the University of Kansas, People for the American Way rolled in a star cast of Hollywood actors for a theatrical reading from the trial, a rendition that made creationist William Jennings Bryan look the fool. A day later, the annual Scopes festival in Dayton, Tennessee, held its play. The local fathers had asked that this year it be more favorable toward Bryan, whose arguments came over as paradigms of reasonableness. "What a study in contrasts," says Scopes historian Edward J. Larson, who spoke at both events.[1]

With the autumn of 2000, the American Association for the Advancement of Science and the Fordham Foundation, which researches education quality, organized a national forum, "Teaching of Evolution in U.S. Schools." It released a report on evolution in state science standards and, seeking balance and amity, featured seventeen speakers expressing several viewpoints. Invariably, the sparks of debate would outglow any agreement achieved at the Washington event.

The intelligent design movement was pushing its cause that year as well. The Discovery Institute had expected Darwin critic Jonathan Wells to be on the AAAS-Fordham program, but when his participation was rejected, Discovery protested. Wells's *Icons of Evolution* had just been published, and though he missed the Washington event, his press release asserting that the Fordham Foundation report on science standards was a "whitewash" got some national media attention. Another intelligent design theorist in the news was William Dembski, who the month before had won his academic freedom dispute at Baylor University over the Polanyi Center, which a group of faculty viewed as a creationist Trojan horse. When the October ruling came down, Dembski was flush with victory; his e-mails declared a defeat of "dogmatic opponents of design" and claimed that his faculty tormentors had "met their Waterloo." The backlash was swift, and a day later the administration charged him with being uncollegial. He was demoted, and soon after the center was dissolved, though Dembski stayed at Baylor under an academic contract.[2]

The controversy was only slightly less heated at the AAAS-Fordham event, at which the centerpiece was the state science standards report by educator Lawrence S. Lerner. The report also made a point of denouncing creationist "arrogance" and warning against a "slicker and trendier intelligent design." Humanities professor Warren Nord reciprocated with comments on the wiles of "scientific fundamentalism" and the "breathtaking act of faith" that evolutionists exhibit in their beliefs. Lisa Graham Keegan, the Arizona superintendent of public instruction, told of a "chilling moment" that year when creationists nearly excised gravity as a "universal force" from the curriculum because it rivaled God. Fighting such creationism, said the school-choice Republican, "is not a compromise-ready issue. It is an issue that must be won."

Barrett Duke, an officer with the Southern Baptist Convention, made note of its sixteen million members and giant budget to emphasize that believers will not accept trampling of their faith in public schools, where he urged tolerance and accommodation. "A winner-take-all mentality will not help anybody," he said, going on to endorse the intelligent design movement as having "an extraordinary currency." Biologist and educator Paul R. Gross had a final word at one session. He spoke of "massive new forces being thrown into the argument" by creationists and was an oracle of escalating clashes. "There's going to be a lot more fighting before it's over," Gross said. "Things are going to get worse before they get better."[3]

This book has tried to show the many places the argument has unfolded, all of them in some sense places where Darwin meets the Bible. As examples from the year 2000 show, the academic and cultural debate can be as explosive as the legal and political conflicts that typically make the newspaper headlines. But what of the future? Will it get worse and then better, or has science morally and technically already won? Will some form of creationism, on the other hand, gain ground in one of the areas covered by this book—schools, textbooks, churches, museums, the science profession, public debates, media coverage, or the study of human nature?

Reading crystal balls is risky business. But speculation may be worth a try in three American contexts that might color the future debate of evolutionists and creationists. First is a new kind of social stratification of America, second the demise of apocalyptic creationism, and third the fall of Darwinism but survival of materialism in biology.

One striking social fact of the past few decades has been how social conservatism has moved into America's suburbs. "The Christian right is not a rural movement, it is a suburban one," says writer Peter Beinart, who investigated the phenomenon in Kansas.[4] The implication, of course, is that creationism has gained social mobility, both financial and educational. This means it also will have its "knowledge class," a class once associated only with elite secular culture. Indeed, according to some new thinking, this entire class structure is in transition and may give a future map of the evolution-creation debate.

While the "knowledge class" theory pitted secular elites such as judges, professors, television executives, and lawyers against ordinary Americans, sociologist Peter Berger has pointed to social and economic changes that call for a "New

Class theory."[5] Simply, the elite knowledge class has grown into a knowledge economy that undergirds half of the American middle class. Today, half of ordinary middle-class Americans have a stake in the same "redistribution" process—redistribution of wealth, ideas, health, rights, information—that once only knowledge class elites had. In other words, half of the middle class makes its living on the growth of government, government experts, welfare, and regulation. Much of this class, Berger says, is career driven, and thus its values tend to align with smaller families and with a more secular outlook—more secular than the other half of the middle class.

That other half constitutes the nation's "producers." They value economic liberty and thus oppose larger government, outside experts, and regulation. Career satisfaction is not as great as in the "knowledge economy," so the producing class gets its satisfaction from its families and religious faith. Berger has not drawn any conclusion for the evolution-creation debate. But the storm center will be the vast middle class, where the knowledge economy might be the vehicle of the evolutionist cause, the producer economy the haven of the creationist cause.

Second, the mainstreaming of the Christian right could be deflating apocalyptic creationism, which has long associated the end of the world with a recent young earth. "Before the 1960s relatively few Americans, including religious fundamentalists, subscribed to such restrictive views of earth history," says historian Ronald Numbers. Yet such a view had caught a rising tide of both political and religious enthusiasm. Both parts, with their roots in the 1970s, had a "militancy that was unknown" by the old Christian right.[6]

What is more, this "restrictive view" had caught the wind of century-end "dispensationalism," a view that the world might end any day, according to prophecy. The best-selling book in 1970s America was Hal Lindsey's *Late Great Planet Earth*, which assured millions of readers that it was "a time of electrifying excitement" and predicted a cosmic upheaval in 1988. The approach of 2000 kept the apocalyptic book-selling excitement going. Next in line was young-earth creationism champion Tim LaHaye, a Baptist minister. For five years beginning in 1997, he broke American book-peddling records by selling seventeen million copies, hardcover and paperback, of an apocalyptic novel series known for the title of its first volume, *Left Behind*. Yet if sales climbed, driven by the market force of repeat buyers, the apocalyptic year 2000 passed uneventfully.[7]

What was most apparent at the turn of the millennium was how, with its relative success, the Christian right had moderated itself. More pluralism came under its tent, prompting a political scientist to say it will "probably become institutionalized as a permanent part of American politics."[8] For these reasons, the less moderate, more sectarian young-earth creationism, which included a large package of other religious mandates, may have to leave center stage. Some observers find this doubtful. The historian Numbers points to the young-earth movement's growth in Canada, South Korea, Australia, and more recently Turkey, where it is linked to the Muslim religious resurgence. Geologist Davis Young, a Christian, lamented before a geology convention in 1999 that "the flood geology movement is disturbing because of its perpetual appeal to the general public." To stop its agenda, he said scientists must hit flood geology with a tide of "persistent challenges" before it will dissipate.[9]

Already, however, this question regarding the earth's age may have fatally undermined the "perpetual appeal" of popular creationism. Young-earth creationism may be on an upswing in South Korea and Turkey, but so is religious fundamentalism. The populist creationist movement in the United States is probably on a down cycle. Indeed, the brightest young-earth researchers, meaning those with doctorates and studies at research universities, have taken to ivory towers and longer term research. There will still be ample sermons on a literal Genesis from creationist pulpits, but that is different from a movement that once captured the political imagination, and in some pockets the scientific imagination, of a nation.

For all these reasons, a type of creationism that ducks the age question, or that works with an ancient earth, and does not rely on apocalyptic backdrops is likely to take prominence in the next few decades. The obvious candidate is the intelligent design movement. But it is still a philosophy movement, and like young-earth philosophy, it must finally be scientifically fruitful to survive—the never-ending challenge to creationism.

Changes are in store not just for the creationists, however. A third major trajectory is seen in the evolutionist camp: a dead end for Darwinism. Michael Crichton is only a science novelist, but he was not far wrong in his *Lost World* when he wrote that "a century and a half after Darwin, nearly all positions on evolution remain strongly contended, and fiercely debated."[10] The way to gauge this crisis in Darwinism is not to ask critics but to ask the most ardent supporters of materialist biological science, people like geneticist Richard Lewontin and philosopher Daniel Dennett.

Besides the materialist premise of evolutionary biology, all the rest is "up for grabs," Lewontin told me when I visited his Harvard laboratory one spring. "Evolutionary genetics wants to study processes in which the forces are very, very weak," he said. "Natural selection is a very weak force. And the time [that it takes to work] is long." And now, he said, "My colleagues are beginning to feel that crisis." Lewontin has rallied science to stop the "divine foot from getting in the door," so his use of the term "crisis" is no concession to theology. But he is criticizing the utopian hubris of science, that it will surely master all knowledge. "If you're a materialist, you're modest about this," Lewontin said. "Human beings are material objects, and there are some things that they can't do."[11]

What Lewontin calls the "divine foot in the door," Dennett refers to as "sky hooks"—the human tendency to find theological answers for what science cannot explain. Confident as ever in the power of science, though, Dennett says that biological problems have always required "cranes"—earthly, material levers. Even if evolutionary biology is lacking new-model cranes to do heavier lifting than ever before, science will provide. "In the history of the controversy over evolution, many skeptics and critics have tried to find sky hooks, and ended up discovering cranes, that is, new mechanisms that actually do enhance the powers of evolution," Dennett tells a university audience. "Evolution turns out to have more tricks up its sleeve than you might initially have imagined."[12]

The point is that materialism is not in a crisis in biology, but the power of Darwinian evolution—mutation and natural selection—to explain all things seems to be. Evolutionist and philosopher Holmes Rolston III put it nicely when he said, "If Darwin is biology's Newton, its Einstein may be still to come."[13] As to which country will give birth to that new genius, it has been noted that the debate over biology in the United States is not nearly as vigorous as in other parts of the world. This is said, moreover, as young people's interest in biological science is in apparent decline, and some seniors in the field are wishing for some fresh air, even a little shake-up, in the field. The physicist Max Planck famously said that "a new scientific truth" does not triumph by making opponents "see the light, but rather because its opponents eventually die."[14]

Oddly, however, it seems that innovation in biology is being stopped not by long life spans or interminable university tenures but by one other thing—fear that creationists will capitalize on disagreements within evolutionary biology. The fear is probably merited, but it may not justify halting scientific ferment, or killing an Einstein for biology in the crib.

This book opened with the observation that, despite the intensity of debate over evolution and creation in America, the public is not better informed of its details as a result. One reason might be that the debate perpetuates certain areas of confusion. To bring clarity, however, will demand some remarkable candor, and some portion of humility, on both sides. Five areas come to mind.

TAUTOLOGY

First is the idea of a tautology, or a circular argument. Both sides accuse the other in this way. British philosopher Anthony Flew has written vividly on how once you have God, you can explain everything (and thus nothing). Lawyer Norman Macbeth says the same about evolution: there is simply nothing it cannot explain, since that which survives, survives. "I regard this [tautology] as a major discovery, a sort of lethal gene," he wrote in 1971.[15]

This power to explain everything, of course, can undermine either idea's usefulness, except as a philosophy, belief, or ideology. It seems reasonable that both sides should put their cards on the table and admit this "explain everything" problem from the outset.

The public might take these confessions of tautology as confessions of relativism—that truth just begins with opinion. But that overlooks how much the stances of evolution and creation, unlike true relativism, agree that there is a real world "out there" that the mind may understand. Nobel chemist Dudley R. Herschbach, speaking at the New York Academy of Sciences in 1996, recalled hearing a claim that myth and poetry have the same value as to science. "A statement like that, of course, is meaningless unless you ask, 'Valuable for what?'" Herschbach said. "If the question is how can we get *Apollo 13* back, obviously myth is not very helpful. If the question is, does God exist, science is irrelevant."[16]

Science philosopher John Leslie, who surveys the cosmology debate, notes that its much-contested "anthropic principle" has been dismissed as a tautology: the universe is fine-tuned for humans to be here, for how else could we be here? A tautology, yes. But still, Leslie says, "it can *encourage* predictions without itself making them."[17] So both evolutionist and creationist tautologies can, in some ways, lead to science. Science's attempt to deny creationism this access has produced a well-known paradox: on one hand it says creationism is tested and proved wrong (failed science) but on the other that it is untestable (not *even* science). What is more, science equally poaches on religious turf by speculating on the theology of God. Increasingly, for example, evolutionists speak about how a Creator (if God existed) would have best designed nature. Thus, they say, nature's imperfections—"errors that no intelligent designer could have committed"— prove evolution's blind work, not the design of God.[18]

Giving creation and evolution equality as tautologies at this time in history benefits the former most, and is protested most by the latter. Herschbach had worried about confusing poetry and science, but in the evolution-creation debate both sides (unlike subjective poetry) are presuming an objective reality "out there." Both in turn have major tautologies. In the end, poetry will not bring back *Apollo 13,* but science still might say something about a Creator—or belief in God say something about nature.

PRACTICAL OR THEORETICAL

Herschbach was addressing a conference on the "future of science," but also at issue was science journalist John Horgan's provocative book *The End of Science* (1996). Horgan argued that science had found all the universal laws within human reach. This had forced scientists down zany paths into "ironic science," which Horgan likened to modern literary criticism and its unverifiable viewpoints. The main rebuttal to Horgan was this: there is much practical science still to do. "You equate science with the search for great universal truths," Cornell University biologist Thomas Eisner said to Horgan at the forum. "To a biologist, aware that most of nature remains to be discovered, the process of discovery . . . is very much a part of science." This includes explanations for how bees, bats, and owls behave. "These are truths that are not universal," said Eisner, "but they are made of the extraordinary process of discovery."[19] The creationist public would cheer such a program of science, which generally is viewed as "practical," not theoretical. But science will not stop there. It pushes ahead with its special "way of knowing," which seems to exceed practical and commonsense science—and leads to the third concern.

EVOLUTION AS ORIGINS

Does study of evolution mean the study of "origins," the ultimate beginning of life itself and of human life in particular? When Supreme Court justice Antonin Scalia dissented in the 1987 ruling on Louisiana's "balanced treatment" law, his favorite term was "origins." Evolutionist Stephen J. Gould cried foul. "Justice Scalia has defined evolution as the search for life's origins—and nothing more," Gould said. Evolution, Gould corrected, is actually limited to the "study of how life changes after it originates." To illustrate this narrow focus of science, Gould turned to the work of the early geologist James Hutton, who never worried about the start and end of the world, just what happened in between.[20]

Should the public be making this careful scientific distinction along with Hutton? Often it seems that science does want to take in everything, not just the in-between part. "The goal of science is to seek naturalistic explanations for phenomena—and the origins of life, the earth, and the universe are, to scientists, such phenomena—within the framework of natural laws and principles and the operational rule of testability," says the National Academy of Sciences. It further counsels, "To teach biology without explaining evolution deprives students of a powerful concept that brings great order and coherence to our understanding of life." When a blue-ribbon panel studied the nation's biology textbooks, its complaint was about too many facts and mechanisms, but a lack of sweeping themes. In September 2001, the PBS documentary *Evolution* aired with the subtitle *Where We're From and Where We're Going*. While Dennett told viewers that a giant natural process, and not skyhooks, produced a world that looks designed, Gould elaborated: "Evolution is about who we are, what we're made of, what our life means in so far as science can answer that question. [So] in many ways, it's the singularly deepest and most discombobulating of all discoveries that science has ever made."[21]

With all of this, the public—and a Supreme Court justice—might be forgiven for seeing science as making fairly absolute and religion-like claims. Unlike Hutton, who looked at a limited phenomenon (cycles of buildup and erosion on the earth's surface), biologists are in search of universal truth, Lewontin suggested. Biologists do not want to face "the real truth about biology, which is there are no general laws, and that you have to worry about the details of every interaction. . . . They want to make universal laws. And if I tell you you've got to study each case on its own, they don't want to hear that. That's not the way to fame and fortune. The way to fame and fortune is to find 'the law' that governs all of life."[22]

The final two areas of candor pose a challenge to religion. In the Supreme Court's 1987 ruling against mandated creationism in science class, it added that "teaching a variety of *scientific* theories about the origins of humankind to school children might be validly done with the clear secular intent of enhancing the effectiveness of science instruction."[23] Remarkably, in the years since, no one in science has proposed what that "variety" of theories might include. One-size-fits-all Darwinian evolution rules. The only contender with a scientific feel, in fact, comes from the creationist side in the idea of "intelligence" or "intelligent information" in nature. Materialists in science also have alluded to such intelligence, pointing to outer space, the abrupt "emergence" of creative new things in nature, or some deeper laws at work in matter. Even Dennett will concede, "We cannot positively and definitively rule out the hypothesis that there has been intelligent tampering with our genomes in the past. After all, we ourselves have in recent years been intelligently tampering with the genomes of different species."[24] Everyone knows, however, that when scientists of a religious persuasion speak of intelligence, they are probably talking about God.

As a church-state issue, it may become one of the more interesting of the future. The courts' means of ruling on what is secular and what is religious is truly a tangled equation, jurists agree. Speaking to some of this legal morass, the intelligent design advocates say that drawing inferences of intelligence from nature is not religion: it is neither a form of worship nor a dogma. They also appeal to new court rulings against "viewpoint discrimination," which have given religion a secular standing as a kind of free speech deserving neutral treatment.[25]

This loathing of viewpoint discrimination and joy at finding "intelligence" in nature may have public appeal. Established science will bitterly oppose it; biologist Kenneth Miller says viewpoint discrimination is what science is all about. Indeed, you cannot teach a flat-earth viewpoint. But this will finally be a legal matter, and that puts success within tenuous reach of creationists.

GOD'S INTERVENTION

Americans may respond to "intelligence," moreover, because they also tell pollsters of high belief rates in prayer and in miracles.[26] In other words, someone or something intelligent out there responds to human appeal. In an age of science, this fifth area poses a great theological and pastoral challenge—unless natural science is simply ignored. Natural science states that no natural law may be broken. Period. Theologians have had to respond by putting God back further and further into realms of transcendent mystery or increasingly under rubrics of doctrinal logic.

The philosopher Walter T. Stace argued darkly in 1948 that, even as science makes the mind corrosively secular and skeptical, religion could survive on a minimum claim of divine purpose.[27] This quest for purpose is both good theology and good metaphysics. But it does not come to grips with the high rates of belief in prayer and miracles—both presumptions that God interferes in nature. It is not an easy issue for theology in a scientific age, and one that has often led to obfuscation at worst, innovation at best. Interestingly, some scientists say theology has "lost its nerve" on this front, so it is believing scientists (who know the ambiguity and soft spots in nature) who are trying to put ideas such as "divine action" back into modern discussions about human reality.

The debate, we have seen, has never given reason to either side to wave a white flag of surrender. Take the new century. When the human genome sequence was completed in June 2000, Caltech president David Baltimore, a Nobelist in medicine, said its proof of "the same humble beginnings" for all creatures was irrefutable. "That should be, but won't be, the end of creationism," he wrote. Indeed, creationists took the same DNA sequence, which turned out to be more complex than once believed, and ask how the coded information got there. With the new century also, evolutionist Niles Eldredge wrote the book *The Triumph of Evolution*, subtitled *And The Failure of Creationism*. Intelligent design "godfather" Phillip Johnson, in turn, issued a progress report—not a victory letter—that evoked Churchillian images. The movement is succeeding, and ten years out it finally made the front page of the *New York Times* with a story from the science desk, he noted. "It's not the beginning of the end, but it is the end of the beginning," Johnson declared.[28]

Wherever Darwin meets the Bible—wherever biblical theism meets evolution—this rivalry is a likely product. The polarization seems inevitable and perennial, but one way to surmount it, at least momentarily, is through a shift in perspective. To evoke an idea that Churchill also surely knew, either side can be portrayed as an underdog. Great cultural arguments often produce a perceived underdog. It is a sympathetic status that evolutionists and creationists as causes and as people both may claim—depending on the circumstances.

The evolutionist's underdog lament is along these lines: "The scientists have lost every battle involving evolution since the Scopes trial in the arena of public opinion," says science professor Walter L. Manger of the University of Arkansas. He is speaking of the failure of society to let biology professors teach their science. Evolutionists are also made underdogs by a broad swath of society that resists the special claims of science. Scientists, esteemed as professionals also, receive appreciation only when they deliver a medical breakthrough or a new creature comfort. They look across the United States and see legions of churches promulgating nonscientific doctrines. They see opinion polls in which more than half of Americans say "we need more faith and less science" and in which 64 percent of teenagers say that if a scientific and a religious explanation disagree, they are "more likely to accept the religious answer." These are the people, scientists know, who have great influence in the statehouses, on school boards, and in national elections.[29]

But creationists are underdogs, too. They see their camp as embattled by a battery of superior social forces, especially the lawyers of the American Civil Liberties Union, surely the most effective anticreationist cadre of all. "They scare everybody," says creationist leader Duane Gish. "They have millions and millions of dollars."[30] Though the ACLU lost the Scopes trial of 1925, it has won the support of judges and the cultural elite and since 1968 has been successful in silencing all attempts to allow criticism of evolution in public schools.

The creationists are further marginalized as the evolutionists alone receive federal dollars. Evolutionists control the accrediting agencies, research centers, and nearly all the universities. The news media and the courts, mavens of the old "knowledge class," are required by definition to rule negatively on creationists and their agenda. Belief in God has also been sidelined at research universities and in the courts, neutering one key argument of all creationism. As the campus saying goes, "Every department can preach except religious studies."

What underdog status produces most in a democracy is public sympathy, which if broad enough can decide the social standing of an American institution. "America loves the underdog," is another way of saying it. "It's a very useful metaphor for both of us," says evolutionist Eugenie Scott, a participant in the debate for a quarter century. "They claim it as consciously as we do. There is truth on both sides."[31]

APPENDIX

Table 1. James H. Leuba's Survey Questions on God and Immortality[1]

A. Concerning the Belief in God
I believe in a God in intellectual and affective communication with mankind, i.e., a God to whom one may pray in expectation of receiving an answer. By "answer" I do not mean the subjective, psychological effect of prayer.

B. Concerning the Belief in Personal Immortality
I believe in immortality for all people . . . [or in] conditional immortality.

The questions, used by James H. Leuba in 1914 and 1933 surveys, were first published in Leuba, *The Belief in God and Immortality: A Psychological, Anthropological and Statistical Study* (Boston: Sherman, French, 1916), 224-25.

1. Possible responses were belief, disbelief, or agnostic (doubt). The survey had secondary questions, such as desire for immortality, but these two questions were the core. "Mankind" was changed to "humankind" in repeat surveys in 1996 and 1998.

Table 2. Natural Scientists on God (A) and Immortality (B)[1]

	A. Belief	Disbelief	Agnostic	B. Belief	Disbelief	Agnostic
RANKING SCIENTISTS[2]						
1914[3]						
Total	41.8	41.5	16.7	50.6	20+/−	27+/−
Biologists	30.5			37		
Physicists	43.9			50.7		
1996						
Total	39.6	45.5	14.9	38	46.9	15
Biologists	42.5	43.5	14	40.5	44.5	14.9
Physicists/ Astronomers	29.1	53.9	17	29.7	53.1	17
Mathematicians	43.6	41.6	14.8	40.93	45.6	13.4
GREATER SCIENTISTS						
1914						
Total	27.7	52.7	20.9	35.2	25.4	43.7
1933						
Total	15	68	17	18	53	29
NATIONAL ACADEMY OF SCIENCES						
1998						
Total	7.1	72.7	20.2	7.9	72.7	19.4
Biologists	5.5	65.3	29.1	7	68.5	23.6
Physicists/ Astronomers	7.47	80.3	12.1	7.47	78.5	14
Mathematicians	14.3	71.4	14.2	15	65.9	19

The 1914 and 1933 surveys were conducted by James Leuba and reported in Leuba, *The Belief in God and Immortality: A Psychological, Anthropological and Statistical Study* (Boston: Sherman, French, 1916); Leuba, "Religious Beliefs of American Scientists," *Harper's*, August 1934, 291–300. The 1996 and 1998 surveys were conducted by Edward J. Larson and Larry Witham and reported in Larson and Witham, "Scientists Are Still Keeping the Faith," *Nature*, April 3, 1997, 435–36; Larson and Witham, "Leading Scientists Still Reject God," *Nature*, July 23, 1998, 313.

1. See Leuba's two questions in Table 1. In 1914 and 1933, Leuba surveyed a random sample of 1,000 biological and physical scientists from *American Men of Science (AMS)* with response rates, respectively, of 70% and 75%. In 1996, Larson and Witham repeated Leuba's survey with a random sample of 1,000 biologists (50%), physicists (25%), and astronomers/mathematicians (25%) from the 1995 edition of *American Men and Women of Science (AMWS)*. In 1998, to compare today's top natural scientists with Leuba's "greater scientists," Larson and Witham

Table 2 continued

surveyed 511 biologists, physicists, astronomers, and mathematicians in the National Academy of Sciences (NAS) in biological and physical sciences. The survey responses were 60% for 1996 and 50% for 1998.

2. A "ranking scientist" is the author's term for those qualified to be listed in *AMS*, used by Leuba for his 1914 and 1933 surveys, and in *AMWS*, used by Larson and Witham in the 1996 survey. A "greater scientist" is Leuba's term for those judged eminent and given a star by their name in the *AMS*, which ended the starring practice in 1941. In this comparison, NAS members are considered "greater scientists."

3. Leuba did not report all his findings, so there are gaps, and in one case his responses are more than 100%.

Table 3. Scientists and Citizens on Evolution-Creation[1]

---

1. Man developed over millions of years from less developed forms of life. God had no part in this process.

        Scientists: 55%   Public: 10%

2. Man developed over millions of years from less advanced forms of life, but God guided this process, including mankind's creation.

        Scientists: 40%   Public: 39%

3. God created man pretty much in his present form at one time within the last 10,000 years.

        Scientists: 5%   Public: 44%

This Gallup poll question was used in 1982, 1991, 1993, 1997, 1999, and 2001 with roughly the same public response; the public response here is from 1997. The scientist response was first reported in Larry Witham, "Many Scientists See God's Hand in Evolution," *Washington Times*, April 11, 1997, A8. For a comparison, that newspaper report used the 1991 poll figures reported in "Many Americans Hold Beliefs in 'Creationism,'" *Emerging Trends*, January 1992, 2. For an overview of public polling, see Deborah Jordan Brooks, "Substantial Number of Americans Continue to Doubt Evolution as Explanation for Origin of Humans," *Gallup News Service*, March 5, 2001, at www.gallup.com/poll/releases/pr010305.asp).

1. In the 1997 public poll, 7% had no opinion. The scientist response comes from a 1996 survey of a random sample of 1,000 biologists (50%), physicists/astronomers (25%), and mathematicians (25%) listed in the 1995 edition of *American Men and Women of Science*. The response was 60%.

Table 4. Theology Schools and Science in America

	ATS Students[2]	ATS Schools	Protestant	Interdenominational/ Nondenom- inational	Catholic
Q. Which view of natural history and human origins predominates at your theological school? Check the closest one.[1]					
Theistic evolution	24.26	35.2	34.3	17.6	51.2
Progressive creation	14.7	8.8	7.29	8.8	12.8
Young-earth creation	7.4	7.6	7.29	17.6	0
Mix of theistic evolution and progressive creation	28.1	26.4	23.9	26.4	29
Mix of progressive creation and young-earth creation	20.2	16.4	20.8	23.5	0
Q. Which seems the most important or appropriate way that theology meets science? Check the closest one to your view.[3]					
Support the biblical account of the human Creation and Fall	14.28	13.4		30.3	2.7
Give overall meaning and purpose of life in a material universe	76.39	77.5		51.5	94.5
Put ethical limits on sciences such as biotechnology	4.34	3.37		9	2.7
Combination/ other/no answer	5	5.8		9.2	0

The survey, conducted in the spring and summer of 1999, was sent to academic deans at the 237 theological schools affiliated with the Association of Theological Schools (ATS). The response was 72% and was representative of all traditions. The results were reported, in part, in Edward J. Larson and Larry Witham, "The God Who Would Intervene," *Christian Century*, October 27, 1999, 1026–27.

Table 4 continued

1. There was 5.3% no answer, and from 6% to 7% in each tradition could not identify a predominant view. The Eastern Orthodox theology schools chose "Mix of theistic evolution and progressive creation." Notes clarifying each stance were, in order: "God guides unbroken evolution over millions of years"; "God specially created new levels of life at points over millions of years"; "God created the universe and humans only thousands of years ago"; "millions of years in common"; and "special creation in common."

2. The student number is based on total "head count" enrollment at the responding ATS schools. A larger percentage of students are creationist than schools are creationist because the more conservative theological seminaries have larger enrollments than the more liberal ones.

3. This answer is the personal view of the academic dean, but in most cases it may be taken as the emphasis at the institution. Faculty views may vary, and student views would be less developed. In answer to another question, only four in ten of the schools had a course "focused mainly on theology/religion and science."

Table 5. Theologians and Scientists

	Belief in a God Who Answers Prayers	Belief in Immortality for All or Some
ATS theologians	91.5	81.3
AMWS scientists	39.3	38
NAS scientists	7.1	7.9

This is the author's comparison of responses to Leuba's two questions (see Table 1). Not all theologians could accept Leuba's "interventionist" God, and various theologies deal differently with the afterlife (or lack of it). In the *AMWS* and NAS survey responses, some scientists suggested Leuba's "personal God" narrowed their option, so they chose disbelief. Some scientists believe in an impersonal God as a cosmic principle or force. See chapter 11.

NOTES

ABBREVIATIONS

AAAS   American Association for the Advancement of Science
ABT   *American Biology Teacher*
AP   Associated Press
ECPE   Daniel Yankelovich Group, *Evolution and Creationism in Public Education: An In-Depth Reading of Public Opinion* (Washington, D.C.: People for the American Way, March 10, 2000)
GNS   Deborah Jordan Brooks, "Substantial Number of Americans Continue to Doubt Evolution as Explanation for Origin of Humans," Gallup News Service, March 5, 2001 (www.gallup.com/poll/releases/pr010305.asp)
JASA   *Journal of the American Scientific Affiliation*
JCST   *Journal of College Science Teaching*
NCSE   National Center for Science Education
NSES   National Research Council, *National Science Education Standards* (Washington, D.C.: National Academy Press, 1996)
NYRB   *New York Review of Books*
NYT   *New York Times*
SC   National Academy of Sciences, *Science and Creationism: A View from the National Academy of Sciences* (Washington, D.C.: National Academy Press, 1984)
SC2   National Academy of Sciences, *Science and Creationism: A View from the National Academy of Sciences*, 2nd ed. (Washington, D.C.: National Academy Press, 1999)
SEI   National Science Board, *Science and Engineering Indicators—1996* (Washington, D.C.: U.S. Government Printing Office, 1996)
TAE   National Academy of Sciences, *Teaching About Evolution and the Nature of Science* (Washington, D.C.: National Academy Press, 1998)
WP   *Washington Post*
WT   *Washington Times*

PREFACE

1. Only 11 percent of scientists have "a great deal of confidence" in the print media, while 22 percent have "hardly any confidence." See Robert O. Wyatt, *Worlds Apart: Gauging the Distance Between Science and Journalism, Survey Responses* (Nashville, Tenn.: First Amendment Center, 1997), 2; in turn, 58 percent of mainline ministers, 70 percent of Catholic priests, and 92 percent of conservative Christian clergy said most religion cover-

age "is biased against ministers and organized religion." See John Dart and Jimmy Allen, *Bridging the Gap: Religion and the News Media* (Nashville, Tenn.: First Amendment Center, 1993), 36.

INTRODUCTION

1. Charles Darwin, *On the Origin of Species* (New York: Penguin, Books, 1968), 435.
2. Letters to the editors, *Scientific American*, January 2000, 6.
3. *SC*, 6; Stephen J. Gould, "President's Address" (presented at the AAAS annual meeting, Washington, D.C., February 19, 2000), author's tape.
4. David Papineau, "Natural Selections," *NYT*, May 14, 1995, sec. 7, 13.
5. Ronald L. Numbers, discussant, "Anti-evolutionism: What Is Changed and Unchanged 20 Years After *McLean v Arkansas*?" (forum at the AAAS annual meeting, San Francisco, February 18, 2001), author's tape.
6. Interview with Ernst Mayr, April 1997; Mayr, "Influence of Darwin on Modern Thought," *Scientific American*, July 2000, 82–83; Mayr, *One Long Argument: Charles Darwin and the Genesis of Modern Evolutionary Thought* (Cambridge, Mass.: Harvard University Press, 1991), ix.
7. John Haught, *God After Darwin: A Theology of Evolution* (Boulder, Colo.: Westview, 2000); Kenneth R. Miller, *Finding Darwin's God: A Scientist's Search for Common Ground Between God and Evolution* (New York: Cliff Street Books, 1999), 267.
8. Darwin, *Origin of Species*, 223.
9. "American Museum of Natural History's Science and Nature Survey" (New York: Louis Harris and Associates, 1994), 61; *ECPE*, 32–35. Similarly, see *GNS* for a February 2001 Gallup poll that found a minority of Americans watched the evolution-creation debate closely: 34 percent were "very informed" on evolution, and 40 percent were "very informed" on creationism; Barna Research Group, "Bible Reading in America," July 15, 1996, news release by Tyndale House Publishers; George Gallup and Jim Castelli, *The People's Religion* (New York: Macmillan, 1989), 60.
10. Gary Wills, *Under God: Religion and American Politics*, 2nd ed. (New York: Simon and Schuster, 1990), 124; Barna, "Bible Reading in America."
11. *SEI*, sec. 7–3. "Americans continue to hold the scientific community in high regard," the reports says, citing two decades of General Social Survey findings that indicate 40 percent of Americans are "very confident" in scientific and medical leaders, the highest for any profession.
12. Edwin O. Reischauer, *The Japanese Today: Change and Continuity* (Cambridge, Mass.: Belknap Press, 1988), 203; George F. Bishop, "What Americans Really Believe," *Free Inquiry* 19 (summer 1999): 38–42. Bishop cites the 1993 International Social Survey's "knowledge" question about human evolution, on which Americans were 44.2 percent "correct," European nations ranged from 60 percent to about 81 percent correct, and 81 percent of Japanese were correct.
13. John Hedley Brooke, *Science and Religion: Some Historical Perspectives* (Cambridge: Cambridge University Press, 1991), 33; James Gilbert, *Redeeming Culture: American Religion in the Age of Science* (Chicago: University of Chicago Press, 1997), 9.
14. Carol Kaesuk Yoon, "Evolutionary Biology Begins Tackling Public Doubts," *NYT*, July 8, 1998, B8; the first month after the April 1998 release of *TAE*, the academy Web site received fifty thousand hits, and when the state of Alabama denied its distribution, an Alabama Academy of Sciences teach-in distributed 875 copes; the updated *SC2* was issued in April 1999; *Biology Textbooks of 1990: The New Generation* (Washington, D.C.: People for the

American Way, 1990); *NSES*, 104; NSTA report in Larry Witham, "49 States Mandate Teaching Evolution," *WT*, April 8, 2000, A3.

15. *The Condition of Education, 1998* (Washington, D.C.: U.S. Department of Education, 1998), see indicator no. 22; Jon D. Miller, "American Attitudes Toward Evolution" (paper presented at the Washington, D.C., Science Writers meeting, January 26, 2000), 6, 7; *Tangipahoa Parish Board of Education v Freiler*, 530 US 1251 (2000). See also *Tangipahoa* 975 FSupp 819 (1997), and *Tangipahoa* 185 F3d 337 (1999). The 1997 ruling by the U.S. District Court for the Eastern District of Louisiana was the first to equate "intelligent design" curriculum with creationism.

16. *The ELSI Research Planning and Evaluation Group Report* (Bethesda, Md.: National Human Genome Research Institute, May 2000), appendix 4a. ELSI is the Ethical, Legal, and Social Implications program of the Human Genome Project, and since the project's start in 1990, ELSI received 3 to 5 percent of the total annual budget to do research on societal topics.

17. John Hanna, "'Science Guy' Goes Ape over Ruling," AP, August 14, 1999; "WGBH/NOVA Science Unit and Clear Blue Sky Productions to Co-produce First American Series on Evolution," press release, December 6, 1999; see Robert Wright, *The Moral Animal: Evolutionary Psychology and Everyday Life* (New York: Pantheon, 1994).

18. William Broad, "Creationists Limit Scope of Evolution Case," *Science*, March 20, 1981, 1331–32; Robert T. Pennock, *Tower of Babel: The Evidence Against the New Creationism* (Cambridge, Mass.: MIT Press, 1999), xvi.

19. Interview with Phillip E. Johnson, July 1996.

20. Interview with Michael Behe, April 1997.

21. Michael Crichton, *The Lost World* (New York: Knopf, 1995), 205; William Provine, "Evolution: Free Will and Punishment and Meaning in Life" (keynote speech at Darwin Day 1998, University of Tennessee, Knoxville, February 12, 1998), author's tape.

22. Bill Gates, *The Road Ahead* (Boulder, Colo.: Blue Penguin, 1996), 228; interview with Gene Myers, February 2001; Tom Abate, "Human Genome Map Has Scientists Talking About the Divine: Surprisingly Low Number of Genes Raises Big Questions," *San Francisco Chronicle*, February 19, 2001, B1. David F. Noble argues that mastery of machines and systems easily evokes religious aspirations; see Noble, *The Religion of Technology: The Divinity of Man and the Spirit of Invention* (New York: Knopf, 1997).

23. Philip Handler, "Public Doubts About Science," *Science*, June 6, 1980, 1093; testimony of Warren Nord in U.S. Commission on Civil Rights, *Schools and Religion: Proceedings* (Washington, D.C.: Government Printing Office, 1999), 59; Larry Witham, "Senate Bill Tackles Evolution Debate," *WT*, June 18, 2001, A4. For final wording, see chapter 7, note 8, and for Senate debate, see *Congressional Record*, June 13, 2001, amendment 779, bill S.1.

24. Interview with David Raup, June 1996.

25. Quoted in Ronald L. Numbers, *The Creationists* (New York: Knopf, 1992), 180; Garrett Hardin, *Nature and Man's Fate* (New York: Mentor Books, 1961), 216.

CHAPTER 1

1. Quoted in Mark Monmonier, *Drawing the Line: Tales of Maps and Cartcontroversy* (New York: Henry Holt, 1995), 164.

2. Quoted in Thomas Glick, ed., *The Comparative Reception of Darwinism* (Austin: University of Texas Press, 1972), vii.

3. On nature and faith, see John Harmon McElroy, *American Beliefs: What Keeps a Big Country and Diverse People Together* (Chicago: Ivan R. Dee, 1999), 111–20. Christopher Toumey

posits three American perceptions of science, saying Protestant "curiosity" and "useful knowledge" integrate well in America, while European pure research is "alien to large parts of the general population"; see Toumey, *Conjuring Science: Scientific Symbols and Cultural Meanings in American Life* (New Brunswick, N.J.: Rutgers University Press, 1996), 25–26. When *Time* chose Einstein as "person of the century," essayist Roger Rosenblatt said "he won his stardom in the only way that Americans could accept—by dint of intuitive, not scholarly, intelligence and by having his thought applied to practical things, such as rockets and atom bombs"; see Rosenblatt, "The Age of Einstein," *Time*, December 31, 1999, 90.

4. Interview with Duncan Porter, April 1996, and follow-up; interview with Peter Graham, April 1996; Porter and Graham, *The Portable Darwin* (New York: Penguin, 1993), xiv; George Gallup Jr. and D. Michael Lindsay, *Surveying the Religious Landscape* (Harrisburg, Pa.: Morehouse, 1999), 36–38. See *GNS* for polls in 1982, 1991, 1993, 1997, 1999, and 2001. The 1997 results (which found that 10 percent of Americans reject a deity behind nature and 7 percent have no opinion) roughly matched those of the other years. Of note, a major 2001 poll found Christian identification had dropped to just below 80 percent.

5. Charles Darwin to Catherine Darwin, April 6, 1834, in Frederick Burkhardt et al., eds., *The Correspondence of Charles Darwin*, vol. 1, *1821–1836* (Cambridge: Cambridge University Press, 1985), 379.

6. William Cabell Rives, *An Address Delivered Before the Alumnus of the University of Virginia*, June 27, 1883, 32. Monograph in the Library of Congress.

7. Emma Rogers and William Sedgwick, eds., *Life and Letters of William Barton Rogers*, vol. 1 (Boston: Houghton Mifflin, 1896), 304–8; Adrian Desmond, *The Politics of Evolution: Morphology, Medicine, and Reform in Radical London*, 2nd ed. (Chicago: University of Chicago Press, 1992), 414.

8. Charles Darwin to Joseph D. Hooker, January 11, 1844, in Frederick Burkhardt et al., eds., *The Correspondence of Charles Darwin*, vol. 3, *1844–1846* (Cambridge: Cambridge University Press, 1988), 2.

9. Ernst Mayr, *One Long Argument: Charles Darwin and the Genesis of Modern Evolutionary Thought* (Cambridge, Mass.: Harvard University Press, 1991), 5.

10. Charles Darwin, *On the Origin of Species* (New York: Penguin, 1968), 296.

11. Rogers and Sedgwick, *Life and Letters*, 1:307.

12. Charles Lyell, *Principles of Geology*, vol. 1, 6th ed. (London, 1840), 112; Charles Darwin to Leonard Horner, August 29, 1844, in Burkhardt et al., *Correspondence of Charles Darwin*, 3:55.

13. Darwin, *Origin of Species*, 460, 132, 452.

14. Emma Rogers and William Sedgwick, eds., *Life and Letters of William Barton Rogers*, vol. 2 (Boston: Houghton Mifflin, 1896), 24, 25.

15. W. M. Smallwood, "The Agassiz-Rogers Debate on Evolution," *Quarterly Review of Biology* 16 (March 1941): 5; Ronald L. Numbers, *Darwinism Comes to America* (Cambridge, Mass.: Harvard University Press, 1998), 33.

16. Rogers and Sedgwick, *Life and Letters*, 2:27–28.

17. See Edward J. Pfeifer's account in *The Comparative Reception of Darwinism*, 176–77, 181; Rogers and Sedgwick, *Life and Letters*, 2:29–30; Smallwood, "Agassiz-Rogers Debate on Evolution," 1–12.

18. Smallwood, "The Agassiz-Rogers Debate on Evolution," 4–11; *Proceedings of the Boston Society of Natural History*, vol. 7, *1859–1861* (Boston, 1866), 231–71.

19. Numbers, *Darwinism Comes to America*, 31.

20. Pfeifer, *Comparative Reception*, 181.

21. Adrian Desmond and James R. Moore, *Darwin: The Life of a Tormented Evolutionist* (New York: Warner Books, 1991), 661–66, 671.

22. Rogers and Sedgwick, *Life and Letters*, 2:387–88. Rogers's religion has been described as, "if anything, a Presbyterian, though he had close ties to Unitarians," in Numbers, *Darwinism Comes to America*, 154.

23. Rives, *An Address*, 28.

24. Rogers and Sedgwick, *Life and Letters*, 2:389–90. See Francis Walker, "Memoir of William Barton Rogers, 1804–1882," in *Biographical Memoirs*, vol. 3 (Washington, D.C.: National Academy of Sciences; 1887), 3–12.

25. Rexmond C. Cochrane, *The National Academy of Sciences: The First Hundred Years, 1863–1963* (Washington, D.C.: National Academy of Sciences, 1978), 136.

26. The rise of theoretical science in America is traced to the 1944 report to the Roosevelt White House, *Science: The Endless Frontier*, by MIT president Vannevar Bush; see Deborah Shapley and Rustum Roy, *Lost at the Frontier: U.S. Science and Technology Policy Adrift* (Philadelphia: ISI Press, 1985), 1–34. The Bush report successfully urged more funding for basic science, even though Americans were "still far from true understanding of the nature of basic research and the fundamental difference between science and technology"; quoted in Rodger W. Bybee and Joseph D. McInerney, eds., *Redesigning the Science Curriculum* (Colorado Springs, Colo.: Biological Sciences Curriculum Study, 1995), 5.

27. Edwin S. Gaustad, *Sworn on the Altar of God: A Religious Biography of Thomas Jefferson* (Grand Rapids, Mich.: Eerdmans, 1996), 51–55, 63–69.

28. *TAE*, viii.

29. Thomas E. Buckley, "The Use and Abuse of Jefferson's Statute: Separating Church and State in Nineteenth-Century Virginia," in *Religion and the New Republic: Faith in the Founding of America*, ed. James H. Huston (Lanham, Md.: Rowman and Littlefield, 2000), 49, 51, 50.

30. Ibid., 50; Peter Lee, "Faith, Freedom and Virtue" (sermon delivered at the Washington National Cathedral, Washington, D.C., September 20, 1998).

31. See Jerry Falwell, *Strength for the Journey* (New York: Simon and Schuster, 1987); William Schneider, "The Republicans in '88," *Atlantic*, July 1987, 80.

32. Interview with Henry M. Morris, July 1996.

33. Edward J. Larson, *Summer for the Gods: The Scopes Trial and America's Continuing Debate over Science and Religion* (New York: Basic Books, 1997), 110, 200.

34. On Baptist populations, see Barry A. Kosmin and Seymour P. Lachman, *One Nation Under God: Religion in Contemporary American Society* (New York: Harmony Books, 1993), 15, 52–55, 88–89.

35. Richard Dawkins, "Acceptance Speech" (presented at the Humanist of the Year Award, American Humanist Association, Atlanta, March 29, 1996), author's tape; Larry Witham, "Humanists Pronounce Movement a Positive Force," *WT*, April 5, 1996, A2.

36. Adrian Desmond, *Huxley: From Devil's Disciple to Evolution's High Priest* (Reading, Mass.: Addison-Wesley, 1997), 480.

37. Interview with Richard Dawkins, March 1996.

38. Desmond, *Huxley*, 482, 463. Desmond notes that Huxley visited on the 1876 centenary "when Americans were seeking measures of national achievements" (482).

39. Interview with Dawkins.

40. Darwin Day flyer, February 12, 1998, author's copy.

41. Eugenie Scott, "A Brief History of the Creation/Evolution Controversy as It Affects Textbooks, Curriculum, and Teachers" (talk presented at Darwin Day, University of Tennessee, Knoxville, February 11, 1998), author's notes from campus reports; interview with Scott, July 1996; Eugenie Scott, "Monkey Business," *The Sciences* 36 (January–February 1996): 24.

42. William Provine, "Evolution: Free Will and Punishment and Meaning in Life" (keynote talk presented at Darwin Day, University of Tennessee, Knoxville, February 12, 1998), author's tape; interview with Provine, February 1998.

CHAPTER 2

1. Roland M. Frye, "The Two Books of God," in *Is God a Creationist? The Religious Case Against Creation-Science*, ed. Roland Frye (New York: Scribner's, 1983), 200.
2. James Gilbert, *Redeeming Culture: American Religion in An Age of Science, 1925–1962* (Chicago: University of Chicago Press, 1997), 25.
3. Interview with William Provine, February 1998.
4. The Millikan pamphlets were discovered by Edward B. Davis, who provided me with a copy of his National Science Foundation research proposal, "Science and Religion in the 1920s"; for Millikan's scientists statement, see Edward LeRoy Long, *Religious Beliefs of American Scientists* (Philadelphia: Westminster Press, 1952), 45–46.
5. George M. Marsden, ed., *The Fundamentals: A Testimony to Truth* (New York, Garland, 1988), xi.
6. Edwin H. Wilson, *The Genesis of a Humanist Manifesto*, ed. Teresa Machiocha (Amherst, N.Y.: Humanist Press, 1995), 103; Julian Huxley, *Religion Without Revelation* (London: Ernest Benn, 1927); Huxley, "The Evolutionary Vision," in *Evolution After Darwin: Issues in Evolution*, ed. Sol Tax, vol. 3 (Chicago: University of Chicago Press, 1960), 260; Ursula Goodenough, *The Sacred Depths of Nature* (New York: Oxford University Press, 1998), xvii, 174. Goodenough (176) traces the recent coinage of "religious naturalism" to Holmes Rolston III, *Science and Religion: A Critical Survey* (New York: Random House, 1987); Bruce Alberts, preface to *SC2*, ix.
7. Gilbert, *Redeeming Culture*, 70–93, 75 (repudiate scientism); Wilson, *The Genesis of the Humanist Manifesto*, 58.
8. Albert Einstein, *Ideas and Opinions*, ed. Carl Seeling (New York: Crown, 1954), 48; "Science and Religion," *Time*, September 23, 1940, 46.
9. Gilbert, *Redeeming Culture*, 75; Mortimer J. Adler, *Platonism and Positivism in Psychology* (New Brunswick, N.J.: Transaction, 1995), 117, 115. This is a republication of Adler's *What Man Has Made of Man* (1937).
10. William G. Pollard, *Transcendence and Providence: Reflections of a Physicist and Priest* (Edinburgh: Scottish Academic Press, 1987), 7; interview with Edward LeRoy Long, October 2000, and correspondence.
11. Gilbert, *Redeeming Culture*, 281, 288; interview with Philip Hefner, June 1998.
12. Wilson, *The Genesis of the Humanist Manifesto*, 59; see also "Institute on Religion in an Age of Science," *Who's Who in Theology and Science* (New York: Continuum, 1996), 586–87.
13. James D. Watson, "Watson on Pauling," *Time*, March 29, 1999, 174.
14. Interview with Ernst Mayr, April 1997.
15. Francisco Ayala and Walter M. Fitch, "Genetics and the Origin of Species: An Introduction," *Proceedings of the National Academy of Sciences* 94 (July 22, 1997): 7692.
16. Michael Ruse, *Monad to Man: The Concept of Progress in Evolutionary Biology* (Cambridge, Mass.: Harvard University Press, 1996), 534; Niles Eldredge, *Reinventing Darwin: The Great Debate at the High Table of Evolutionary Theory* (New York: Wiley, 1995), 23; Garland E. Allen. "The Morgan Lab," in *The Evolution of Theodosius Dobzhansky*, ed. Mark B. Adams (Princeton, N.J.: Princeton University Press, 1994), 91; interview with Richard Bambach, April 1996.
17. Interview with Provine.

18. Keith E. Yandel, "Protestant Theology and Natural Science in the Twentieth Century," in *God and Nature: Historical Essays on the Encounter Between Christianity and Science*, ed. David C. Lindberg and Ronald L. Numbers (Berkeley: University of California Press, 1986), 448.

19. C. Luther Fry, "The Reported Religious Affiliations of the Various Classes of Leaders in 'Who's Who,' 1930–31 Edition," in *Yearbook of American Churches, 1933*, ed. Herman C. Weber (New York: Round Table Press, 1933), 314; T. Vaughan, D. Smith, and G. Sjoberg, "The Religious Orientation of American Natural Scientists," *Social Forces* 44 (June 1966): 519.

20. Reinhold Niebuhr, "The Providence of God," in *American Sermons: The Pilgrims to Martin Luther King Jr.*, ed. Michael Warner (New York: Library of America, 1999), 822.

21. Interview with Long.

22. Theodosius Dobzhansky, *The Biology of Ultimate Concern* (New York: New American Library, 1967), 4–5, 96, 134–37; Ruse, *Monad to Man*, 408; interview with Richard Lowentin, April 1997 (crazy); Peter Medawar, "Remarks by the Chairman," in *Mathematical Challenges to the Neo-Darwinian Interpretation of Evolution*, ed. Paul S. Moorhead and Martin Kaplan, 2nd ed. (New York: Alan R. Liss, 1985), xi (bunk).

23. Charles Birch, "My Damascus," in *Spiritual Evolution: Scientists Discuss Their Beliefs*, ed. John Templeton and Kenneth Giniger (Philadelphia: Templeton Foundation Press, 1998), 7; Ruse, *Monad*, 404.

24. Interview with Mayr; David Lack, *Evolutionary Theory and Christian Belief: The Unresolved Conflict* (London: Methuen, 1957), 79.

25. Gilbert, *Redeeming Culture*, 128 (proselytizing), 122–145; Ronald L. Numbers, *The Creationists* (New York: Knopf, 1992), 73–101.

26. Mark Kalthoff, "The Harmonious Dissonance of Evangelical Scientists: Rhetoric and Reality in the Early Decades of the American Scientific Affiliation," *Perspectives in Science and Christian Faith* 43 (December 1991): 264, 265; see also H. Harold Hartzler, "The American Scientific Affiliation: 30 Years," *JASA* 24 (March 1972): 23–27. The man summoned to critique flood geology was Columbia University geologist Lawrence J. Kulp, a convert to fundamentalism whom historian George M. Marsden said "contributed more than any other scientist to splitting conservative Protestants into self-consciously separate camps of 'evangelicals' and 'fundamentalists.'" See Marsden, *Reforming Fundamentalism: Fuller Seminary and the New Evangelicalism* (Grand Rapids, Mich.: Eerdmans, 1987), 36; Bernard Ramm, *The Christian View of Science and Scripture* (Grand Rapids, Mich.: Eerdmans, 1954), 29, 229; Graham cited in Marsden, *Reforming Fundamentalism*, 158. Says Marsden, "By the next summer Ramm's book had indeed caused the largest stir in fundamentalism since the [Revised Standard Version Bible] controversy" (159).

27. Henry M. Morris, *History of Modern Creationism*, 2nd ed. (Santee, Calif.: Institute for Creation Research, 1993), 210.

28. John C. Whitcomb and Henry M. Morris, *The Genesis Flood: The Biblical Record and Its Scientific Implications* (Grand Rapids, Mich.: Baker, 1998). First issued by Presbyterian and Reformed Publishing, 1961; Davis A. Young, *Creation and the Flood* (Grand Rapids, Mich.: Baker Book House, 1977), 7 (stunning).

29. Numbers, *The Creationists*, 129, 132.

30. Marvin Lubenow, *Bones of Contention: A Creationist Assessment of Human Fossils* (Grand Rapids, Mich.: Baker Books 1992), 245.

31. Claude E. Stipe, "Does the ASA Take a 'Position' on Controversial Issues?" *JASA* 29 (March 1977): 4; Richard H. Bube, "Christian Responsibilities in Science," *JASA* 21 (March 1969): 8; interview with Walter Hearn, November 1996; interview with Edward B. Davis, June 1999;

Francis Collins, "Genetics and the Image of God" (talk presented at a "Science and Religion Workshop," Washington [D.C.] Theological Union, June 19, 1999).

32. Jerome Lawrence and Robert E. Lee, *Inherit the Wind* (New York: Bantam Books, 1960), v.

33. Lewis S. Feuer, *The Scientific Intellectual: The Psychological and Sociological Origins of Modern Science* (New York: Basic Books, 1963), 2; Thomas Kuhn, *The Structure of Scientific Revolutions*, 3rd ed. (Chicago: University of Chicago Press, 1970).

34. See David Hull, "The Use and Abuse of Sir Karl Popper," *Biology and Philosophy* 14 (October 1999): 481–504.

35. Ian G. Barbour, *Issues in Science and Religion* (Englewood Cliffs, N.J.: Prentice-Hall, 1966). For Barbour's other important categories in this field, see his 1989 Gifford Lectures, Barbour, *Religion in an Age of Science* (New York: Harper and Row, 1990).

36. Lynn White Jr., "The Historical Roots of Our Ecological Crisis," *Science*, March 10, 1967, 1206.

37. R. Scott Appleby, "Exposing Darwin's 'Hidden Agenda': Roman Catholic Responses to Evolution, 1875–1925," in *Darwin's Reception: The Role of Place, Race, Religion, and Gender*, ed. Ronald L. Numbers and John Stenhouse (New York: Cambridge University Press, 1999); Donald Attwater, ed., *The Catholic Encyclopedia Dictionary* (New York: Macmillan, 1931), 190; for *Humani Generis*, see Claudia Carlen, ed. *The Papal Encyclicals, 1938–1958* (Wilmington, N.C.: McGrath, 1981), 181; Pope John Paul II, "Message to Pontifical Academy of Sciences on Evolution," *Origins*, December 5, 1996, 414–16; T. Vaughan, D. Smith, and G. Sjoberg, "The Religious Orientation of American Natural Scientists," *Social Forces* 44 (June 1966): 519.

38. Interview with Robert Brungs, May 1999.

39. Brent Dalrymple, "Anti-evolutionism: What Is Changed and Unchanged 20 Years After *McLean v Arkansas*?" (forum at the AAAS annual meeting, San Francisco, February 18, 2001), author's tape; Dalrymple, *The Age of the Earth* (Stanford, Calif.: Stanford University Press, 1991); Eldredge, *Reinventing Darwin*, 104.

40. Moorhead and Kaplan, *Mathematical Challenges*; interview with David Berlinski, November 1996.

41. Niles Eldredge and Stephen J. Gould, "Punctuated Equilibrium: An Alternative to Phyletic Gradualism," in *Models of Paleobiology*, ed. T. J. M. Schopf (San Francisco: Freeman, Cooper, 1972), 82–115; Michael Ruse, *The Darwinian Paradigm: Essays on Its History, Philosophy and Religious Implications* (London: Routledge, 1989), 122.

42. Stephen J. Gould, "Is a New and General Theory of Evolution Emerging?" *Paleobiology* 6 (1980): 119–30.

43. Niles Eldredge, *Unfinished Synthesis: Biological Hierarchies and Modern Evolutionary Thought* (New York: Oxford University Press, 1985); Eldredge, *Reinventing Darwin*, xi, 4; Eldredge, "Outlooks on the Evolution of Life" (talk presented at the "Epic of Evolution" conference, Field Museum of Natural History, Chicago, November 12, 1997), author's tape.

44. Interview with Joel Cracraft, February 1998.

45. Interview with David Raup, June 1996; Roger Lewin, "Evolutionary Theory Under Fire," *Science*, November 21, 1980, 883–87; letters to the editor, *Science*, February 20, 1981, 770, 773–74.

46. G. L. Stebbins and F. Ayala, "Is a New Evolutionary Synthesis Necessary?" *Science*, August 28, 1981, 967–71; Robert G. B. Reid, *Evolutionary Theory: The Unfinished Synthesis* (London: Croom Helm, 1985), 10; interview with Raup.

47. Lynn Margulis, *Symbiosis in Cell Evolution* (San Francisco: Freeman, 1981); Margulis, "Kingdom Animalia: The Zoological Malaise from a Microbial Perspective," *American Zoologist* 30 (1990): 867; *TAE*, 4; interview with Raup.

48. International Human Genome Sequencing Consortium, "Initial Sequencing and Analysis of the Human Genome," *Nature*, February 15, 2001, 860–921; interview with Arthur Peacocke, March 2001.

49. Interview with Carl Sagan in Larry Witham, "Sagan's Faith in Science Pays Off," *WT*, June 5, 1996.

50. E. O. Wilson, *Consilience: The Unity of Knowledge* (New York: Knopf, 1999), 262; Wilson, *Naturalist* (Washington, D.C.: Island Press, 1994), 360.

51. Quoted in "An Environmental Agenda for the World's Faiths," *NYT*, October 24, 1998, A15.

CHAPTER 3

1. Everett Mendelsohn, "The Biological Sciences: Studies on the Students of Life" (talk presented at the AAAS conference "Cosmology and Teleology," Seattle, February 13, 1997); Nancey Murphy, discussant, "Cosmology and Teleology" conference, author's tape. The four-part distinction used in this chapter (evolution, theistic evolution, progressive creation, young-earth creation) has been employed by Protestants for decades but is derided as a "conservative Protestant" template. Scientists use the scheme in *SC*2, 7. Others, such as Mark Railey, have divided evolution and creation stances by several nuanced differences. The most common taxonomy for God and nature, however, has a different aim: to show how science and religion interact. This began with Ian Barbour's *Issues in Science and Religion* (1966) with the religion-science modes being conflict, independence, dialogue, and integration. Since then, John Haught, Arthur Peacocke, James B. Miller, John Polkinghorne, Ted Peters, and Philip Hefner all have proposed a series of epistemic categories.

2. I became familiar with "boundary theory" in Jon R. Stone, *On the Boundaries of American Evangelicalism: The Postwar Evangelical Coalition* (New York: St. Martin's, 1997), 43–48.

3. Ronald L. Numbers, *Darwinism Comes to America* (Cambridge, Mass.: Harvard University Press, 1998), 29; Neal C. Gillespie, *Charles Darwin and the Problem of Creation* (Chicago: University of Chicago Press, 1979), 147.

4. John A. Moore, "Science as a Way of Knowing—Evolutionary Biology," *American Zoologist* 24 (1984): 467–525. Moore explains an eight-year initiative (470), aimed mostly at college education, to tie the "way of knowing" with eight areas of biology: evolution, genetics, developmental biology, biology and human affairs, cells, physiology, the animal kingdom, and ecology; "RENO! NABT National Convention Information," *ABT*, October 1998, 624–29.

5. Adam S. Wilkins, "Evolutionary Processes: A Special Issue," *BioEssays*, December 2000, 1051. In his 1999 presidential address to the American Society of Naturalists, Peter Grant made the same point: "Not all biologists who would call themselves naturalists pay attention to [evolutionary theory] or even feel the need to. An ecologist's world can make perfect sense, in the short term at least, in the absence of evolutionary considerations." See Grant, "What Does It Mean to Be a Naturalist at the End of the Twentieth Century?" *American Naturalist* 155 (January 2000): 9; *American Zoologist: Final Program and Abstracts, Annual Meeting, December 26–30, 1995* 35 (1995); John E. Repetski, ed., *Sixth North American*

Paleontological Convention Abstract of Papers, Special Publication no. 8, June 9–12, 1996; Carolyn J. Boyd and Michael S. Strauss, eds., *Exploring Frontiers—Expanding Opportunities: 1998 AAAS Annual Meeting* (Washington, D.C.: AAAS, 1998).

6. Paul Thompson, "Evolutionary Ethics: Its Origins and Contemporary Face," *Zygon* 34 (September 1999): 475.

7. Ernst Mayr, *One Long Argument* (Cambridge, Mass.: Harvard University Press, 1991), 15; interview with Mayr, April 1997. Embryologists who theorized jumps in biology were not invited to the Darwin Centennial by the purists; see Vassiliki Betty Smocovitis, "The 1959 Darwin Centennial Celebration in America," *Osiris* 14 (1999): 296.

8. Stephen J. Gould, "The Return of Hopeful Monsters," *Natural History* 86 (June–July 1977): 22–30.

9. Science writer Robert Wright dubbed Gould "evolutionist laureate"; Stephen J. Gould, "Darwinian Fundamentalism," *NYRB*, June 12, 1997, 34; Gould, "Evolution: The Pleasures of Pluralism," *NYRB*, June 26, 1997, 47; John Maynard Smith, "Genes, Memes, and Minds," *NYRB*, November 30, 1995, 46. Dennett also has criticized Gould, suspicious of his "religious yearnings" because of his literary use of the Bible, in Dennett, *Darwin's Dangerous Idea: Evolution and the Meanings of Life* (New York: Simon and Schuster, 1995), 309.

10. Interview with Stephen J. Gould, April 1997.

11. Stephen J. Gould, "Wallace's Fatal Flaw," in *Scientists Confront Creationism*, ed. Laurie Godfrey (New York: Norton, 1983), 69.

12. Lynn Margulis, "Gaia Is a Tough Bitch," in *The Third Culture*, ed. John Brockman (New York: Touchstone, 1995), 131, 133, 140.

13. Brian Goodwin, "Biology Is Just a Dance," in *The Third Culture*, ed. John Brockman, 101; Goodwin, *How the Leopard Changed Its Spots* (New York: Scribner's, 1994). Even though Gould favors German biology's structural view of organisms over British biology's focus on tiny steps in evolution, he would find Goodwin's "fields" a bit mystical.

14. Stuart Kauffman, *At Home in the Universe* (New York: Oxford University Press, 1995), 112; Michael Denton, *Nature's Destiny: How the Laws of Biology Reveal Purpose in the Universe* (New York: Free Press, 1998), xvii.

15. Denton, *Nature's Destiny*, 269, xvii, xviii.

16. Murphy, discussant, "Cosmos and Teleology."

17. Martin Marty, *Modern American Religion*, vol. 1 (Chicago: University of Chicago Press, 1986), 36. Marty quotes the Reverend Newell Dwight Hillis of Plymouth Congregational Church in Brooklyn from 1910.

18. Margaret Wertheim, "Faith and Reason," New River Media, Washington, D.C., one-hour PBS documentary, September 11, 1998; interview with Robert Russell, June 1998; interview with Templeton officer Charles Harper, June 1998; Sharon Begley, "Science Finds God," *Newsweek*, July 20, 1998, 46–52; Stephen J. Gould, "Rocks of Ages: Science and Religion in the Fullness of Life" (talk presented at the Smithsonian Associates program, Washington, D.C., March 18, 1999), author's tape.

19. Interview with John Polkinghorne, June 1998.

20. John Polkinghorne, *Scientists as Theologians* (London: SPK, 1996); interview with Polkinghorne; Arthur Peacocke, "Science and Religion: The Challenges and Possibilities for Western Theism" (paper presented at the Science and the Spiritual Quest Conference, Berkeley, California, June 7, 1998).

21. John Polkinghorne, *Belief in God in an Age of Science* (New Haven, Conn.: Yale University Press, 1998), 58, xi; Polkinghorne, "So Finely Tuned a Universe of Atoms, Stars, Quanta and God," *Commonweal*, August 16, 1996, 16.

22. Pope John Paul II, "Message to Pontifical Academy of Sciences on Evolution," *Origins*, December 5, 1996, 414–16; letter to the author from Lawrence C. Brennan, academic dean of Kenrick-Glennon Seminary in St. Louis, March 19, 1999.

23. Kenneth Miller and Joseph Levin, *Biology*, 2nd ed. (New York: Prentice Hall, 1993), 658; interview with Miller, June 1996.

24. Numbers, *Darwinism Comes to America*, 50, 51, 29. Edward J. Larson cites an earlier use of "the creationist" in a Darwin notebook of 1837; see Larson, *Evolution's Workshop: God and Science on the Galapagos Islands* (New York: Basic Books, 2001), 78, 260 n. 61.

25. Richard H. Bube, "We Believe in Creation," *JASA* 23 (December 1971): 121–22; Keith Miller, "Theological Implications of an Evolving Creation," *Perspectives in Science and Christian Faith* 45 (September 1993): 150–60.

26. Francis Crick, *What Mad Pursuits* (New York: Basic Books, 1988), 138; Michael Shermer, *How We Believe: The Search for God in an Age of Science* (New York: Freeman, 1999), 84.

27. William A. Dembski, ed., *Mere Creation: Science, Faith and Intelligent Design* (Downers Grove, Ill.: InterVarsity Press, 1998), 9; George Johnson, "Science and Religion: Bridging the Great Divide," *NYT*, June 30, 1998, F4; interview with Stephen Meyer, November 1996.

28. Robert C. Newman, "Progressive Creationism," in *Three Views on Creation and Evolution*, ed. J. P. Moreland and John Mark Reynolds (Grand Rapids, Mich.: Zondervan, 1999), 106; see papers in Dembski, *Mere Creation*.

29. Denton comments from author's notes; Dembski, *Mere Creation*, 15.

30. William A. Dembski, *The Design Inference* (New York: Cambridge University Press, 1999); Dembski, *Mere Creation*, 16, 21.

31. Walter L. Bradley and Roger Olsen, "The Trustworthiness of Scripture in Areas Related to Science," in *Hermeneutics, Inerrancy, and the Bible*, ed. Earl D. Radmacher and Robert D. Preus (Grand Rapids, Mich.: Zondervan, 1984), 285, 288; interview with Duane Gish, May 1999; Henry M. Morris, "A Response," in *Hermeneutics, Inerrancy, and the Bible*, 347.

32. Robert E. Walsh, ed., *Technical Symposium Sessions: Proceedings of the Fourth International Conference on Creationism, 1998* (Pittsburgh: Creation Science Fellowship, 1998); interviews with Walsh and Kurt Wise, August 1998.

33. Danny Falkner, "The State of Creation Astronomy" (talk presented at the "Fourth International Conference on Creationism," Pittsburgh, August 6, 1998), author's tape; Duane Gish, *Evolution? The Fossils Say No!* 2nd ed. (San Diego: Creation Life Publishers, 1973), 42 (italics in original).

34. Falkner, "The State of Creation Astronomy."

35. Interview with Kurt Wise, April 1996.

36. Wise, "Religion and Public Life: Seventy Years After the Scopes Trial" (talk presented at the Vanderbilt University conference, November 2, 1995); interview with Ronald Numbers, November 1995.

37. *ECPE*, 32–34; see GNS for a similar finding in a February 2001 Gallup Poll: only 34 percent of respondents were "very informed" on evolution, and 40 percent were "very informed" on creationism.

38. Interview with Mark Noll, July 1998; interview with Edward LeRoy Long, October 2000.

39. Interview with Philip Hefner, June 1998; this view of Lutherans, which is solely mine, was aided by interviews with Duane H. Larson of the Evangelical Lutheran Church in America and Joel Lehenbauer of the Lutheran Church–Missouri Synod.

40. Philip Handler, "Public Doubts About Science," *Science*, June 6, 1980, 1093; Robert A. Frosch, "What's Next," *American Scientist* 86 (May–June 1998): 210.

41. *ECPE*, 48, 49; Edward J. Larson and Larry Witham, "Inherit an Ill Wind," *The Nation*, October 4, 1999, 29.

42. Paul R. Gross, *Politicizing Science Education* (Washington, D.C.: Fordham Foundation, 2000), 13.

43. George Gallup Jr. and D. Michael Lindsay, *Surveying the Religious Landscape* (Harrisburg, Pa.: Morehouse, 1999), 23–25; see also George Barna, *The Index of Spiritual Indicators* (Dallas: Word, 1996). Barna found that a quarter of believers opt for God as higher consciousness (11 percent), full realization of personal potential (8 percent), many gods (3 percent), and everyone as their own god (3 percent). Bill Marvel, "List of 100 Best Spiritual Books Includes Works of Popes, Malcolm X, Tolkien," *Dallas Morning News*, November 4, 1999, A1; Deepak Chopra, "Address to the National Press Club Luncheon," transcript, April 30, 1997, 3.

44. Ronald Numbers, *The Creationists* (New York: Knopf, 1992), 6.

CHAPTER 4

1. The PBS *Firing Line* debate, which first aired on December 19, 1997, had four people on each side. PBS researcher John Fuller said that while 540,000 households watched the debate's first showing, over the week the cumulative viewership reached more than 2 million. Still, the first-showing *Firing Line* debate drew only half the viewership of an average daytime PBS program (994,000 households per minute) and only a quarter of the prime-time PBS audience, which averages 2 million households.

2. Interview with Eugenie Scott, July 1996.

3. *SEI*, 72; a June 1999 poll by CNN/USA Today/Gallup Poll found that 68 percent of respondents "favor" teaching "creationism along with evolution," cited in *NYT*, August 15, 1999, sec. 4, 1.

4. William V. Mayer, "Evolution and the Law," *ABT*, March 1973, 145

5. In a 1978 University of Texas dissertation, "Chronology and Analysis of Regulatory Actions Relating to the Teaching of Evolution in Public Schools," Richard David Wilhelm identified twenty-four creationism bills in twelve states between 1971 and 1977; Henry M. Morris, "No. 26, Resolution for Equitable Treatment of Both Creation and Evolution," *ICR Impact Series*, 1975; Wendell R. Bird, "Freedom of Religion and Science Instruction in Public Schools," *Yale Law Journal* 87 (January 1978): 515–70.

6. Duane Gish, "Creation, Evolution, and the Historical Evidence," *ABT*, March 1973, 132–40; Joan C. Creager, "Freedom in Science Teaching," *ABT*, January 1975, 11.

7. Interview with Walter Hearn, November 1996.

8. Wayne A. Moyer, "The Problem Won't Go Away," *BioScience* 39 (March 1980): 147; interview with Moyer, March 1999.

9. Robert Booth Fowler, "The Failure of the Religious Right," in *No Longer Exiles: The Religious New Right in American Politics*, ed. Michael Cromartie (Washington, D.C.: Ethics and Public Policy Center, 1993), 72.

10. Alex Heard, "Creationism Debate Planned for TV and State Legislatures," *Education Week*, October 5, 1981, 5; Doolittle and Gish quoted in Philip J. Hilts, "Science Loses One to Creationism," *WP*, October 15, 1981, A1.

11. Quoted in Patrick McQuaid, "Evolution Supporters Develop Strategy to Counter Creationism," *Education Week*, November 9, 1981, 6.

12. John Walsh, "At AAAS Meeting, a Closing of Ranks," *Science*, January 22, 1982, 380; interview with Walter Hearn, November 1996.

13. Quoted in Alex Heard, "Scientists Urged to Continue Their Efforts," *Education Week*, January 12, 1982, 13.

14. *SEI*, appendix tables 2–26, 2–28; *Statistical Abstract of the United States, 1998* (Washington, D.C.: U.S. Census Bureau), 616.

15. "A Statement Affirming Evolution as a Principle of Science," *The Humanist* 37 (January–February 1977): 4–5.

16. Eugenie Scott, "Creationism Lives," in letters, *Nature*, September 24, 1987, 282; Kevin Padian, comments at the forum "Only a Theory: Presenting Evolution to the Public," held by the NCSE at the annual AAAS convention, San Francisco, January 16, 1989, author's tape; Henry M. Morris, *History of Modern Creationism*, 2nd ed. (Santee, Calif.: Institute for Creation Research, 1993), 389; ICR documents on the case, author's copy.

17. Paul R. Gross, comments at "A Symposium on the Unity of Knowledge," Woodrow Wilson International Center for Scholars, Washington, D.C., April 27, 1998, author's tape.

18. Interview with Phillip E. Johnson, July 1996.

19. Interview with Stephen Meyer, November 1996.

20. Interview with Owen Gingerich, February 1997.

21. Interview with Jonathan Wells, July 2001 (sharp). The "wedge" concept has been used to describe an early group who worked with Johnson but also as a metaphor for splitting off "metaphysical naturalism from science" or splitting apart an obstacle; Johnson comment from 1996, five years before he retired.

22. Stephen J. Gould, "Impeaching a Self-Appointed Judge," *Scientific American*, July 1992, 118, 119.

23. Interview with Kurt Wise, April 1996.

24. Interview with David Raup, June 1996.

25. Interview with Duane Gish, May 1999.

26. Phillip Johnson, *The Wedge of Truth: Splitting the Foundations of Naturalism* (Downers Grove, Ill.: InterVarsity Press, 2000); comments by Johnson and John G. West Jr. at the "Life After Materialism" conference, Biola University, Los Angeles, December 3, 1999.

27. John Mark Reynolds, "God of the Gaps," in *Mere Creation*, ed. William Dembski (Downers Grove, Ill.: InterVarsity, 1998), 318, 325.

28. "NABT Unveils New Statement on Teaching Evolution," *ABT*, January 1996, 61.

29. Ira Rifkin, "Teachers Change Evolution Wording," Religion News Service, in *Cleveland Plain Dealer*, October 16, 1997, E10; Laurie Goodstein, "Christians and Scientists: New Light for Creationism," *NYT*, December 21, 1997, sec. 4, 1; statement by Massimo Pigliucci; interview with Wayne Carley, February 1998; statement by Eugenie Scott.

30. Statement by Phillip Johnson; Johnson, "Is Man the Measure of All Things? One Scientist's Reply," *WT*, May 24, 1998, B8.

31. Eugenie Scott, "'Science and Religion,' 'Christian Scholarship,' and 'Theistic Science': Some Comparisons," *NCSE Reports* 18 (March–April 1998): 30–33; Scott, "The 'Science and Religion Movement': An Opportunity for Improved Public Understanding of Science?" *Skeptical Inquirer* 23 (July–August 1999): 29–31.

32. Interview with Niles Eldredge, July 1997.

CHAPTER 5

1. Joseph McInerney, "Voting in Science: Raise Your Hand If You Want Humans to Have 48 Chromosomes," *ABT*, March 1993, 132–33.

2. Interview with Joseph McInerney, April 1996, and follow-up.

3. Vassiliki Betty Smocovitis, "The 1959 Darwin Centennial Celebration in America," *Osiris* 14 (1999): 294.

4. "Coalition to Educate Health Professionals About Genetics Names Executive Director," NHGRI press release, October 2000; "Remarks by the President," et al., transcript, White House press office, June 26, 2000; correspondence with McInerney, September 1, 2000.

5. William V. Mayer, "Biological Education in the United States During the 20th Century," *Quarterly Review of Biology* 61 (December 1986): 490; Constance Holden, "Science Education Axed," *Science*, March 20, 1981, 1330; "Reagan Budget Would Reshape Science Policies," *Science*, March 27, 1981, 1399–1400.

6. Arnold B. Grobman, *The Changing Classroom: The Role of the Biological Sciences Curriculum Study* (Garden City, N.Y.: Doubleday, 1969), 62.

7. Carl Sagan, *The Demon Haunted World* (New York: Random House, 1995), 295–306.

8. Ibid., 278.

9. John Simmons, *The Scientific 100: A Ranking of the Most Influential Scientists, Past and Present* (New York: Carol Publishing, 1997), 305; interview with Ernst Mayr, April 1997.

10. Interview with William Provine, February 1998; Daniel Otte and John A. Endler, *Speciation and Its Consequences* (Sunderland, Mass.: Sinauer Associates, 1989).

11. Mayr quoted in *Evolution After Darwin: Issues in Evolution After Darwin*, vol. 3, ed. Sol Tax (Chicago: University of Chicago Press, 1960), 124. One reviewer of Mayr's *What Evolution Is* (New York: Basic Books, 2001), said at career's end he was willing to "acknowledge" theories such as non-geographic speciation but remained polemical on genetic theories. See Menno Schilthuizen, "A Grand Old Synthesizer's Overview," *Science*, January 4, 2002, 50.

12. Ernst Mayr, *This Is Biology: The Science of the Living World* (Cambridge, Mass.: Harvard University Press, 1997), 64–67.

13. Interview with Richard Lewontin, April 1997.

14. Interview with Michael Ruse, May 1998.

15. "Decision of the Court," in *Science and Creationism*, ed. Ashley Montagu (New York: Oxford University Press, 1984), 380. Stephen J. Gould praised the judge for elucidating science: "Judge Overton's definitions of science are so cogent and clearly expressed that we can use his words as a model for our own proceedings." See in Gould, *Bully for Brontosaurus* (New York: Norton, 1991), 431.

16. Philip L. Quinn, "The Philosopher of Science as Expert Witness," in *But Is It Science? The Philosophical Question in the Creation/Evolution Controversy*, ed. Michael Ruse (Buffalo, N.Y.: Prometheus Books, 1988), 369.

17. Larry Laudan, "The Demise of the Demarcation Problem," in *But Is It Science?*, 338.

18. Michael Ruse, *Mystery of Mysteries: Is Evolution a Social Construction?* (Cambridge, Mass.: Harvard University Press, 1999), 32–33.

19. Michael Ruse letter to author, February 3, 2000; Ruse, *The Darwinian Revolution: Science Red in Tooth and Claw* (Chicago: University of Chicago Press, 1979).

20. Michael Ruse, comments at "The New Antievolutionists" symposium, AAAS annual meeting, Boston, February 13, 1993, author's transcript; Arthur M. Shapiro, "Did Michael Ruse Give Away the Store?" *NCSE Reports* 3 (spring 1993): 20–21; interview with Niles Eldredge, July 1997.

21. Ruse letter to author, February 3, 2000; Ruse, "Booknotes," *Biology and Philosophy* 8 (1993): 353, 354.

22. Interview with Mayr.

23. Interview with Jeffrey Schloss, November 1996.

24. Michael Ruse, *Monad to Man: The Concept of Progress in Evolutionary Biology* (Cambridge, Mass.: Harvard University Press, 1996), 539.

25. For Francisco Ayala's testimony and "observer" comment, see Langdon Gilkey, *Creationism on Trial: Evolution and God at Little Rock* (Minneapolis, Minn.: Winston Press, 1985),

139–40; for reference to "Lysenkoism" at the trial, see James Gorman, "Judgment Day For Creationism," *Discover*, February, 1982, 17.

26. For more on Allan Wilson and "mitochondrial Eve," see Michael H. Brown, *The Search for Eve* (New York: Harper and Row, 1990), 24–27, 42–57; also see Ann Gibbons, "Looking for the Father of Us All," *Science*, January 25, 1991, 378–80.

27. Ayala quoted in Reuters, "'Eve' Theory Takes Hit from Scientists on DNA Finding," *WT*, December 22, 1995, A6; Francisco Rodriguez-Trelles, Rosa Tarrio, and Francisco Ayala, "Erratic Overdispersion of Three Molecular Clocks," *Proceedings of the National Academy of Sciences*, 98 (September 25, 2001): 11405.

28. Interview with Francisco Ayala, July 1996.

29. Telephone interview with William Provine, February 1998.

30. Pope John Paul II, "Message to Pontifical Academy of Sciences on Evolution," *Origins*, December 5, 1996, 414–16.

31. Francisco Ayala, "Darwin and the Teleology of Nature" (paper presented at the "Cosmology and Teleology" conference of the AAAS, Seattle, February 13, 1997), author's tape.

32. Interview with Eldredge.

33. Interview with Mayr.

34. Gould quoted in John Brockman, ed., *The Third Culture* (New York: Touchstone, 1995), 90.

35. Niles Eldredge, *The Unfinished Synthesis: Biological Hierarchies and Modern Evolutionary Thought* (New York: Oxford University Press, 1985).

36. Interview with Richard Dawkins, March 1996.

37. When the New York State Board of Regents in 1978 wanted to revise the biology syllabus, a Bureau of Science Education official urged Luther D. Sunderland to make his case by interviewing paleontologists on the fossil record. Sunderland's own summary of five interviews, including Eldredge, was published in 1986. See Sunderland, *Darwin's Enigma: Ebbing the Tide of Naturalism* (Green Forest, Ark.: Master Books, 1988).

38. Niles Eldredge, "Evolution Theory Stands Tall Without Reservation," *NYT*, June 29, 1987, A16. The original editorial was "Louisiana's Fig Leaf for Bad Science," *NYT*, June 23, 1987, A30.

39. Niles Eldredge, *Dominion* (Berkeley: University of California Press, 1997), 175–76.

40. David Raup, *The Nemesis Affair: A Story of the Death of Dinosaurs and the Ways of Science*, rev. ed. (New York: Norton, 1999), 1.

41. Interview with David Raup, June 1996.

42. Raup, *Nemesis*, 23.

43. Luis W. Alvarez, Walter Alvarez, Frank Asaro, and Helen V. Michel, "Extraterrestrial Cause for the Cretaceous-Tertiary Extinction," *Science,* June 6, 1980, 1095–1108.

44. Raup, *Nemesis*, 152; David M. Raup and J. John Sepkoski, "Mass Extinctions in the Marine Fossil Record," *Science*, March 19, 1982, 1501–2. As with any scientific consensus, the one around mass extinction was eventually challenged, as in Richard A. Kerr, "Mass Extinctions Face Downsizing, Extinction," *Science*, August 10, 2001, 1037.

45. Raup, *Nemesis*, 18.

46. Ibid., 154.

47. "A Death-Star Theory Is Born: Nemesis," *Newsweek*, March 5, 1984, 85.

48. Raup quoted in Ruse, *Monad to Man*, 494.

49. Steven Stanley, *Children of the Ice Age: How a Global Catastrophe Allowed Humans to Evolve* (New York: Harmony Books, 1996), 179–87.

50. Raup quoted in Ruse, *Monad to Man*, 494.

51. David M. Raup, "Conflicts Between Darwin and Paleontology," *Bulletin of the Field Museum of Natural History* 50 (January 1979): 22–29.

52. Laurie R. Godfrey, "Scientific Creationism: The Art of Distortion," in *Science and Creationism*, ed. Ashley Montagu (New York: Oxford University Press, 1984), 176–77.

53. Mike Foote, "On the Probability of Ancestors in the Fossil Record," *Paleobiology* 22 (spring 1996): 141–51.

CHAPTER 6

1. Interview with Kurt Wise, April 1996, in Dayton, and follow-up by telephone.

2. Kurt Wise, "The Importance of the Young-Earth Creation Model" (talk presented at the "Fourth International Conference on Creationism," Pittsburgh, August 8, 1998), author's tape.

3. Interviews with John Wiester, July 1996 and February 1998.

4. Theodosius Dobzhansky, *Genetics and the Origin of Species* (reprint, New York: Columbia University Press, 1982), 12.

5. Kenneth Miller and Joseph Levin, *Biology*, 2nd ed. (New York: Prentice Hall, 1993), 29 (reference section).

6. *Science Framework for California Public Schools: Kindergarten to Twelve* (Sacramento: California Department of Education, 1990).

7. The document was drafted by the eleven-member Committee on Science and Creationism, which included four lawyers because it began in 1982 as a legal brief. See *SC*, 3.

8. Committee for Integrity in Science Education, *Teaching Science in a Climate of Controversy: A View from the American Scientific Affiliation*, 4th ed. (Ipswich, Mass.: American Scientific Affiliation, 1993).

9. William J. Benetta, "Scientists Decry a Slick New Packaging of Creationism," *Science Teacher*, May 1987, 36–43.

10. Miller and Levine, *Biology*, 658.

11. "NABT Unveils New Statement on Teaching Evolution," *ABT*, January 1996, 61–62. The statement had been circulating since September 1995.

12. Eugenie Scott quoted in e-mail to John Wiester.

13. *SC*, 6.

14. *SC2*, ix.

15. John Wiester, *What's Darwin Got to Do with It? A Friendly Conversation About Evolution* (Downers Grove, Ill.: InterVarsity Press, 2000).

16. Interview with Henry M. Morris, July 1996.

17. Dorothy Nelkin, *The Creation Controversy: Science of Scripture in the Schools* (New York: Norton, 1982), 51, 127–35; interview with Dorothy Nelkin, 1999.

18. In comparison to zoologists and physicists, for example, chemical engineers were more inclined to religion according to T. Vaughan, D. Smith, and G. Sjoberg, "The Religious Orientation of American Natural Scientists," *Social Forces* 44 (June 1966): 524; mathematicians were more religious than biologists and physicists in Edward J. Larson and Larry Witham, "Scientists Are Still Keeping the Faith," *Nature*, April 3, 1997, 435–36.

19. Rexmond C. Cochrane, *The National Academy of Sciences: The First Hundred Years, 1863–1963* (Washington, D.C.: NAS, 1978), 572; Norman L. Geisler and J. Kerby Anderson, *Origin Science: A Proposal for the Creation-Evolution Controversy* (Grand Rapids, Mich.: Baker Book House, 1987).

20. Ronald L. Numbers, *The Creationists* (New York: Knopf, 1992); Numbers, discussant, "Antievolutionism: What Is Changed and Unchanged 20 Years After *McLean v Arkansas*?" (forum at the AAAS annual meeting, San Francisco, February 18, 2001), author's tape.

21. Wendell R. Bird, "Freedom of Religion and Science Instruction in Public Schools," *Yale Law Journal* 87 (January 1978): 515–70.

22. Whitcomb quoted in Numbers, *The Creationists*, 246.

23. Interviews with Howard Van Till, January 1996 and February 1997.

24. Davis A. Young, *The Biblical Flood: A Case Study of the Church's Response to Extrabiblical Evidence* (Grand Rapids, Mich.: Eerdmans, 1995), 272.

25. Henry Morris, *History of Modern Creationism,* 2nd ed. (Santee, Calif.: Institute for Creation Research, 1993), 179.

26. Howard Van Till, Davis A. Young, and Clarence Menninga, *Science Held Hostage: What's Wrong with Creation Science and Evolution?* (Downers Grove, Ill.: InterVarsity Press, 1988).

27. The form stipulated adherence to the Canons of Dort, the Heidelberg Catechism, and the Belgic. The Synod of Dort, which founded the Dutch Church, condemned Jacobus Arminius and was the "highwater mark of Calvinist creed making," said Wilston Walker, *Church History,* 3rd ed. (New York: Scribner's, 1970), 400.

28. For quote and more on McCosh, see George M. Marsden, *The Soul of the American University: From Protestant Establishment to Established Disbelief* (New York: Oxford University Press, 1994), 203. See James McCosh, *Christianity and Positivism: A Series of Lectures to the Times on Natural Theology and Apologetics* (New York: R. Carter, 1871), 38.

29. Howard Van Till, "Science After Kuhn: Values, Theory Evaluation, and Folk Science" (talk presented at the "Following Christ: Shaping Our World" conference, Chicago, December 30, 1998), notes provided to author.

30. Howard Van Till, "The Creation: Intelligently Designed or Optimally Equipped?" *Theology Today* 55 (October 1998): 344–64.

31. John J. O'Connor, "Putting 'Cosmos' into Perspective," *NYT,* December 14, 1980, sec. 2, 36; Carl Sagan, *Cosmos* (New York: Random House), 4.

32. Interview with Owen Gingerich, February 1997.

33. Owen Gingerich, "The Copernican Quinquecentennial and Its Predecessors: Historical Insight and National Agendas," *Osiris* 14 (1999): 37, 54.

34. Sergei Shargorodsky, "Copernicus Tempts Thieves Worldwide," AP, February 14, 2000.

35. Quoted in Chris Floyd, "Eyes Wide Open: The Case for a Coherent Cosmos," *Science and Spirit* 10 (November–December 1999): 19, 38.

36. Lecture reprinted in Timothy Ferris, *The World Treasury of Physics, Astronomy, and Mathematics* (Boston: Little, Brown, 1991).

37. Owen Gingerich, "Is There Design and Purpose in the Universe?" (talk presented at the "Cosmology and Teleology" conference of the AAAS, Seattle, February 13, 1997), author's tape.

38. *Evolution,* PBS documentary, seven parts, September 24–28, 2001.

39. Owen Gingerich, "Let There Be Light: Modern Cosmology and Biblical Creation," in *Is God a Creationist? The Religious Case Against Creation-Science,* ed. Roland M. Frye (New York: Scribner's, 1983), 119–40.

40. Michael Behe, "Breaking Ranks with Darwinian Orthodoxy" (talk presented at the Ethics and Public Policy Center, Washington, D.C., April 30, 1996); interview with Michael Behe, April 1997.

41. Interview with Kenneth Miller, June 1996.

42. Daniel C. Dennett, "The Case of the Tell-Tale Traces: A Mystery Solved, a Skyhook Grounded" (speech presented to the forum "Darwin's Black Box," University of Notre Dame, April 5, 1997), author's tape.

43. Michael Behe, "Reply to My Critics: A Response to Reviews of Darwin's Black Box," *Biology and Philosophy* 16 (2001): 685–709.

44. Dennett, "The Case"; Tom Woodward, "Meeting Darwin's Wager," *Christianity Today*, April 28, 1997, 17.

45. Ivan Amato, "Johnson vs. Darwin," *Science*, July 26, 1991, 379.

46. Charles Darwin to Asa Gray, May 22, 1860, in Frederick Burkhardt et al., eds., *The Correspondence of Charles Darwin*, vol. 8, *1860* (Cambridge: Cambridge University Press, 1993), 224.

47. Francis H. Crick and Leslie E. Orgel, "Directed Panspermia," *Icarus* 19 (1973): 341.

48. John Maddox, "Down with the Big Bang," *Nature*, August 10, 1989, 425; interview with Behe. During his visit to Virginia Tech, Behe gave the keynote address on "Darwin's Black Box" to the Veritas '98 Quest for Truth Conference, October 27, 1998.

CHAPTER 7

1. Adrian Desmond, *The Politics of Evolution: Morphology, Medicine, and Reform in Radical London* (Chicago: University of Chicago Press, 1989), 413.

2. "Fairfax County Characteristics, 1994" (County Office of Management and Budget brochure).

3. Vassiliki Betty Smocovitis, "The 1959 Darwin Centennial Celebration in America," *Osiris* 14 (1999) : 293–94; *NSES*, 104, 119.

4. Larry Witham, "U.S. Scientists Seek Formula for Image Boost," *WT*, March 26, 1996, A2; interview with Jon Miller, March 1996.

5. Dorothy Nelkin, *The Creation Controversy: Science or Scripture in the Schools* (New York: Norton, 1982), 51, 127–35.

6. National Science Foundation press release, March 27, 1997; Michael Zimmerman, "The Science Budget Must Be Insulated from the Creationist Threat," *Chronicle of Higher Education*, February 8, 1989, B2. Zimmerman spoke in January 1989.

7. Constance Holden, "Science Education Axed," *Science*, March 20, 1981, 1330; "Reagan Budget Would Reshape Science Policies," *Science*, March 22, 1981, 1399–1400; "Gore and Bush Offer Their Views on Science," *Science*, October 13, 2000, 262–69.

8. Larry Witham, "Senate Bill Tackles Evolution Debate," *WT*, June 18, 2001, A4. Due to House opposition, the Senate language was revised and moved to the "Joint Explanatory Statement of the Committee of Conference," part A, title I, item 78. The statement that "science education should prepare students to distinguish the data and testable theories of science from religious and philosophical claims that are made in the name of science" retains the Senate language. However, the focus on evolution alone was revised by the words, "Wherever topics are taught that may generate controversy (such as biological evolution), the curriculum should help students to understand the full range of scientific views that exist, why such topics may generate controversy, and how scientific discoveries can profoundly affect society." The legal implications of the language are unclear. For the Senate debate, see *Congressional Record*, June 13, 2001, amendment 779, bill S.1.

9. "JFK Spoke of Attraction from the Sea at 1962 Dinner," AP, July, 22, 1999.

10. Reagan remarks in "Republican Candidate Picks Fight with Darwin," *Science*, September 12, 1980, 1214; "Carter on Creation," *Science*, October 3, 1980, 35; telephone news conference with Jimmy Carter, November 12, 1996, author's transcript.

11. Interview with Stephen J. Gould, April 1997; interview with Lynn Nofziger, former Reagan press secretary, 1997; Transnational Association of Christian Colleges and Schools, *Foundational Standards*, 1 (see www.tracs.org/foundstandards.pdf).

12. Albert Gore, *Earth in the Balance: Ecology and the Human Spirit* (New York: Houghton Mifflin, 1992), 63, 229, 240, 241, 245.

13. Hanna Rosin, "Gore Avoids Stance Against Creationism," *WP*, August 27, 1999, A8; Steve Kraske, "Creationism Evolves into Campaign Topic," *Kansas City Star*, September 3, 1999, A1.

14. International Human Genome Sequencing Consortium, "Initial Sequencing and Analysis of the Human Genome," *Nature*, February 15, 2001, 860–921; J. Craig Venter et al., "The Sequence of the Human Genome," *Science*, February 16, 2001, 1304–51; Helen Briggs, "Dispute over Number of Human Genes," *BBC News Online*, July 7, 2001; Malcolm Ritter, "Scientists Disagree on How Many Genes Make a Person," AP, December 17, 2000.

15. Larry Witham, "Labs Work Nonstop on Human Puzzle," *WT*, March 23, 1997, D8; Witham, "Genome Map Challenges Long-Held Evolutionary Theories," *WT*, February 26, 2001, A3; interview with Mark Bloom, February 2001.

16. David Baltimore, "Our Genome Unveiled," *Nature*, February 15, 2001, 816.

17. John Maddox, *What Remains to Be Discovered* (New York: Free Press, 1998), 263.

18. Larry Witham, "Scientists Plan Gigantic Task," *WT*, March, 8, 1998, D8.

19. Joel Cracraft, "Strategies for Building Global Systematics Research" (talk presented at the AAAS annual meeting, Philadelphia, February 14, 1998). The PBE's eight-page report, *Teaming With Life* (c. 1998), requested $1 billion over five years. At the "Assembling the Tree of Life" conference in 2002, E. O. Wilson said species estimates range from 3.6 million to 100 million. Favoring a high number, he cited many new projects worth funding. For context, in the 1990s, the annual U.S. research and development budget was roughly $80 billion. The National Science Foundation portion (about $4 billion a year) is dispensed in thousands of grants with an average value of $85,000.

20. Larry Witham, "Theologians Not Upset by Possibility of Life on Mars," *WT*, August 17, 1998, B8; David Perlman, "NASA Institute's Metaphysical Mission," *San Francisco Chronicle*, May 19, 1999, A4. The news story marked the appointment of a first director, a biochemist and Nobel laureate, and NASA head Daniel S. Goldin gave a speech outlining the institute's ambitious goals. For a history of the founding of the institute, see *NASA Astrobiology Institute: Annual Science Report, July 1999–June 2000* (Moffett Field, Calif.: Ames Research Center, 2000), 6.

21. Neal Lane, "Science and the American Dream: Healthy or History?" (talk presented at the AAAS annual meeting, Baltimore, February 9, 1996); Michael Ruse, *Monad to Man: The Concept of Progress in Evolutionary Biology* (Cambridge, Mass.: Harvard University Press, 1996), 438; Niles Eldredge comment on George C. Williams in *The Third Culture*, ed. John Brockman (New York: Touchstone, 1995), 49.

22. Deborah Shapley and Rostum Roy, *Lost at the Frontier: U.S. Science and Technology Policy Adrift* (Philadelphia: ISI Press, 1985), 45. This summary is based on *Survey of Federal Funds for Research and Development*, vol. 46 (Arlington, Va.: National Science Foundation, 1996). See table 8 for funding from 1951 to 1998; table 62 for percentage of funds to colleges and universities going to biological research; table 64 for what part of biological research funds goes to "applied" research. The "nearly a fifth" comes from rounding the 17.6 percent that universities/colleges received ($12,340,627) from a total annual spending of $69,972,891, according to the latest figures in 1996.

23. Interview with Richard Bambach, April 1996; interview with Richard Lewontin, April 1997.

24. Interview with Francisco Ayala, July 1996; Jonathan Wells, *Icons of Evolution: Science or Myth? Why Much of What We Teach About Evolution Is Wrong* (Washington, D.C.: Regnery, 2000), 240–44.

25. Evolution Working Group, *Evolution, Science, and Society : Evolutionary Biology and the National Research Agenda* (New Brunswick, N.J.: Rutgers University Office of Print and Electronic Communications, 1999), 27, 43.

26. Ibid., 18, 36, 43.

27. Interview with Mark Noll, July 1998; interview with David Raup, June 1996.

28. Interview with James Guth, August 1997; John Green et al., "Who Elected Clinton? A Coalition of Values," *First Things* 75 (August–September 1997): 37, 38. The pattern continued in the 2000 Bush-Gore contest, though Gore sought a larger share of traditionalists by selecting Senator Joseph Lieberman of Connecticut, an Orthodox Jew, as his running mate.

29. John O. McGinnis, "The Origin of Conservatism," *National Review*, December 22, 1997, 31; Patricia Wen, "WGBH Swaps Lists with Democrats," *Boston Globe*, May 8, 1999, A1; Anne E. Kornblut, "PBS Chief Resigns," *Globe*, September 10, 1999, A1.

30. John Hanna, "Board's Evolution Debate Shows Conservatives Still a Force," AP, August 22, 1999.

31. Edward J. Larson and Larry Witham, "Inherit an Ill Wind," *The Nation*, October 4, 1999, 25–29; David Miles, "Staff to Rewrite Science Standards for State Board," AP, October 13, 1999; John W. Fountain, "Kansas Puts Evolution Back into Public Schools," *NYT*, February 14, 2001, A18.

32. Interviews with William Nowers, Mark H. Emery, Mark Sickles; Emery, "Creationism in the Science Classroom," press release, October 1995; Robert O'Harrow Jr., "Creationism Issue Evolves in Fairfax School Election," *WP*, October 21, 1995, A1; "Science and Values in Fairfax," editorial, *WP*, October 28, 1995, A26; Larry Witham, "Creation-Evolution Furor Hits Fairfax School Board Races," *WT*, November 5, 1995, A3; Roberta Holland, "Board Chairman Wins Full Term with 53% of Vote," *Fairfax Journal*, November 8, 1995, A1; Jeremy Redmon, "School Board Vote Backs Mainstream," *Journal*, November 8, 1995, A1; Les Fettig, "Tuesday's Tepid Turn-Out," *Journal*, November 9, 1999, editorial page; Larry Witham, "'Radical Right' Rumors Unproved," *WT*, November 10, 1995, C8.

CHAPTER 8

1. *Edwards v. Aguillard*, 482 US 578, 591 (1987).

2. U.S. Commission on Civil Rights, *Schools and Religion: Proceedings* (Washington, D.C.: Government Printing Office, 1999), 60. The 1998 hearings took place on May 20 in Washington, June 12 in New York City, and June 23 in Seattle.

3. Ibid., 52.

4. Ibid., 54.

5. *Epperson v. Arkansas*, 393 US 97, 116, 103, 108 (1968).

6. The 1987 ruling against "balanced treatment" of creation science and evolution science used the same establishment of religion logic as the 1968 ruling. See *Edwards*, 482 US at 583, for Justice William Brennan's description of the Lemon test: "First, the legislature must have adopted the law with a secular purpose. Second, the statute's principal or primary effect must be one that neither advances nor inhibits religion. Third, the statute must not result in an excessive entanglement of government with religion."

7. *Schools and Religion*, 53, 54, 60.

8. Ibid., 59.

9. Ibid., 1.

10. Poll cited in George Johnson, "It's a Fact: Faith and Theory Collide over Evolution," *NYT*, August 15, 1999, sec. 4, 1. The June 1999 poll by CNN/USA Today/Gallup Poll asked what "would you generally favor or oppose" in public schools? The result: teaching creationism *along with* evolution (yes, 68 percent; no, 29 percent; no opinion, 3 percent); or teaching

creationism *instead of* evolution (yes, 40 percent; no, 55 percent; no opinion, 5 percent); *ECPE*, 48–49.

11. *Edwards*, 482 US at 597, 593, 594.

12. *ECPE*, 13, 15.

13. *Schools and Religion*, 218, 223, 226.

14. David K. DeWolf, Stephen C. Meyer, and Mark E. DeForrest, *Intelligent Design in Public School Science Curricula: A Legal Guidebook* (Richardson, Tex.: Foundation for Thought and Ethics, 1999), 16, 22, 24–25. See also DeWolf, Meyer, and DeForrest, "Teaching the Origins Controversy: Science, or Religion, or Speech?" *Utah Law Review* 39 (2000): 39–110.

15. *NSES*, 36, 201.

16. Interview with Stephen J. Gould, April 1997.

17. "High School Graduation Requirements" (Denver, Colo.: Education Commission of the States, November 1998). The report says that among the forty-four states that have unit requirements for high school graduation, the average is 20 units. Of the forty-six states that have a social studies requirement for high school graduation, the average is 2.8 units. Forty-two states require students to take electives; the state requirements range from 2 to 10 units, averaging 6.5 units nationally.

18. Interview with Charles Haynes, November 1995; "A Response by Warren A. Nord," *Phi Delta Kappan* 81 (February 2000): 467; Warren Nord and Charles Haynes, *Taking Religion Seriously Across the Curriculum* (Alexandria, Va.: Association for Supervision and Curriculum and Development, 1998).

19. William V. Mayer, "Biological Education in the United States During the Twentieth Century," *Quarterly Review of Biology* 61 (December 1986): 481–82.

20. Arnold B. Grobman, "National Standards," *ABT* 60 (October 1998): 562.

21. Gerald D. Skoog, prepared remarks provided to the author; Larry Witham, "49 States Mandate Teaching Evolution," *WT*, April 8, 2000, A3.

22. Thomas Lord and Suzanna Marino, "How University Students View the Theory of Evolution, *JCST* 22 (May 1993): 353–54; Ganga Shankar and Gerald Skoog, "Emphasis Given Evolution and Creationism by Texas High School Biology Teachers," *Science Education* 77 (1993): 224, 228; Eugenie Scott, "Problem Concepts in Evolution: Cause, Purpose, Design, and Chance," in *The Evolution-Creation Controversy II: Perspectives on Science, Religion, and Geological Education*, ed. Walter L. Manger, *Paleontological Papers* 5 (October 1999): 179 (a third); D. Aguillard, "Evolution Education in Louisiana Public Schools," *ABT* 61 (1999): 182–88.

23. Michael Clough, "Reducing Resistance to Evolution Education," in *Investigating Evolutionary Biology in the Laboratory*, ed. William F. McComas (Reston, Va.: National Association of Biology Teachers, 1994), 15; Anton E. Lawson and William A. Worsnop, "Learning About Evolution and Rejecting a Belief in Special Creation," *Journal of Research in Science Teaching* 29 (February 1992): 144, 164–65.

24. *TAE*, 59; Scott, comment at the forum "Only a Theory: Presenting Evolution to the Public," held by the NCSE at the AAAS annual convention, San Francisco, January 16, 1989, author's tape.

25. Edward J. Larson, *Trial and Error: The American Controversy over Creation and Evolution*, rev. ed. (New York: Oxford University Press, 1989), 14.

26. Quoted by Robert M. May, "Creation, Evolution, and High School Texts," in *Science and Creationism*, ed. Ashley Montagu (New York: Oxford University Press, 1984), 307.

27. Gerald Skoog, "Creationism Has No Legitimate Place in the Science Curriculum of Public Schools," *Academy* 42 (1986): 38. For an overview, see Skoog, "The Coverage of Evolution in Secondary School Biology Textbooks, 1900–1989," in *The Textbook Controversy: Issues, Aspects and Perspectives*, ed. John G. Herlihy (Norwood, N.J.: Ablex, 1992), 71–87.

28. The Association of American Publishers report on book sales, 1998; Hermann J. Muller, "One Hundred Years Without Darwinism Are Enough," *School Science and Mathematics* 59 (April 1959): 316.

29. For Texas, see Larson, *Trial and Error*, 139; Shankar and Skoog, "Emphasis Given Evolution," 221; Wayne A. Moyer, "How Texas Rewrote Your Textbooks," *Science Teacher* 52 (January 1985): 23–27.

30. Introduction to "Statements by Scientists in the California Textbook Dispute," *ABT*, October 1972, 411.

31. Quoted in Larson, *Trial and Error*, 139–43.

32. Raymond Eve and Francis B. Harold, *The Creationist Movement in Modern America* (Boston: Twayne, 1991), 157 (italics added).

33. Ibid., 166. The 1987 survey found that 83 percent of college students who graduated from California high schools learned evolution "with creation," but it was less so with Texas students (33 percent learned both) and Connecticut students (31 percent). However, nearly all California students were taught evolution (91 percent), while more than two in ten Texas (24 percent) and Connecticut (22 percent) students were not.

34. Lawrence S. Lerner, *Good Science, Bad Science: Teaching Evolution in the States* (Washington, D.C.: Thomas B. Fordham Foundation, 2000), xiv, 12, 14; Chester E. Finn Jr. and Michael J. Petrilli, *The State of State Standards, 2000* (Washington, D.C.: Thomas B. Fordham Foundation, 2000), 34; see also *Science Framework for California Public Schools: Kindergarten to Twelve* (Sacramento: California Department of Education, 1990).

35. Interviews with Jack Greene and Nowers; author's notes and documents; parents' letter, March 17, 1997.

36. Interview with Gerald Skoog, December 1998.

37. Gerald Skoog, "The Coverage of Evolution in High School Biology Textbooks Published in the 1980s," *Science Education* 68 (1984): 127; Shankar and Skoog, "Emphasis Given Evolution," 221; *Biology Textbooks of 1990: The New Generation* (Washington, D.C.: People for the American Way, 1990).

38. Interview with Norris Anderson, December 1998; "A Mockery of Science That Trivialized Religion," editorial, *Mobile Press Register*, December 8, 1995, A14.

39. Norris Anderson, "The Alabama Insert: A Call for Impartial Science," position paper, May 15, 1996, 1.

40. Ibid., 8.

41. Interview with Norris Anderson; Anderson, "Education or Indoctrination? Analysis of Textbooks in Alabama" (Alabama Textbook Commission, December 17, 1995), 2, 4, 3.

42. Anderson, "Education or Indoctrination?" 9; Anderson, "The Alabama Insert," 9.

43. Textbook quoted in Robin Marantz Henig, "Evolution Called a 'Religion,' Creationism Defended as 'Science,'" *BioScience* 29 (September 1979): 514; and in Gerald Skoog, "Does Creationism Belong in the Biology Curriculum?" *ABT*, January 1978, 25.

44. See Henry M. Morris, *Scientific Creationism* (Green Forest, Ark.: Master Books, 1974, 1985), iii–v.

45. Interview with Jon Buell, December 1998; Todd Pruzan, "The Secret Creator," *NYT Magazine*, August 29, 1999, sec. 6, 18.

46. Jonathan Wells, *Icons of Evolution: Science or Myth? Why Much of What We Teach About Evolution Is Wrong* (Washington, D.C.: Regnery, 2000), 249–58. The examples, or "icons," Wells assessed were the Miller-Urey experiment; Darwin's tree; homology of vertebrate limbs; E. Haeckel's embryos; the "bird" skeleton of an archaeopteryx; peppered moths; Darwin's Galapagos finches; four-winged fruit flies; directional evolution of horses; and

human evolution. James Glanz, "Biology Text Illustrations More Fiction Than Fact," *NYT*, April 8, 2001, A18.

47. Helen Alexander, "For a Complete Education, Students Must Learn Evolution," *Kansas City Star*, August 7, 1999, B6; Stephen L. Carter, *God's Name in Vain: The Wrongs and Rights of Religion in Politics* (New York: Basic Books, 2000), 172.

48. Statement by Wayne Carley, August 19, 1999.

CHAPTER 9

1. Vassiliki Betty Smocovitis, "The 1959 Darwin Centennial Celebration in America," *Osiris* 14 (1999): 278, 282.

2. *AAUP Bulletin* 8 (October 1922): 405. The committee and its watch on the antievolution problem appear episodically in the *Bulletin* up to October 1929.

3. Thomas Lord and Suzanna Marino, "How University Students View the Theory of Evolution," *JCST* 22 (May 1993): 355; *SEI*, appendix tables 2–26, 2–28. These tables list natural science degrees: associate (4,000); bachelor's (70,000); master's (13,000); doctorate (10,000).

4. The Darwin Centennial marked a period when a university education in science was an indicator of likely agnosticism or atheism. See Rodney Stark, "On the Incompatibility of Religion and Science: A Survey of American Graduate Students," *Journal for the Scientific Study of Religion* 3 (fall 1963): 3–21.

5. "One Voice in the Cosmic Fugue," episode 2 of *Cosmos*. See also Carl Sagan, *Cosmos* (New York: Random House, 1980), 38.

6. Interview with Dean Kenyon, July 1996.

7. Dean Kenyon, "Teaching a Balanced View of Biologic Origins in a Secular University" (talk presented at the "Fourth International Conference on Creationism," Pittsburgh, August 6, 1998), author's tape.

8. Dean Kenyon, "Memorandum" to Jan Gregory, January 19, 1993, 1; letter from science dean James C. Kelley to Kenyon, October 21, 1992; chairman also quoted in Jan Gregory, *Punctuated Equilibrium: A Report of the Academic Freedom Committee* (San Francisco State University, June 4, 1993), 5.

9. Letter from Gregory to Kenyon, June 4, 1993; Gregory, *A Report of the Academic Freedom Committee*, 8, 13; letter from AAUP's Jonathan Knight to science dean James Kelley, November 2, 1993; Constance Holden, "'Intelligent Design' at San Francisco State," *Science*, December 24, 1993, 1977.

10. When Professor Stephen Meyer, a philosopher of science and intelligent design advocate, defended Kenyon in an op-ed piece, Discovery Institute president Bruce Chapman tapped Meyer to start up the Center for the Renewal of Science and Culture. See Meyer, "A Scopes Trial for the 90s," *Wall Street Journal*, December 6, 1993, A14.

11. Interview with Duncan Porter, April 1996.

12. Interview with David West, April 1996.

13. Richard Bambach, "The Meaning of Biotic Succession," in *The Evolution-Creation Controversy II: Perspectives on Science, Religion, and Geological Education*, ed. Walter L. Manger, *Paleontological Papers* 5 (October 1999): 23–46.

14. Interview with Richard Bambach, April 1996.

15. J. John Sepkoski et al., "Phanerozoic Marine Diversity and the Fossil Record," *Nature*, October 8, 1981, 435–37.

16. Malcolm W. Browne, "Mass Extinction of Permian Era Linked to a Gas," *NYT*, July 30, 1996, C1; Tom Waters, "Death by Seltzer," *Discover*, January 1997, 54.

17. Davis A. Young, *Christianity and the Age of the Earth* (Grand Rapids, Mich.: Zondervan, 1982), 66; Richard Bambach, "Responses to Creationism," *Science*, May 20, 1983, 851–53.

18. Interview with Paul Sattler, April 1996.

19. Quentin Shultze, "The Two Faces of Fundamentalist Higher Education," in *Fundamentalisms and Society*, ed. Martin Mary and Scott Appleby (Chicago: University of Chicago Press, 1993), 507.

20. Michael Isikoff, "Creationism in the Schools?" *WP*, May 20, 1982, B1.

21. Jerry Falwell, "LBC: The Dream and the Dilemma," *Fundamentalist Journal*, October 1984, 8.

22. Jerry Falwell, "Should Science or Religion Be Taught in the Public Schools?" (debate on CNN *Crossfire*, August 17, 1999), author's transcript.

23. Interview with Charles Detwiler, April 1996.

24. David Wells, *God in the Wasteland: The Reality of Truth in a World of Fading Dreams* (Grand Rapids, Mich.: Eerdmans, 1994), 192–93, 233n. a, b.

25. A. Pierre Guillermin, "Creationism and Biology at LBC," *Fundamentalist Journal*, October 1984, 12.

26. Interview with Vincent Synan, July 1996. Author's survey of eight of eleven Regent University theology faculty found two emphasizing "no official position" and others giving personal stances: one theistic evolution, two progressive creation, one young-earth creation, and two "mixed" of theistic evolution and progressive creation.

27. William E. Evenson, "Evolution," in *Encyclopedia of Mormonism*, ed. Daniel E. Ludlow, vol. 2 (New York: Macmillan, 1992), 478; Ronald L. Numbers, *The Creationists* (New York: Knopf, 1992), 308; interview with Gordon Hinckley, December 1996. Duane Jeffery, a zoology professor at Brigham Young University, estimated that about 40 percent of students in the required biology class were opposed to the idea of evolution, in Hannah Wolfson, "Science or Religion? Teaching Evolution the Mormon Way," AP, October 4, 1999; interview with Jeffery, May 2001. He said Mormons hold to an eternal universe, so "ex nihilo" creation is rejected, but an ancient earth is accepted.

28. Philip Sloan, "Response to Harold Morowitz, November 9, 1997, Washington D.C.," *Origins*, February 12, 1998, 568.

29. A 1996 survey of 1,000 of the 78,431 people in *American Men and Women of Science* under biology, math, physics, and astronomy produced 5 percent who agreed that "man was created 10,000 years ago." That amounts to 3,921 scientists. Survey reported in Larry Witham, "Many Scientists See God's Hand in Evolution," *WT*, April 11, 1997, A8. See also Appendix, table 3; interview with Kenyon.

30. Interview with Duane Gish, July 1999.

CHAPTER 10

1. *SEI*, 7–8; Barna Research Group, "Bible Reading in America," press release for Tyndale House Publishers, July 15, 1996; George Gallup and Jim Castelli, *The People's Religion* (New York: Macmillan, 1989), 60.

2. Larry Witham, "Churches Not Ready for New Welfare Role," *WT*, October 23, 1999, A1; interview with Martin Marty, October 1999.

3. Lutheran Church–Missouri Synod, "To Reaffirm Our Position on Creation, Fall, and Related Subjects," *Book of Resolutions: Theological Matters*, adopted in 1967; United Methodist Church, "The Natural World," *Social Principles*, 33, adopted in 1976.

4. Interview with Marty.

5. Author's tape of *Galapagos*, March 21, 2000.

6. Zahava D. Doering, *Who Attends Our Cultural Institutions?* (Washington, D.C.: Smithsonian Institution, May 1995), 4.

7. Ernst Mayr, "Museums and Biological Laboratories," in *Toward a New Philosophy of Biology: Observations of an Evolutionist* (Cambridge, Mass.: Belknap Press, 1988), 289. Only a quarter of Americans in 1994 appreciated Mayr's call for museums to be research centers. Nearly seven in ten, however, agreed with museums' educational role. These finding are in "American Museum of Natural History's Science and Nature Survey" (New York: Louis Harris and Associates, 1994), 87.

8. Mary Winsor, *Reading the Shape of Nature: Comparative Zoology at the Agassiz Museum* (Chicago: University of Chicago Press, 1991), 9–10; Owen quoted in British Museum's Web site history.

9. Interview with Mac West, May 2000.

10. "Natural History and Natural Science Museums," in *The Official Museum Directory*, 28th ed., vol. 1 (New Providence, N.J.: National Register Publishing, 1998), 1950–53. The 1995 Smithsonian study *Who Attends* found forty-one million adults who made at least one visit to a science and technology center. See also *Yearbook of Science-Center Statistics, 1997* (Washington, D.C.: Association of Science-Technology Centers Incorporated, 1997). It points to a forty million attendance range (see "Types," 7; figures 24, 21).

11. See Harris and Associates, *Survey*, which found that 65 percent of the public said that "real artifacts" interested them most, followed by models/simulations, video presentations, and computer displays (81). After animals, bones, and plants, the public ranks dinosaurs as the fourth most expected item to see at a natural history museum—with an evolution exhibit ranking fifth (86). The public most distinguishes natural history museums for their "focus on the past" (23 percent), focus on "natural occurrences" (11 percent), and the "history of life/mankind" (5 percent) (83).

12. Douglas J. Preston, *Dinosaurs in the Attic: An Excursion into the American Museum of Natural History* (New York: St. Martin's, 1986), 66.

13. Interview with Peter Crane, January 1996; Crane, "Public Understanding of Evolution" (talk presented at the AAAS annual meeting, Washington, D.C., February 21, 2000).

14. Heather MacDonald, "Revisionist Lust: The Smithsonian Today," *New Criterion* 15 (May 1997): 17–31. The article surveys the "new museology," which, like postmodernism in universities, is "honoring multiple ways of interpreting the world." Under new "gender and race equity" guidelines, Smithsonian exhibits offensive for racism, to include the Africa Hall and Human Evolution exhibit, got "dilemma labels," since funds were not available to alter exhibits.

15. Lon Tuck, "Natural Selections: 'Evolution' Comes to the Smithsonian," *WP*, May 18, 1979, C1.

16. *Crowley v. Smithsonian Institution*, 462 F. Supp. 725 (DC 1978); *Crowley*, 636 F2d 738, 743 (2nd Cir 1980). Two organizations joined Dale Crowley Jr. as plaintiffs, the National Foundation for Fairness in Education and National Bible Knowledge, Inc.; interview with William Dannemeyer, September 1996.

17. Committee for Integrity in Science Education, *Teaching Science in a Climate of Controversy: A View from the American Scientific Affiliation*, 4th ed. (Ipswich, Mass.: American Scientific Affiliation, 1993), 52. See "Addendum: Classroom Exercises," 49–63.

18. Author's notes on horse display.

19. Author's visits and tape; Sharon Begley, "Science Contra Darwin," *Newsweek*, April 8, 1985, 81.

20. Nicholas Wade, "Dinosaur Battle Erupts in British Museum," *Science*, January 2, 1981, 35–36.

21. Charles Hodge, *What Is Darwinism? And Other Writings on Science and Religion*, ed. Mark Noll and David N. Livingstone (Grand Rapids, Mich.: Baker Books, 1994), 20.

22. Stephen J. Gould, *Wonderful Life: The Burgess Shale and the Nature of History* (New York: Norton, 1989), 240–63.

23. Author's notes from New York and Washington museums.

24. Harry L. Shapiro, "The Role of the American Museum of Natural History in 20th Century Paleoanthropology," in *Ancestors: The Hard Evidence*, ed. Eric Delson (New York: Liss, 1985), 6, 7.

25. Eric Delson, "The Ancestors Project: An Expurgated History," in *Ancestors: The Hard Evidence*, 1.

26. "Old Bones Week," *Discover*, June 1984, 69; Virginia Morell, *Ancestral Passions: The Leakey Family and the Quest for Humankind's Beginnings* (New York: Simon and Schuster, 1995), 533.

27. "Darwin's Death in South Kensington," *Nature*, February 28, 1981, 735; "Darwin's Survival," correspondence, *Nature*, March 12, 1981.

28. George Gallup Jr. and D. Michael Lindsay, *Surveying the Religious Landscape* (Harrisburg, Pa.: Morehouse, 1999), 15, 95.

29. Edward J. Larson and Larry Witham, "The God Who Would Intervene," *Christian Century*, October 27, 1999, 1026–27. See Appendix, table 4, for detail. Of the 170 responses from 237 theology schools (72 percent), nearly all answered all six questions. Some theologians objected to these natural history categories: A Lutheran theologian said they sounded too "conservative Christian"; a Jesuit said they were too simple.

30. "Science and Religion Workshop," Washington [D.C.] Theological Union, June 28 to July 2, 1999. Haught spoke on the opening and closing days. Miller spoke on the last day.

31. Talks by Haught and Miller. When Americans were asked by a 1997 poll, "Is it against God's will to clone human beings?" nearly three in four said yes. See poll in Jeffrey Kluger, "Will We Follow the Sheep?" *Time*, March 10, 1997, 67–73.

32. Interviews with Protestant scholar Michael E. Dixon and Catholic liturgist Dennis McManus.

33. Lawrence Goodman, "Evolution Revolution: O'C Says Adam and Eve Possibly 'Lower Animals,'" *New York Daily News*, November 25, 1996, national edition, C3. Cardinal spokesman Joe Zwilling provided a fuller quote, "perhaps that earth," as reported in Larry Witham, "Prelate Expands on Papal Message," *WT*, November 26, 1996, A3.

34. Interdicasterial Commission, *Catechism of the Catholic Church* (Washington, D.C.: United States Catholic Conference, 1994), 98; interview with Kenneth Miller, June 1996.

35. Whitlock sermon at Biola University chapel, November 17, 1996, author's notes.

36. James L. Guth et al., *The Bully Pulpit: The Politics of Protestant Clergy* (Lawrence: University Press of Kansas, 1997), 112–13; interview with John C. Green, 2000.

37. The 1998 National Congregations Study found 18 percent self-described independents. See M. Chaves, *How Do We Worship* (Bethesda, Md.: Alban Institute, 1999), 53; the 2001 U.S. Congregational Life Survey ranks 12 percent unaffiliated. See C. Woolever and D. Bruce, *A Field Guide to U.S. Congregations* (Louisville, Ky.: Westminster John Knox, 2002), 18.

38. Lee Strobel, "Evolution or Creation" (sermon at Willowcreek Community Church, April 27, 1996), tape series no. 4.

39. Interview with Hugh Ross, November 1996; on time dilation, see Max Jammer, *Einstein and Religion* (Princeton, N.J.: Princeton University Press, 1999), 196; and Gerald L. Schroeder, *Genesis and the Big Bang* (New York: Bantam Books, 1992), 42–48.

40. Writings by Ken Ham that use the terms "Rossism" and "Rossist" are found on the Answers in Genesis Web site; Ham, "Dinosaurs and the Bible" (Florence, Ky.: Answers in Genesis USA, 1993), 1, 8, 12.

41. Ibid., 19–20.

42. *ECPE*, 5, 21; see *GNS* for an overview of Gallup polls.

43. Interview with Philip Hefner, June 1998; Hefner, "Why I Don't Believe in Miracles," *Newsweek*, May 1, 2000, 61.

44. Telephone news conference with Louis Palau, March 25, 1999; Palau with Steve Halliday, *Where Is God When Bad Things Happen? Finding Solace in Time of Trouble* (New York: Doubleday, 1999).

45. David R. Griffin, "Process Theology and the Christian Good News: A Response to Classic Free Will Theism," in *Searching for an Adequate God: A Dialogue Between Process and Free Will Theists*, ed. John B. Cobb Jr. and Clark H. Pinnock (Grand Rapids, Mich.: Eerdmans, 2000), 23; John F. Haught, *Science and Religion: From Conflict to Conversation* (New York: Paulist Press, 1995), 160, 200. Haught uses the theological terms "kenotic" for self-emptying and "apophatic" for silent awe.

46. Interview with William Hasker, August 1999; Hasker, "A Philosophical Perspective," in *The Openness of God: A Biblical Challenge to the Traditional Understanding of God*, ed. Clark Pinnock et al. (Downers Grove, Ill.: InterVarsity Press, 1994), 126–54.

47. John Wilson, "Your Darwin Is Too Large," *Christianity Today*, May 22, 2000, 54; Adrian Rodgers, "From the Chairman of the Committee on the Baptist Faith and Message," June 2000.

48. Notes written by deans on survey responses, February–August, 1999 (see full survey in Appendix, tables 4 and 5).

CHAPTER 11

1. Edward J. Larson and Larry Witham, "Scientists Are Still Keeping the Faith," *Nature*, April 3, 1997, 435–36; Michael Shermer, *How We Believe: The Search for God in an Age of Science* (New York: Freeman, 1999), 73; Atkins quoted in Roger Highfield, "Disbelief Proves to Be a Constant Among Scientists," *Daily Telegraph*, April 3, 1997, 4.

2. "NABT Unveils New Statement on Teaching Evolution," *ABT*, January 1996, 61; *TAE*, 58; James A. Leuba, *The Belief in God and Immortality: A Psychological, Anthropological and Statistical Study* (Boston: Sherman, French, 1916), 229; NAS physicist's comment in 1998 survey response (see full survey in Appendix, table 2).

3. Richard J. Goss, "The Riddle of the Religious Scientist," *American Rationalist* 39 (May–June 1994): 105; John Brockman, *The Third Culture* (New York: Touchstone, 1995); Benjamin Beit-Hallahmi and Michael Argyle, *The Psychology of Religious Behavior, Belief and Experience* (London: Routledge, 1997), 178; Albert Einstein, *Ideas and Opinions*, ed. Carl Seeling (New York: Crown, 1954), 40; on Edgar Anderson, who was a Quaker, see Vassiliki Betty Smocovitis, "The 1959 Darwin Centennial Celebration in America," *Osiris* 14 (1999): 277; Wilson first used "epic of evolution" in his 1978 book, *On Human Nature;* E. O. Wilson, *Consilience: The Unity of Knowledge* (New York: Knopf, 1998), 264.

4. Edward LeRoy Long, *Religious Beliefs of American Scientists* (Philadelphia: Westminster Press, 1952), 145; Rodney Stark, Laurence R. Iannaccone, and Roger Finke, "Religion, Science, and Rationality," *American Economic Review* 86 (May 1996): 436. Stark cites the Carnegie Commission 1969 Survey of American Academics, which found that more than four in ten faculty in math, physical science, and life science "attend regularly" a house of worship. Three-quarters of scientists surveyed belonged to a religious body in T. Vaughan, D. Smith, and G. Sjoberg, "The Religious Orientation of American Natural Scientists," *Social Forces* 44 (June 1966): 520; Einstein quoted in Max Jammer, *Einstein and Religion* (Princeton, N.J.: Princeton University Press, 1999), 49.

5. Beit-Hallahmi and Argyle, *Psychology of Religious Behavior*, 180.

6. Henry M. Morris, *Men of Science—Men of God* (San Diego: Master Books, 1982); Eric C. Barrett and David Fisher, eds., *Scientists Who Believe* (Chicago: Moody Press, 1984); Sheldon Glashow comments in "Does Ideology Stop at the Laboratory Door?" *NYT*, October 22, 1989, sec. 4, 24; Allan Sandage, "Reductionism vs. Holism," comments at the "Science and the Spiritual Quest" (SSQ) conference, Berkeley, June 9, 1998, author's tape.

7. Robert Jastrow, *God and the Astronomers* (New York: Norton, 1978), 116; John Maddox, *What Remains to Be Discovered* (New York: Free Press, 1998), 25; see also Maddox, "Down with the Big Bang," *Nature*, August 10, 1989, 425.

8. Arthur Stanley Eddington, *The Nature of the Physical World* (Cambridge: Cambridge University Press, 1928), 350. The Bridgman article, "The New Vision of Science," *Harper's*, March 1929, is collected with other writings in Percy W. Bridgman, *Reflections of a Physicist* (New York: Philosophical Library, 1955), 189.

9. Ronald W. Clark, *Einstein: The Life and Times* (New York: World Publishing, 1971), 19 (malicious, dice); Jammer, *Einstein and Religion*, 26–27 (sons); Einstein, *Ideas and Opinions*, 38 (cosmic); Jammer, *Einstein and Religion*, 124 (emotion).

10. Jammer, *Einstein and Religion*, 49 (Spinoza); Einstein, *Ideas and Opinions*, 48 (abandon, interfering, imbued).

11. Charles Darwin, *The Autobiography of Charles Darwin, 1809-1182*, ed. Nora Barlow (New York: Norton, 1963), 87.

12. Ibid., 90, 91. For a discussion of Darwin's belief, see Kenneth R. Miller, *Finding Darwin's God: A Scientist's Search for Common Ground Between God and Evolution* (New York: Cliff Street Books, 1999), 286–92. On the debate over Darwin's disbelief, see James R. Moore, "Of Love and Death: Why Darwin 'Gave Up Christianity,'" in *History, Humanity, and Evolution: Essays for John C. Greene*, ed. James R. Moore (Cambridge: Cambridge University Press, 1989), 195–229; Neal C. Gillespie, *Charles Darwin and the Problem of Creation* (Chicago: University of Chicago Press, 1970), 137. For the topic of evolution and evil, see Cornelius G. Hunter, *Darwin's God: Evolution and the Problem of Evil* (Grand Rapids, Mich.: Brazos Press, 2001).

13. Adrian Desmond, *The Politics of Evolution: Morphology, Medicine, and Reform in Radical London* (Chicago: University of Chicago Press, 1989), 412–13; Darwin, *Autobiography*, 93.

14. Moore, "Of Love and Death," 197, 209, 224; Steven Ozment, *Protestants: The Birth of a Revolution* (New York: Doubleday, 1992), 167.

15. Charles Darwin to Asa Gray, May 22, 1860, in Frederick Burkhardt et al., eds., *The Correspondence of Charles Darwin*, vol. 8, *1860* (Cambridge: Cambridge University Press, 1993), 224. Gertrude Himmelfarb observed this willingness of Darwin to explain religion by science in Himmelfarb, *Darwin and the Darwinian Revolution* (Chicago: Elephant Paperback, 1996), 386.

16. Darwin, *Autobiography*, 95; James R. Moore, *The Post-Darwinian Controversies: A Study of the Protestant Struggle to Come to Terms with Darwin in Great Britain and America, 1870-1900* (Cambridge: Cambridge University Press, 1979), 102–3; Jon H. Roberts, *Darwinism and the Divine in America: Protestant Intellectuals and Organic Evolution* (Madison: University of Wisconsin Press, 1988), 136.

17. Ronald L. Numbers, *Darwinism Comes to America* (Cambridge, Mass.: Harvard University Press, 1998), 24–48, 137–59. Of the eighty naturalists, Numbers could ascertain the beliefs of fifty-three. Thirteen were agnostic or atheists—about a fourth—and the rest covered eleven denominations, "Christian," and "theist."

18. Cyprian L. Drawbridge, *The Religion of Scientists* (London: Ernest Benn, 1932), 29, 59, 81.

19. Edwin T. Dahlberg, "Science and Religion at the Crossroads," in *Science and Religion: Twenty-three Prominent Churchmen Express Their Opinions*, ed. John Clover Monsma (New York: Putnam's, 1962), 17; David F. Noble, *The Religion of Technology: The Divinity of Man and the Spirit of Invention* (New York: Knopf, 1997), 125–42. Noble notes that the first project to launch a man in space, begun in 1959, was called Project Adam (125); Monsma, *Science and Religion*, 11; John C. Monsma, ed., *The Evidence of God in an Expending Universe: Forty American Scientists Declare Their Affirmative Views on Religion* (New York: Putnam's, 1958). The forty essayists were diverse, middle-ranking scientists: 10 at state universities; 13 at Christian colleges, and 17 in government, industry, or private sectors; 16 were in life and biological science, 3 in medicine, 3 in technology, 11 in math/physics, and 7 in chemistry; Gerald R. Bergman, "Religious Beliefs of Scientists: A Survey of the Research," *Free Inquiry* 16 (summer 1996): 41.

20. Stephen Sargent Visher, *Scientists Starred, 1903–1943, In American Men of Science* (Baltimore, Md.: Johns Hopkins University Press, 1947).

21. Leuba, *Belief in God*, 224–25, 247; Leuba, "Religious Beliefs of American Scientists," *Harper's*, August 1934, 291.

22. Leuba, *Belief in God*, 281; Leuba, "Religious Beliefs," 291, 297.

23. George F. Bishop, "Are Scientists Still Keeping the Faith?" *Free Inquiry* 19 (spring 1999): 11–12. Bishop complained that the 1914 and 1996 samples were different, the first having 40 percent greater scientists and the second having no such designations. It may be argued that the modern levels of education, professionalization, and secularization in the 1996 *AMWS* sample put it on a par with the 1914 Leuba sample, which had 40 percent "eminent" scientists.

24. Marginal note on 1996 survey response; academy member's e-mail response to Larson and Witham, "Leading Scientists Still Reject God," *Nature*, July 23, 1998, 313.

25. This spectrum came across in Long's book, even though his two-pronged approach looked at scientists who move from science to belief, and at how scientists who held a religious creed reconciled it with science; Long, *Religious Beliefs*, 150, 121.

26. Ursula Goodenough, "Theism and Non-theism," June 10, 1998, essay in response to a lecture by Pauline Rudd at the SSQ conference, Berkeley, June 8, 1998; Nathaniel Carleton, letter with the survey response, August 27, 1996.

27. Joel Primack, "Is There Common Ground in Practice and Experience of Science and Religion" (talk presented at SSQ, June 8, 1998); Jocelyn Bell-Burnell, "Science and Morality" (talk presented at SSQ, June 9, 1998).

28. Charles H. Townes, "The Distinctiveness of Being Human" (talk presented at SSQ, June 10, 1998); Allan Sandage, "Reductionism Versus Holism" (talk presented at SSQ, June 9, 1998).

29. Interview with Henry F. Schaefer III, November 1996.

30. The Galton work is summarized in Derek J. De Solla Price *Little Science, Big Science . . . and Beyond* (New York: Columbia University Press, 1986), 31–33. In 1874 Francis Galton measured eminence by Royal Society membership, a university chair, a medal, office-holding, and membership in an elite club. He found 180 men of this rank and generalized to 300, thus his 1 in 100,000 general population; Galton, *Hereditary Genius* (New York: Meridian Books, 1962); Galton, *English Men of Science* (London, 1874).

31. Price, *Little Science*, 33, 55.

32. Leuba, *Belief in God*, 254; Larson and Witham, "Leading Scientists Still Reject God," 313.

33. Anthony F. C. Wallace, *Religion: An Anthropological View* (New York: Random House, 1966), 264–65.

34. Stark et al., "Religion, Science, and Rationality," 434; Beit-Hallahmi and Argyle, *Psychology of Religious Behavior*, 183; interview with Richard Dawkins, March 1996.

35. Rees quoted in Highfield, "Disbelief," 4; interview with Ernst Mayr, April 1997.
36. Beit-Hallahmi and Argyle, *Psychology of Religious Belief*, 181–83; Leuba, *Belief in God*, 241.
37. Rodney Stark, "On the Incompatibility of Religion and Science: A Survey of American Graduate Students," *Journal for the Scientific Study of Religion* 3 (fall 1963): 8; Stark et al., "Religion, Science, and Rationality," 435.
38. Stark et al., "Religion, Science, and Rationality," 436; interview with Rodney Stark, 1999, also quoted in Edward J. Larson and Larry Witham, "Scientists and Religion in America," *Scientific American*, September 1999, 81.
39. Leuba, *Belief in God*, 250; Visher, *Scientists Starred*, 2; Visher, "Distribution of the Psychologists Starred," *American Journal of Psychology* 52 (April 1939); 280.
40. Visher, *Scientists Starred*, 4 n. 1; Bruce Alberts, "Science and Human Needs" (president's address to the NAS annual meeting, Washington, D.C., May 1, 2000), transcript.
41. Interview with Michael Ruse, June 1998; Herman C. Weber, "The 1932–1933 Edition of Who's Who and Organized Religion," in *Yearbook of American Churches, 1933*, ed. Herman C. Weber (New York: Round Table Press, 1933), 135.
42. Larson and Witham, "Leading Scientists Still Reject God," 313; Vaughan et al., "The Religious Orientation of American Natural Scientists," 524; Hewish quoted in Highfield, "Disbelief," 4.
43. Wolpert quoted in Highfield, "Disbelief," 4; Leuba, "Religious Beliefs," 299; Leuba, *Belief in God*, 279.
44. Bruce Alberts, press conference, Washington, D.C., April 9, 1999, author's tape; Larson and Witham, "Scientists and Religion," 81.
45. Long, *Religious Beliefs*, 146; interview with Long, October 2000, and correspondence; Matt Cartmill, "Oppressed by Evolution," *Discover*, March 1998, 83.
46. Interview with Arthur Peacocke, June 1998.

CHAPTER 12

1. Henry M. Morris, *History of Modern Creationism*, 2nd ed. (Santee, Calif.: Institute for Creation Research, 1993), 375.
2. Interview with Kenneth R. Miller, June 1996; Miller, *Finding Darwin's God: A Scientist's Search for Common Ground Between God and Evolution* (New York: Cliff Street Books, 2000); interview with Michael Behe after he read *Finding*, September 2000.
3. Charles Darwin, *The Autobiography of Charles Darwin, 1809–1182*, ed. Nora Barlow (New York: Norton, 1963), 106; Ronald L. Numbers, ed., *Creation-Evolution Debates* (New York: Garland, 1995), ix.
4. Toby A. Appel, *The Cuvier-Geoffroy Debate: French Biology in the Decades Before Darwin* (New York: Oxford University Press, 1987), 6, 5.
5. W. M. Smallwood, "The Agassiz-Rogers Debate on Evolution," *Quarterly Review of Biology* 16 (March 1941): 4.
6. Adrian Desmond, *Huxley: From Devil's Disciple to Evolution's High Priest* (Reading, Mass.: Addison-Wesley, 1997), 276, 278, 279, 280.
7. Interview with Arthur Peacocke, June 1998.
8. Smallwood, "Agassiz-Rogers Debate," 8, 11.
9. Numbers, *Creation-Evolution Debates*, x–xi; "S.F. Debate on Evolution Ends in 'Tie,'" *San Francisco Examiner*, June 15, 1925, 13.
10. Interview with Ernan McMullin, April 2000; interview with Richard B. Lewontin, April 1997; Chris Kazan, "'Great Debate' Draws 1,600," *Arkansas Gazette*, June 29, 1966, A1, 2; "Second Debate Draws 700," *Gazette*, June 30, 1966, A3.

11. Richard B. Lewontin, "Billions and Billions of Demons," *NYRB*, January 9, 1997, 28; interview with Lewontin.
12. Interview with Henry Morris, July 1996.
13. Interview with Duane Gish, July 1999. Gish's debate history was compiled in August 1999 by his assistant Connie Horn. Of 247 debates: 135 at U.S. colleges or universities; 30 in foreign countries; 28 debates on radio, television, or in magazine/newspaper format; 22 in churches; 11 in high school auditoriums; 9 in civic centers/community centers; 12 with unrecalled locations.
14. Duane Gish, "Creation, Evolution, and the Historical Evidence," *ABT*, March 1978, 132–40.
15. Interview with Gish; Walter L. Bradley and Roger Olsen, "The Trustworthiness of Scripture in Areas Related to Science," in *Hermeneutics, Inerrancy, and the Bible*, ed. Earl D. Radmacher and Robert D. Preus (Grand Rapids, Mich.: Zondervan, 1984), 311.
16. Interview with Gish; Marvin Lubenow, *From Fish to Gish* (San Diego, Calif.: Creation Life, 1983), 24, 256–57.
17. David H. Milne, "How to Debate with Creationists—and 'Win,'" *ABT*, May 1981, 237, 245.
18. Interview with Miller; Lubenow, *From Fish to Gish*, 230; Eugenie Scott, "Monkey Business," *The Sciences* 36 (January–February 1996): 25.
19. Dobson comment and debate in "Gish/Ross Creation Timescale Debate," transcript of the August 12, 1992, *Focus on the Family* debate. Before this, Dobson had Gish on his program in December 1985 and Ross on in 1991; interview with Hugh Ross, November 1996, and follow-up.
20. Charles B. Thaxton, Walter L. Bradley, and Roger L. Olsen, *The Mystery of Life's Origin: Reassessing Current Theories*, 2nd ed. (Dallas: Lewis and Stanley, 1992), 200; Jon Buell and Virginia Hearn, eds., *Darwinism: Science or Philosophy? Proceedings* (Richardson, Tex.: Foundation for Thought and Ethics, 1994). See also Phillip E. Johnson, "The Rhetorical Problem of Intelligent Design," *Rhetoric and Public Affairs* 1 (winter 1998): 587–91.
21. Interview with Jon Buell, December 1998.
22. Interview with Charles Thaxton, January 2001. See also Michael Polanyi, "Life Transcending Physics and Chemistry," *Chemical and Engineering News* 75 (1967): 54–66.
23. Leslie E. Orgel, *The Origins of Life: Molecules and Natural Selection* (New York: Wiley, 1973), 189.
24. Fred Hoyle, *The Intelligent Universe* (New York: Holt, Reinhart, and Winston, 1984); AP, "Space Face," in *Fort Worth Star-Telegram*, July 8, 1988, sec. 1, p. 12.
25. Interview with William Dembski, May 2000.
26. Interview with Stephen Meyer, November 1996; of the eight *Firing Line* participants, Eugenie Scott, Michael Ruse, Barry Lynn, and Kenneth Miller sided with evolution against Phillip Johnson, Michael Behe, host William Buckley, and mathematician David Berlinski. Waggish moderator Michael Kinsley closed with, "Well, I'm glad that's settled."
27. Interviews with Padian, Meyer, Behe, and Scott, December 1997; Larry Witham, "Evolving Debate on Evolution Redraws Scopes Battle Lines," *WT*, December 31, 1997, A2.
28. Larry Witham, "Contesting Science's Anti-religious Bias," *WT*, December 29, 1999, A2; interview with Dembski.
29. "Nature of Nature: An Interdisciplinary Conference on the Role of Naturalism in Science," conference program, Baylor University, April 12–15, 2000 (at www.baylor.edu/polanyi/confsched.htm); Ron Nissimov, "Baylor Professors Concerned Center Is Front for Promoting Creationism," *Houston Chronicle*, July 3, 2000, A1; Beth McMurtrie, "Committee Backs Baylor U. Center That Studies 'Intelligent Design' of Universe," *Chronicle of Higher Education*, October 27, 2000, A14.

30. "Evolution Versus Creation: John Roach Straton and Charles Francis Potter," in Numbers, *Creation-Evolution Debates*, 21–132. This anthology is comprised of debate transcripts.

31. "The San Francisco Debates on Evolution: Maynard Shipley, Francis D. Nichol, and Alonzo L. Baker," in Numbers, *Creation-Evolution Debates*, 191–362.

32. "A Debate: William Bell Riley and Harry Rimmer," in Numbers, *Creation-Evolution Debates*, 393–425.

33. Bradley and Olsen, "Trustworthiness of Scripture," 285–311; Henry M. Morris, "A Response," in Radmacher and Preus, *Hermeneutics, Inerrancy, and the Bible*, 337–47.

34. Gish-Saladin debate, May 10, 1988, author's tape; Kenneth Saladin, "Saladin-Gish Debate," *Creation/Evolution Newsletter* 8 (November–December 1988): 11, 14.

35. Johnson-Provine debate, "Darwinism: Science or Naturalistic Philosophy," April, 30, 1994, study guide and video, 1995.

36. Behe-Dennett debate at the University of Notre Dame, April 5, 1997, author's tapes.

37. Witham, "2 Physicists Debate God, Good, Evil and What If," *WT*, April 16, 1999, A1, author's notes.

38. Gould-Falwell debate, "Should Science or Religion Be Taught in the Public Schools?" on CNN *Crossfire*, August 17, 1999, author's transcript.

39. Milne, "How to Debate," 245.

40. Interview with Stephen J. Gould, April 1997; interview with Phillip Johnson, July 1996.

41. Interview with Francisco Ayala, July 1996; interview with Michael Ruse, May 1998.

CHAPTER 13

1. "He Called It as He Saw It," *Baltimore Sun*, August 13, 1999, 1E.

2. Edward J. Larson, *Summer for the Gods: The Scopes Trial and America's Continuing Debate over Science and Religion* (New York: Basic Books, 1997); Michael Williams, "At Dayton, Tennessee," Commonweal, July 22, 1925, 262; "Weird Adventures of 200 Reporters at Tennessee Trial," *Editor and Publisher*, July 18, 1925, 1.

3. The fundamentalist retreat is described in George M. Marsden, *Fundamentalism and American Culture: The Shaping of Twentieth-Century Evangelicalism, 1870–1925* (New York: Oxford University Press, 1982), 184–95; the *New Republic's* Greg Easterbrook coined the "IQ test" notion at "Can the Kansas School Board Decision Be Defended?" a forum at the Ethics and Public Policy Center, Washington, D.C., September 21, 1999; Ellen Goodman, "The New Creationists," *WP*, April 15, 1980, A17.

4. Quoted in Cynthia A. McCune, "Framing Reality: Shaping News Coverage of the 1996 Tennessee Debate on Evolution" (master's thesis, San Jose State University, 1998), 72; Lou Boccardi, "Can Journalism Be Impartial?" (remarks at a Columbia University journalism forum, December 4, 1997), see www.cjr.org/html/session1.html.

5. Robert O. Wyatt, *Worlds Apart: Gauging the Distance Between Science and Journalism, Survey Responses* (Nashville, Tenn.: First Amendment Center, 1997). Scientists think creationism should get "no" priority (69 percent) or "low" priority (25 percent) in coverage, but they expect evolution to receive "moderate" (52 percent) or "high" (22 percent) coverage. News editors and science writers give low priority to creationism, but few (3 percent) give "high" priority to evolution compared with medical research (57 percent), the environment (51 percent), genetics (25 percent), and space exploration (18 percent).

6. Edward Caudill, "A Content Analysis of Press Views of Darwin's Evolution Theory, 1860–1925," *Journalism Quarterly* 64 (winter 1987): 786.

7. Edward Caudill, "The Roots of Bias: An Empiricist Press and Coverage of the Scopes Trial," *Journalism Monographs*, no. 114 (July 1989): 2, 32, 18, 22 (this study combed nine

newspapers and six magazines and journals); Caudill, *Darwinism in the Press: The Evolution of an Idea* (Hillsdale, N.J.: Erlbaum, 1989), 110.

8. Vassiliki Betty Smocovitis, "The 1959 Darwin Centennial Celebration in America," *Osiris* 14 (1999): 286, 285, 298.

9. Ibid., 314, 312, 321–22.

10. *Epperson v Arkansas*, 393 US 97, 103 (1968). While concurring in *Epperson* because the Arkansas law was too "vague" to continue, Justice Hugo L. Black argued that antievolution did indeed promote neutrality: "Since there is no indication that the literal Biblical doctrine of the origin of man is included in the curriculum of Arkansas schools, does not the removal of the subject of evolution leave the State in a neutral position toward these supposedly competing religious and anti-religious doctrines?" See *Epperson*, 393 US at 113. For the legal politics of the case, see Edward J. Larson, *Trial and Error: The American Controversy over Creation and Evolution* (New York: Oxford University Press, 1989), 114–19.

11. Larson, *Trial and Error*, 119–20.

12. John Dart and Jimmy Allen, *Bridging the Gap: Religion and the News Media* (Nashville, Tenn.: First Amendment Center, 1993), 47, 3.

13. Peter Gwynne and Gerald C. Lubenow, "'Scopes II' in California," *Newsweek*, March 16, 1981, 67; Philip J. Hilts, "Evolution on Trial Again in California," *WP*, March 2, 1981, A1. A Nexus search of available databases from 1981 found that the total news coverage amounted to thirty-four separate stories, twenty from morning and evening cycles of the Associated Press and United Press International.

14. Philip J. Hilts, "Fundamentalists Drop Wide Attack on Schools' Teaching of Evolution," *WP*, March 5, 1981, A2.

15. Gwynne and Lubenow, "'Scopes II,'" 67.

16. Hilts, "Evolution on Trial," A1; Hilts, "Fundamentalists Drop Wide Attack," A2; Larson, *Trial and Error*, 141.

17. "Seventy-five News Organizations Register to Cover Trial," *Arkansas Gazette*, December 20, 1981, A2; Philip J. Hilts, "Creationist Tells of Belief in UFO's, Satan, Occult," *WP*, December 12, 1981, A11; Charles A. Taylor and Celeste M. Condit, "Objectivity and Elites: A Creation Science Trial," *Critical Studies in Mass Communication* 5 (December 1988): 297, 303–4. Science writer Hilts called it the first U.S. court setting that "allowed a head-on collision" between science and religion, in Hilts, "Religion Influenced Ark. Legislator Who Wrote Creationism Law," *WP*, December 14, 1981, A3.

18. Marcel C. La Follette, "Creationism in the News: Mass Media Coverage of the Arkansas Trial," in *Creationism, Science, and the Law: The Arkansas Case*, ed. Marcel C. La Follette (Cambridge, Mass.: MIT Press, 1983), 192, 194, 199.

19. Thomas F. Gieryn, George M. Bevins, and Stephen C. Zahr, "Professionalization of American Scientists: Public Science in the Creation/Evolution Trials," *American Sociological Review* 50 (1985): 392–409; Wayne Frair, "Effects of the 1981 Arkansas Trial on the Creationist Movement" (talk presented at the "Fourth International Conference on Creationism," Pittsburgh, August 7, 1998), author's tape.

20. W. Neuman, M. Just, and A. Crigler, *Common Knowledge: News and the Construction of Political Meaning* (Chicago: University of Chicago Press, 1992), 111.

21. William A. Gamson and Gadi Wolfsfeld, "Movements and Media as Interacting Systems," *Annals of the American Academy of Politics and Social Science* 528 (July 1993): 117–19 (events, deciding); Gamson, "News as Framing," *American Behavioral Scientist* 33 (November–December 1989): 157 (central).

22. La Follette, "Creationism in the News," 203.

23. Interview with Jon Buell, December 1998, Stuart Taylor Jr., "High Court Voids Curb on Teaching Evolution Theory," *NYT*, June 20, 1987, A1; Al Kamen, "7–2 Ruling Deals Blow to Fundamentalists," *WP*, June 20, 1987, A1, 13; Scalia quote in *Edwards v Aguillard*, 482 US 578, 634 (1987).

24. Lionel Van Deerlin, "Creationists Set the Stage for a New 'Monkey Trial' in Vista," *San Diego Union-Tribune*, January 1, 1993, B5.

25. AP, "Board Backs Teaching of Evolution to Avoid Legal War of Words," in *Chicago Tribune*, January 23, 1993, A3.

26. Michael Granberry, "Vista Board OKs Teaching of Creationism," *Los Angeles Times*, August 14, 1993, A1.

27. Randy Dotinga, "Wronging the Right," *Columbia Journalism Review* 33 (March–April 1995): 17–18. Dotinga reported for the *Times Advocate* of Escondido, California.

28. Constance Holden, "Alabama Schools Disclaim Evolution," *Science*, November 24, 1995, 1305; AP, "Schoolbooks to Carry Evolution Caveat," in *WT*, November 11, 1995, A2.

29. Tad Szulc, *Pope John Paul II* (New York: Scribner, 1995), 440–41.

30. Larry Witham, "Prelate Expands on Papal Message," *WT*, November 26, 1996, A3; "Evolution Compatible with Faith, Pope Rules," *Baltimore Sun*, November 25, 1996, A1.

31. McCune, "Framing Reality," 64. The two front-page headlines read, "70 Years After Scopes Trial, Creation Debate Lives," *NYT*, March 10, 1996, and "Scopes Revisited: South Puts Creationism into Classroom," *Christian Science Monitor*, March 8, 1996.

32. McCune, "Framing Reality," 87, 5, 54. McCune looked at sixty-two articles in Tennessee's three largest papers and a small daily: *Commercial Appeal of Memphis, Knoxville News Sentinel, Tennessean of Nashville*, and *Herald Citizen* of Cookeville. She interviewed eleven people, from advocates to reporters.

33. Ibid., 76, 52.

34. John Corry, *TV News and the Dominant Culture* (Washington, D.C.: Media Institute, 1986), 1; Laurence I. Barrett, "The 'Religious Right' and the Pagan Press," *Columbia Journalism Review* 32 (July–August 1993): 33; Gary Wills, *Under God: Religion and American Politics*, 2nd ed. (New York: Simon and Schuster, 1990), 101–7. Wills says Mencken made his name in 1908 with *The Philosophy of Friedrich Nietzsche*, a book that joined the "will to power" with natural selection.

35. Caudill, "A Content Analysis," 782.

36. John L. Petterson and John A. Dvorak, "Lawmakers Consider Move to Eliminate Kansas Board of Education," *Kansas City Star*, August 18, 1999, B1; William K. Piotrowski, "The Kansas Compromise," *Religion in the News* 2 (fall 1999): 10–12; interviews with Piotrowski and Mark Silk, January 2000; John Rennie, "A Total Eclipse of Reason," *Scientific American*, October 1999, 124; Stephen J. Gould, "Dorothy, It's Really Oz," *Time*, August 23, 1999, 59.

37. Jim McLean, "Evolution, Creation Should Be Taught," *Capital-Journal*, August 23, 1999, opinion page; John Hanna, "'Science Guy' Goes Ape over Ruling," AP, August 14, 1999.

38. For a national poll after the Kansas vote, see chapter 8, note 10; Melissa Bruener, CNN's *Talkback Live*, August 16, 1999. Bruener was anchor for CNN affiliate WIBU in Topeka, Kansas, author's transcript.

39. Mark Silk, *Unsecular Media: Making News of Religion in America* (Urbana: University of Illinois Press, 1995), 74; interview with Silk; James C. Petersen and Gerald E. Markle, ed. "Controversies in Science and Technology," in *Science Off the Pedestal: Social Perspectives on Science and Technology*, ed. Daryl E. Chubin and Ellen W. Chu (Belmont, Calif.: Wadsworth, 1989), 14.

40. Interview with Tim Graham, January 2000.

41. Jonathan Wells, "Ridiculing Kansas School Board Easy, but It's Not Good Journalism," reprinted in Wells, *Icons of Evolution: Science or Myth? Why Much of What We Teach About Evolution Is Wrong* (Washington, D.C.: Regnery, 2000), 330.

42. Frances X. Clines, "Creationist Captain Sees Battle 'Hotting Up,'" *NYT*, December 1, 1999, A15; interview with Phillip Johnson, December 1999. He said that "at the very moment that Robert Wright was blasting Stephen J. Gould for giving so much ammunition to [intelligent design advocate and biochemist Michael] Behe and Johnson, the *New York Times* was telling its readers that nothing has changed, this is just a re-run of 'Inherit the Wind'"; James Glanz, "Darwin vs. Design: Evolutionists' New Battle," *NYT*, April 8, 2001, A1; Phillip Johnson, "The Wedge: A Progress Report," memo, April 16, 2001.

43. Correspondence with Joseph McInerney, September 1, 2000; Robert Wright, "The Accidental Creationist: Why Stephen J. Gould Is Bad for Evolution," *New Yorker*, December 13, 1999, 56.

44. John Yemma, "Survival of the Theorists: Professors Battle over Darwin's Concept of Evolution," *Boston Globe*, October 3, 1997, A1; Sharon Begley, "Science Contra Darwin: Evolution's Founding Father Comes Under Attack," *Newsweek*, April 8, 1985, 80. Duane Gish used the *Newsweek* article in a 1988 debate with Kenneth Saladin, who objected to the news angle, author's tape.

45. Chandler Burr, "The Geophysics of God," *U.S. News & World Report*, June 16, 1997, 55–58.

46. Silas Deane, "Science and Religion Media Coverage for SSQ and AAAS Cosmic Questions." The three-page document was provided January 17, 2000, by Deane, who managed press relations for the two conferences.

47. Judy Smith, "Setting Record Straight on Kansas Board of Education," *Las Vegas Review-Journal*, August 22, 1999, 7D. A Nexus search for all 1999 stories on the Kansas school board decision found about 120 separate news reports, columns, or editorials. The total would be higher if two Kansas papers were on Nexus, as was the larger *Kansas City Star*, which ran just over 20 stories; *ECPE*, 32, 37, 26.

48. David Lack, *Evolutionary Theory and Christian Belief: The Unresolved Conflict* (London: Methuen, 1957), 110.

CHAPTER 14

1. *Edwards v. Aguillard*, 482 US 578 (1987). Oral arguments of December 10, 1986, on author's tape.

2. Texas School Board quoted in "Liberty and Learning in Schools" (Washington, D.C.: American Association of University Professors, 1986), 9; *TAE*, 57; William Jennings Bryan, "The Most Powerful Argument Against Evolution Ever," *Skeptic* 4, no. 2 (1996): 93.

3. Francis Fukuyama, "The Great Disruption: Human Nature and the Reconstitution of Social Order," *Atlantic*, May 1999, 76; Eugene B. Habecker, "Parenting in the Scriptures," *Record*, June–July 1999, 1.

4. Charles Darwin to Joseph D. Hooker, July 13, 1856, in Frederick Burkhardt et al., eds., *The Correspondence of Charles Darwin*, vol. 6, *1856–1857* (Cambridge: Cambridge University Press, 1990), 178; Loren Eiseley, *Darwin's Century: Evolution and the Men Who Discovered It* (Garden City, N.Y.: Anchor, 1961), 348; Darwin, *The Descent of Man* (Amherst, N.Y.: Prometheus, 1998), 119, 126–27.

5. Thomas H. Huxley, "Evolution and Ethics," in *Evolution and Ethics and Other Essays* (New York: AMS Press, 1970), 81, 82, 83.

6. Henry Fairfield Osborn, "Evolution and Religion," *NYT*, March 5, 1922, 16–30, in *Creation-Evolution Debates*, ed. Ronald L. Numbers (New York: Garland, 1995), 13.

7. Gertrude Himmelfarb, *Darwin and the Darwinian Revolution* (Chicago: Elephant Paperback, 1996), 409; Peter J. Bowler, *The Non-Darwinian Revolution: Reinterpreting a Historical Myth* (Baltimore: Johns Hopkins University Press, 1988), 192, 187.

8. Guenter Lewy, *Why America Needs Religion: Secular Modernity and Its Discontents* (Grand Rapids, Mich.: Eerdmans, 1996), 121.

9. George Gaylord Simpson, *The Meaning of Evolution*, 2nd ed. (New Haven, Conn.: Yale University Press, 1966), 297, 310–11.

10. Author's tour of the museum in Santee, California, July 2, 1996.

11. Interview with Philip Kitcher, July 1996.

12. Philip Kitcher, *Abusing Science: The Case Against Creationism* (Cambridge, Mass.: MIT Press, 1982), 197, 201, 198; Kitcher, *Vaulting Ambition: Sociobiology and the Quest for Human Nature* (Cambridge, Mass.: MIT Press, 1985).

13. E. O. Wilson, *On Human Nature* (Cambridge, Mass.: Harvard University Press, 1978), ix; Wilson, *Naturalist* (Washington, D.C.: Island Press, 1994), 349; Wilson, *Consilience: The Unity of Knowledge* (New York: Knopf, 1998), 266.

14. Kitcher, *Vaulting Ambition*, 435.

15. Steven Pinker, "Why They Kill Their Newborns," *NYT Magazine*, November 2, 1997, sec. 6, 52; Randy Thornhill and Craig T. Palmer, "Why Men Rape," *The Sciences* 40 (January–February 2000): 30–36.

16. "Humanist Manifesto 2000," *Free Inquiry* 19 (fall 1999): 9.

17. Richard Dawkins, *The Selfish Gene* (Oxford: Oxford University Press, 1976), 215; interview with Dawkins, March 1996.

18. William Hamilton, "The Genetical Evolution of Social Behavior I," *Journal of Theoretical Biology* 7 (July 1964): 1–16; the legendary comment was by J. B. S. Haldane; Robert Trivers, *Social Evolution* (Menlo Park, Calif.: Benjamin Cummings, 1985), 47.

19. George C. Williams, *Adaptation and Natural Selection: A Critique of Some Current Evolutionary Thought* (Princeton, N.J.: Princeton University Press, 1966); David Sloan Wilson, "The Rise and Fall and Rise and Fall and Rise of Altruism in Evolutionary Theory" (talk presented at the "Empathy, Altruism, and Agape" [EAA] conference, Boston, October 1, 1999); Williams, *Adaptation*, 438; Michael Ruse and E. O. Wilson, "The Evolution of Ethics," *New Scientist*, October 17, 1985, 52; Robert Wright, *The Moral Animal: The New Science of Evolutionary Psychology* (New York: Pantheon, 1994), 216; Michael T. Ghiselin, *The Economy of Nature and the Evolution of Sex* (Berkeley: University of California Press, 1974), 207.

20. Jeffrey Schloss, "Science and Human Nature" (talk presented at the "Life After Materialism" conference, Biola University, Los Angeles, December 3, 1999).

21. Frans de Waal, "The Communication of Emotions and the Possibility of Sympathy in Monkeys and Apes" (talk presented at the EAA conference, Boston, October 1, 1999).

22. Frans de Waal, *Good Natured: The Origins of Right and Wrong in Humans and Other Animals* (Cambridge, Mass.: Harvard University Press, 1996), 2.

23. Elliot Sober, "The ABC's of Altruism" (talk presented at the EAA conference, Boston, October 1, 1999). See also Mary Midgley, *The Ethical Primate: Humans, Freedom and Morality* (London: Routledge, 1994).

24. Interview with Ernst Mayr, April 1997; Mayr, *This Is Biology: The Science of the Living World* (Cambridge, Mass.: Harvard University Press), 262–65.

25. James Q. Wilson, *The Moral Sense* (New York: Free Press, 1993), 24, 44.

26. Ruth Benedict, considered a founder of "cultural relativism," spoke of it early on as "the fundamental and distinctive cultural configurations that pattern existence and condition the thoughts and emotions of the individuals who participate in those cultures." Quoted from Benedict, *Patterns of Culture* (Boston: Houghton Mifflin, 1959), 55; Richard Rorty, *Contingency, Irony, and Solidarity* (Cambridge: Cambridge University Press, 1989), 189.

27. Interview with Jeffrey Schloss, 2000. See also Schloss, "Seeing Things Whole," *Science and Spirit* 10 (September–October 1999): 17, where he discusses the "emergent property" research of evolutionary psychologist Henry Plotkin.

28. Robert Wright, *Nonzero: The Logic of Human Destiny* (New York: Pantheon, 2000), xii.

29. Nancey Murphy, "Evolution, Neuroscience and Human Nature" (talk presented at the "Epic of Evolution" conference, Field Museum of Natural History, Chicago, November 13, 1997), author's tape.

30. Warren S. Brown, Nancey Murphy, and H. Newton Malony, eds., *Whatever Happened to the Soul? Scientific and Theological Portraits of Human Nature* (Minneapolis: Fortress Press, 1997), xiii.

31. Ibid., 127.

32. J. P. Moreland and Scott B. Rae, *Body and Soul: Human Nature and the Crisis in Ethics* (Downers Grove, Ill.: InterVarsity Press, 2000), 9, 10, 49–85.

33. Ibid., 115. For a concise survey in support of dualism see Howard M. Ducharme, "The Image of God and the Moral Indentity of Persons," in Richard O'Dair and Andrew Lewis, eds., *Law and Religion: Current Legal Issues, vol. 4* (Oxford: Oxford University Press, 2001).

34. Karl E. Peters, "Biocultural Evolution and Human Moral Ambivalence" (draft paper for the Religion and Science Group of the American Academy of Religion, November 21, 1999).

35. Dan Batson, "Addressing the Altruism Question Experimentally" (abstract of talk presented at the EAA conference, Boston, October 3, 1999).

36. Kristen Monroe, "Explicating Altruism" (paper presented at the EAA conference, Boston, October 3, 1999), 4, 5, 7, 15.

37. Donald Browning, "Human Development, Attachment, and Love" (paper presented at the EAA conference, Boston, October 2, 1999), 2, 5, 4; see also Browning, ed., *From Culture Wars to Common Ground: Religion and the American Family Debate* (Louisville, Ky.: Westminster/John Knox, 1997), 113–24.

38. Adam Smith, *The Theory of Moral Sentiments* (Oxford: Clarendon, 1976). James Reichley explains how Smith drew on the "moral sentiments" concept from Scottish colleague Francis Hutcheson and on other individualist theories, in Reichley, *The Values Connection* (Lanham, Md.: Rowman and Littlefield, 2001), 18–21. To speak of human and animal sympathy, Darwin cites Smith's *Moral Sentiments*, especially his "first and striking chapter," and David Hume's *An Enquiry Concerning the Principles of Morals*; see in Darwin, *The Descent of Man* (Amherst, N.Y.: Prometheus, 1998), 109, 109 n. 21, 112 n. 22.

39. Interview with William Galston, October 1999. For the utilitarian value of religion, see James H. Hutson, *Religion at the Founding of the American Republic* (Washington, D.C.: Library of Congress, 1998). The preface notes how George Washington held that religion, as a source of morality, was "a necessary spring of popular government" (xii), and Hutson writes of how Thomas Jefferson "attended church services in Congress" in order to "send to the nation the strongest symbol possible that he was a friend of religion" (93).

40. *ECPE*, 49; Morris, *Scientific Creationism*, 14, 15.

41. Lewy, *Why America Needs Religion*, 87–88. On the problem of religion as mere utility, H. Richard Niebuhr said, "The instrumental value of faith for society is dependent upon faith's conviction that it has more than instrumental value. Faith could not defend us if it

believed that defense was its meaning," in Niebuhr, *The Kingdom of God in America* (New York: Harper and Row, 1937), 12.

42. Interview with Kitcher.
43. Interview with Mayr.
44. Wilson, *Consilience*, 264, 265.
45. Interview with Dawkins.
46. "Humanist Manifesto 2000," 9.
47. William J. Bennett, *The Index of Leading Cultural Indicators: American Society at the End of the Twentieth Century* (New York: Broadway Books, 1999); David G. Myers, *The American Paradox: Spiritual Hunger in an Age of Plenty* (New Haven, Conn.: Yale University Press, 2000). Both authors look at negative social trends, but Myers, denoting "the best of times, the worst of times," also extols social progress in economics, civil rights, medicine, comfort, education, and mobility.
48. Charles Murray, "Deeper into the Brain," *National Review*, January 24, 2000, 46.
49. James Q. Wilson, "Religion and Public Life," *Brookings Review* 17 (spring 1999): 38–39. Wilson develops his argument from Michael Oakeshott, *Religion, Politics, and the Moral Life*, ed. Timothy Fuller (New Haven, Conn.: Yale University Press, 1993).
50. Lewy, *Why America Needs Religion*, 114, 112, 135, 132.
51. Maynard Smith quoted in Robin McKie, "Darwin's Theories Move to Psychology," London Observer Service, in *WT*, April 20, 1998, A2; "Humanist Manifesto 2000," 10.
52. Interview with Eugenie Scott, July 1996.
53. William Jennings Bryan, "God and Evolution," *NYT*, February 26, 1922, 1–15, in *Creation-Evolution Debates*, ed. Ronald L. Numbers (New York: Garland, 1995), 1, 5.

CHAPTER 15

1. Larry Witham, "Scopes Trial Re-enactments Show Controversies Linger," *WT*, July 6, 2000, A11; interview with Edward J. Larson, July 2000.
2. In a July 18, 2000, letter, Fordham Foundation president Chester E. Finn Jr. wrote that "Dr. Wells sounds like he might be a fine choice for one of our sessions," but finally he was not included. Ron Nissimov, "'Intelligent Design' Leader Demoted," *Houston Chronicle*, October 20, 2000, A31. Dembski was demoted October 19, two days after a faculty committee acknowledged the legitimacy of the center and a day after he posted a victorious Internet news release: "Dogmatic opponents of design who demanded that the Center be shut down have met their Waterloo." As the center was dissolved, he issued another press notice: "Intellectual McCarthyism has, for the moment, prevailed at Baylor." He still had a four-year contract.
3. Public addresses, "Teaching of Evolution in U.S. Schools: Where Politics, Religion and Science Converge," Washington, D.C., September 26, 2000, author's notes.
4. Peter Beinart, "Battle for the 'Burbs," *New Republic*, October 19, 1998, 26.
5. Peter L. Berger, *A Far Glory: The Quest for Faith in an Age of Credulity* (New York: Free Press, 1992), 51, 53, 56.
6. Ronald L. Numbers, introduction, *Antievolutionism Before World War I*, ed. Ronald L. Numbers (New York: Garland, 1995), vii; Steve Bruce, Peter Kivisto, and William H. Swatos Jr., eds., *The Rapture of Politics: The Christian Right as the United States Approaches the Year 2000* (New Brunswick, N.J.: Transaction, 1995), 140.
7. Clyde Wilcox, *Onward Christian Soldiers? The Religious Right in American Politics* (Boulder, Colo.: Westview Press, 1996), 149.

8. Hal Lindsey, *The Late Great Planet Earth* (New York: HarperCollins, 1992), 43, 47; Dinitia Smith, "Apocalyptic Potboiler Is Publisher's Dream," *NYT*, June 8, 2000, A1.

9. Davis Young, "The Biblical Flood as a Geological Agent: A Review of Theories," in *The Evolution-Creation Controversy II: Perspectives on Science, Religion, and Geological Education*, ed. Walter L. Manger, *Paleontological Papers* 5 (October 1999): 130.

10. Michael Crichton, acknowledgments, *The Lost World* (New York: Knopf, 1995), 395.

11. Interview with Richard Lewontin, April 1997.

12. Daniel C. Dennett, "The Case of the Tell-Tale Traces: A Mystery Solved, a Skyhook Grounded" (talk presented at the "Darwin's Black Box" forum, University of Notre Dame, April 5, 1997), author's tape.

13. Holmes Rolston III, "Science and Christianity," in *A New Handbook of Christian Theology*, eds. Donald W. Musser and Joseph Price (Nashville: Abingdon Press, 1992), 432.

14. Planck is quoted from his *Scientific Autobiography* of 1949 by Thomas Kuhn, *The Structure of Scientific Revolutions*, 3rd ed. (Chicago: University of Chicago Press, 1970), 151.

15. Anthony Flew, "Theology and Falsification," in *New Essays in Philosophical Theology*, ed. Anthony Flew and Alasdair MacIntyre (New York: Macmillian, 1964), 96–99; Norman Macbeth, *Darwin Retried: An Appeal to Reason* (Boston: Gambit, 1971), 47.

16. Dudley R. Herschbach, comments at the forum "A Billion Seconds of *The Sciences*," New York Academy of Sciences, November 11, 1996, author's tape.

17. John Leslie, "The Anthropic Principle Today," in *Modern Cosmology and Philosophy*, ed. John Leslie (Amherst, N.Y.: Prometheus, 1998), 301 (italics in original).

18. Kenneth R. Miller, "Life's Grand Design," *Technology Review* 26 (February–March 1994): 29 (errors). For a criticism of evolutionist use of theology, see Paul Nelson, "The Role of Theology in Current Evolutionary Reasoning," in *Intelligent Design, Creationism and Its Critics*, ed. Robert Pennock (Cambridge, Mass.: MIT Press, 2001), 677–704.

19. John Horgan, *The End of Science: Facing the Limits of Knowledge in the Twilight of the Scientific Age* (Reading, Mass.: Addison-Wesley, 1996), 7; Thomas Eisner's comments came at the "A Billion Seconds of *The Sciences*" forum, author's tape.

20. Stephen J. Gould, "Justice Scalia's Misunderstanding," in *Bully for Brontosaurus* (New York: Norton, 1992), 458.

21. *SC*, 26; *TAE*, 3; "Big Biology Books Fail to Convey Big Ideas, Reports AAAS's Project 2061," press release, Project 2061, Washington, D.C., June 27, 2000; *Evolution*, PBS documentary, September 24–28, 2001. Dennett and Gould appear in part 1.

22. Interview with Lewontin.

23. *Edwards v. Aguillard*, 482 US 578, 594 (1987) (italics added).

24. Dennett, "The Case" (talk presented at the University of Notre Dame, April 5, 1997).

25. David K. DeWolf, Stephen C. Meyer, and Mark E. DeForrest, *Intelligent Design in Public School Science Curricula: A Legal Guidebook* (Richardson, Tex.: Foundation for Thought and Ethics, 1999), 16, 22, 24–25.

26. Ken Woodward, "What Miracles Mean," *Newsweek*, May 1, 2000, The *Newsweek* poll found that 84 percent of adult Americans believed God performed miracles; 48 percent testified to personal cases; three-fourths of Catholics prayed for a miracle, as did 43 percent of non-Christians and nonbelievers. Gallup polls typically find, through a sample, that more than 90 percent of Americans say they pray, and 75 percent do so daily. See George Gallup Jr. and D. Michael Lindsay, *Surveying the Religious Landscape* (Harrisburg, Pa.: Morehouse, 1999), 45–47.

27. Walter T. Stace, "Man Against Darkness," *Atlantic*, September 1948, 53–58.

28. David Baltimore, "50,000 Genes, and We Know Them All (Almost)," *NYT*, June 25, 2000, sec. 4, 17; Niles Eldredge, *The Triumph of Evolution: and the Failure of Creationism* (New York: Freeman, 2000); Phillip Johnson, "Progress Report," memo, April 16, 2001.

29. Walter L. Manger, "Examining the Creation-Evolution Issue as a Humanities Course," in Manger, *The Evolution-Creation Controversy II*, 239; G. Gallup and A. Gallup, "Teen-agers Favor Religious Explanations over Science," *Baton Rouge Advocate*, March 27, 1999, 2E; *SEI*, appendix table 7–20.

30. Interview with Duane Gish, May 1999.

31. Interview with Eugenie Scott, July 1996.

INDEX

Aristotle, 44, 49, 77, 254
Arkansas, 21, 151, 269
   balanced treatment act, 60, 61
   creation science trial (*McLean*), 61,
      84–85, 88, 90, 141, 164, 170, 180
   debates in, 215–16
   Epperson case, 230
   news media on, 232, 233
Association of Theological Schools, 190
astrobiology, 140
atheism, 57, 64, 87, 221, 229, 257
   Darwin and, 12, 201–2
   evolution and, 8, 23, 68, 71–72, 116, 126,
      187
   scientists and, 50, 198, 200, 204, 207–10
   in universities, 123, 177, 208, 209
Atkins, Peter, 198, 208,
Ayala, Francisco, 30, 39, 141
   on debates, 225–26
   life and ideas of, 88–93

Baltimore, David, 139, 269
Baltimore Sun, 227, 236, 238
Bambach, Richard, 141, 168–71
Baptists, 55, 69, 104, 114, 230
   and Bible, 120, 137, 196
   and creation-evolution, 20, 26, 33, 109,
      172, 190, 223
   history of, 19, 21, 26
   Northern, 26, 28, 93, 214
   and politics, 20, 60, 144–45, 185
   Southern, 69, 113, 137, 171, 196, 223, 262
Barth, Karl, 31
Batson, Dan, 253
Baylor University, 223, 261, 309 n.28, 316
   n.2
Behe, Michael, 8, 212
   Darwin's Black Box, 8, 127, 132, 222
   and debates, 222, 224
   life and ideas of, 127–32
Bell-Burnell, Jocelyn, 206
Benedict, Ruth, 250, 315 n.26
   Patterns of Culture, 250
Bennett, William J., 258
   Leading Cultural Indicators, 258
Berger, Peter, 262–63
Bergman, Gerald R., 203
Berlinski, David, 37
Berry, Mary Frances, 149

Bible, the
   authority of, 51, 176, 193
   cultural role of, 5–6
   literacy in, 5, 179
   as New Testament, 252–54
   as Old Testament, 12, 20, 103, 118
   societies for, 114, 243
   Ten Commandments in, 179, 237, 243
biodiversity, 25, 93, 96, 137, 139
Biola University, 50, 69, 108, 115, 192
Biological Sciences Curriculum Study, 58,
   74–79, 138, 151
   blue version textbook by, 76, 79, 156, 157
biophilia, 40
Bird, Wendell, 64, 116
Boccardi, Lou, 228
Bohr, Niels, 79, 82
Boston Society for Natural History, 17, 214
boundary theory, 42, 287 n.2
Bowler, Peter, 244
Bradley, Walter L., 51, 67, 217, 220, 224
   Mystery of Life's Origin, The, 67, 220
Brigham Young University, 176–77, 302
   n.27
British Museum of Natural History, 66,
   187, 189
Browning, Donald, 254
Bryan, William Jennings, 9, 157, 204, 216
   on human evolution, 26, 242, 244, 259
   and Scopes trial, 26, 229, 231, 261
Bryan College, 52–53, 103–4
Bube, Richard, 49
Buell, Jon, 220–21
Burhoe, Ralph, 28–29, 35
Bush, George, 134
Bush, George W., 136, 137, 140
Bush, Vannevar, 283 n. 26

California
   board of education, 63, 154, 155
   court battles, 231–32
   and creationists, 154
   and evolution in science standards, 155
   natural history museum, 185, 187
   science framework, 63, 109, 154, 231
   textbooks, 154, 155
Calvin College, 72, 117–20
Calvin, John, 55
Calvin, Melvin, 163

Cambrian explosion, 159, 169, 187–88, 190
    creationist view of, 50, 217
Campus Crusade for Christ, 69, 206, 216
Carley, Wayne W., 71, 161
Carlton, Nathaniel, 205
Carnegie, Andrew, 245
Carnegie Museum of Natural History, 181
Carter, Jimmy, 19, 137, 230
Carter, Stephen, 161
Cartmill, Matt, 211
catastrophism, 20, 112, 158
    evolution by, 18, 99–100, 184
    of the Flood, 105–6, 224
Catholicism, Roman
    anti-evolutionists in, 36, 116, 130, 145
    and encyclicals on evolution, 177
    on evolution, 36–37, 55, 177, 192, 236, 254
    neo-Thomism in, 27–28, 48–49
    scientists in, 8, 78, 91, 129–30, 165, 173,
        202–3, 212, 215
    and theistic evolution, 48–49
    views at seminaries in, 190
    See also Humani Generis; papacy
Cattell, J. McKeen, 203, 206–7
Caudill, Edward, 228–29, 241
Center for the Renewal of Science and
    Culture, 69, 165, 301 n.10
Center for Theology and the Natural Sci-
    ences, 35, 47
Chambers, Robert, 46
    Vestiges of the Natural History of Cre-
        ation, 46
chaos theory, 39, 46, 48, 222
Chicago Statement on Biblical Hermeneu-
    tics, 51–52
Chopra, Deepak, 56
Christian Heritage College, 115
Christianity, 17, 35, 42, 187, 199, 204
    Darwin and, 201
    See also Catholicism; Protestant
Church of Jesus Christ of Latter-day
    Saints. See Mormon
Clinton, William J., 75, 134, 140, 148
Cloud, Preston, 109–10
Committees on Correspondence, 60, 62
Conference on Macroevolution, 38, 99
Conference on Science, Philosophy and
    Religion, 27
Congregationalists, 31, 55, 193, 288 n.17

Conklin, Edwin G., 28
constitutionality, 232
    and anti-evolution, 154, 230, 237
    and creationism, 61, 64, 242
    as neutrality, 116
constructivism, 56
Copernicus, Nicholas, 84, 92, 123–24
Cosmic Questions conference, 225, 241
Cosmology and Teleology conference, 47
Cosmos, 122, 163, 215
Cracraft, Joel, 38, 140
Crane, Peter R., 183–84
Creation Research Society, 33, 55, 115, 160,
    170, 216
creation science, 20, 42, 52, 79, 137, 158, 185
    in schools, 43, 59, 64, 85, 149, 231
creationism
    Darwin on, 49
    day-age, 32, 33, 50, 223, 224
    definition of, 42, 49–53
    and dispensationalism, 27, 263, 115
    gap theory, 32
    and physical death by sin, 52, 53, 194
    progressive, 33, 49–52, 54, 121, 190, 223,
        224
    as special creation, 14, 46, 47, 51, 120, 158,
        173, 190–91
    speed of light in, 53
    as theistic evolution, 47, 49–50, 62, 126
    in theology schools, 190–91
    and time dilation, 194
    See also flood geology; young earth cre-
        ationism
Crichton, Michael, 8, 264
    Jurassic Park, 182
Crick, Francis, 50, 131
Cuvier, Georges, 99, 213

Dana, James Dwight, 153
Dannemeyer, William, 185
Darrow, Clarence, 9, 26, 34, 229
Darwin, Charles, 4, 5
    on creationists, 49
    Descent of Man, The, 18, 243, 255
    Essay of 1844, 14–15
    finches and, 32, 108, 241
    and HMS Beagle, 14, 15, 100, 201
    letters by, 11, 12, 15, 202, 213
    modern image of, 12–13, 201

molecular clock in, 89, 139
peripatric speciation in, 80
population genetics and, 29–30, 101
as punctuated equilibrium, 37–38,
94–95, 173
as a religion, 27, 67, 148
stasis in, 37, 93, 95
theology schools and, 190
ultra-Darwinism and, 38, 95
as unfinished synthesis, 39, 95
See also genetics; natural selection; New
Synthesis; species; teleology
Evolution, Science, and Society, 141–42
evolutionary psychology, 7, 250, 254

fairness, 58, 148–49, 290 n.3
See also Arkansas, balanced treatment
act
Falkner, Danny, 52–53
Fall (original sin), 52, 53, 191, 192, 243, 254
Falwell, Jerry, 19–20, 114, 140
creationist advocacy by, 60, 61, 172
and debates, 60, 225
Federalist Papers, 255
Field Museum of Natural History
(Chicago), 38, 101, 183–84
bulletin of, 101
evolution in, 184
Finkelstein, Louis, 27
Firing Line, 57, 70, 222
Flew, Anthony, 221, 265
flood geology, 20, 33–34, 50, 116, 264
Fosdick, Henry Emerson, 26
Foundation for Thought and Ethics, 220
Frair, Wayne, 232
Fukuyama, Francis, 243
Fuller Theological Seminary, 42, 219, 252
fundamentalism, 4, 8, 31, 69, 92, 189,
233–35
Fundamentals, The, 26–27
history of, 26, 32, 116, 137, 227
leaders of, 115, 172, 223
and science, 26, 54, 112, 148, 263

Galapagos Islands, 14, 32, 108, 180
Galilei, Galileo, 9, 122, 156, 236
Galston, William, 255
Galton, Francis, 17, 206–7
Hereditary Genius, 206

Gamson, William, 17, 206–7
Gans, Herbert, 233
Gates, Bill, 9
Genesis, 4, 5
church battles over, 51–52, 113, 118
in church worship, 191–92
as creation myth, 97, 259
"days," 32, 50, 223, 224
on natural events, 33, 103, 106, 113, 118,
173–74, 176, 215
in schools, 194, 230, 235, 236, 255
genetics (DNA)
as beanbag, 29
design in, 9, 75, 126, 129, 131, 159, 160, 221,
224, 269
electrophoresis used in, 30, 39, 129, 171
as evidence of evolution, 93, 129, 186, 210
and evolution of phenotypes, 7, 40, 82,
139
and evolutionist careers, 30
gene diversity in, 29, 89, 91, 139, 159
gene frequency in, 4
gene transfer in, 76
and hopeful monsters, 44
human genome and, 75, 88, 240
Human Genome Project and, 138–39
mathematics and, 37
and medicine, 78
molecular biology in, 13
mutations in, 29, 37
population genetics in, 29–30, 101
public understanding of, 7, 179
religion and, 7, 75, 269
specified complexity in, 221
George, Robert, 147–48
Ghiselin, Michael, 248
Gillespie, Neal C., 43, 201
Gingerich, Owen, 67
life and ideas of, 122–26
Gish, Duane, 51–52, 64, 69, 178, 270
debates of, 59–63, 89, 216–19, 224
God of the gaps, 9, 68, 220, 224, 291 n.27
Godfrey, Laurie R., 101
Geoffroy Saint-Hilaire, Etienne, 213
Goldschmidt, Richard, 44
Goodenough, Ursula, 205
Goodwin, Brian, 46
Gore, Albert, 137
Earth in the Balance, 137

Gould, Stephen J., 85, 97, 104, 150, 238
  on contingency, 45, 240
  on debates, 172–73, 225
  on origins, 267
  and Phillip E. Johnson, 67, 68, 72
  and punctuated equilibrium, 37–38,
    94–95, 100
  on religion, 4, 47, 88, 137, 236
  Rocks of Ages, 47
  on sociobiology, 87
  Wonderful Life, 45, 188
Graham, Billy, 33
Graham, Peter, 12–13
  Portable Darwin, The, 13
Graham, Tim, 239
Gray, Asa, 17, 126, 159, 202, 214
  on Christian Darwinism, 43, 47
Green, John, 193
Griffin, David R., 195
Grobman, Arnold B., 151
Gross, Paul R., 56, 65, 262
Guth, James, 142–43

Haeckel, Ernst, 161
Ham, Kenneth, 194
Hamilton, William, 247
Hasker, William, 196
Haught, John, 190–91, 195–96
Haynes, Charles, 150
Hearn, Walter, 60, 62, 110
Hefner, Philip, 28, 55, 195
Hennig Society, 187
Herschbach, Dudley R., 266
Hewish, Anthony, 210
Himmelfarb, Gertrude, 244
Hobbes, Thomas, 244
Hodge, Charles, 187
Hoyle, Fred, 123, 221
  Intelligent Universe, The, 221
human evolution, 100, 137, 139, 209, 224,
  242
  Ancestors exhibit on, 188, 190
  in museums, 188, 242
  Pleistocene epoch and, 248
  and science standards, 151
  in textbooks, 153, 230
Human Genome Project, 34, 88, 138–40
human nature, 241, 242–47, 249–55,
  257–58

Founders' view of, 255
Humani Generis, 36, 177
humanism
  and ethics, 247, 257
  manifestos of, 27, 157, 247
  religious, 27
  secular, 142, 185, 231
Humanist Manifesto 2000, 247, 257
Hume, David, 202, 255
Hutton, James, 267
Huxley, Julian, 27, 114, 163, 229
Huxley, Thomas Henry, 21–22, 228
  battle with church, 16–17, 211
  debates of, 61, 212–14
  on ethics and morals, 244, 246–47, 249
  on jumps in evolution, 44

immortality, 204–5, 207, 252
Institute for Creation Research, 20, 194,
  219
  accreditation of, 117, 172
  description of, 53, 63, 112, 117
  legislative work of, 59, 116, 160, 234
Institute for Theological Encounter with
  Science and Technology, 36
Institute on Religion in an Age of Science,
  28, 86
intelligent design, 126, 240, 264, 266
  as anti-Darwinian, 8, 23, 111, 160–61
  Christian criticism of, 47, 104–5, 121–22,
    124
  conferences on, 50, 52, 69, 86, 101, 239
  and explanatory filter, 51
  and irreducible complexity, 51, 127
  as a movement, 67, 69, 219–23, 261–62,
    269
  in schools, 149–50, 165, 262, 268
  as science, 51, 67, 70, 127, 149–50, 268
International Conference on Creationism, 52, 103

Jastrow, Robert, 200
Jefferson, Thomas, 14, 18–19, 133, 255
Jesus Christ, 48, 57, 70, 256
Jewish Theological Seminary, 27
Johnson, Phillip E., 86, 101, 131, 239–40,
  269
  Darwin on Trial, 65, 67–68, 131, 220, 225
  and debates, 57, 220, 224, 225

life and ideas of, 65–72
on naturalism, 8, 70, 73
and Stephen J. Gould, 67, 68, 73
Journal of Molecular Evolution, 128–29

Kansas, 55, 105
news media on, 7, 227, 238–39
school board vote, 137–38, 143–44, 161, 168
science standards of, 143–44
Kauffman, Stuart, 46
Kennedy, John F., 136
Kenyon, Dean, 162, 169, 221
academic freedom case of, 163–66, 177
Biochemical Predestination, 163
Of Pandas and People, 160, 164, 234
Kinsley, Michael, 57, 309 n.26
Kitcher, Philip, 245–47, 256
Abusing Science, 245–46
Vaulting Ambition, 245–46
Kuhn, Thomas, 34, 39, 84
Structure of Scientific Revolutions, The, 84
Kulp, Laurence J., 285 n.26

La Follette, Marcel, 232–33
Lack, David, 79
Darwin's Finches, 241
on religion, 32, 241
LaHaye, Tim, 115, 263
Lane, Neal, 140
Larson, Edward J., 153, 230, 261
Leakey, Mary, 189
Leakey, Richard, 137
Lee, Peter, 19
Leibnitz, Gottfried, 206
Lerner, Lawrence, 261
Leuba, James, 197, 203–5, 207–10
Lewontin, Richard, 83, 90
and creationists, 215–16
on evolution's future, 264, 267
on science funding, 141
Lewy, Guenter, 258–59
Liberty University, 144, 166
debates at, 60
science at, 171–75
Lindsey, Hal, 263
Late Great Planet Earth, The, 263
Long, Edward LeRoy, 28, 31, 55

on beliefs of scientists, 199, 203, 205, 211
Louisiana, 43, 116, 242
disclaimer case, 7
Edwards v. Aguillard, 64
Supreme Court rulings on, 7, 64, 148, 149, 233–34, 267
Luther, Martin, 55, 202
Lutherans, 55, 81, 158, 180, 185, 190, 195
Lyell, Charles, 15, 99
Principles of Geology, 15

Macbeth, Norman, 265
Maddox, John, 83–84, 132, 139, 200
Malthus, Thomas, 15
Man: A Course of Study, 135
Manger, Walter L., 269
Margulis, Lynn, 39, 45–46
Marsh, Frank, 33
Evolution, Creation, and Science, 33
Marty, Martin, 180, 189, 196
Massachusetts Institute of Technology, 16–18, 214, 232, 240, 246
Mayer, William, 58–60, 151
Mayr, Ernst, 87, 94, 97, 108, 181
on cladistics, 94
and ethics, 249–50, 256
life and ideas of, 79–84
and New Synthesis, 29, 30, 101
on religion, 5, 32, 44, 208, 256
McCosh, James, 120
McCune, Cynthia, 237
McGinnis, John O., 143
McInerney, Joseph D., 240
life and ideas of, 74–79
McLean v. Arkansas. See Arkansas, creation science trial
McMullin, Ernan, 215
Mencken, H. L., 7, 227, 238
Mendelsohn, Everett, 42
Mere Creation conference, 50–51, 69, 192
Meyer, Stephen C., 50, 66–68, 149, 221–22
Miller, James B., 191
Miller, Jon, 135
Miller, Kenneth R., 127, 212, 268
Biology, 111
and debates, 212, 218–19
Finding Darwin's God, 5, 212
on religion, 5, 49, 192
Miller, Stanley, 163

naturalistic fallacy, 257
Nature of Nature conference, 223
Nelkin, Dorothy, 112–13, 135
Nelson, Paul, 178, 222
Nemesis, 100
Neuhaus, Richard John, 236
neuroscience, 251–52, 260
New Age, 53, 56, 117
New Class theory, 262
new Christian right, 19–20, 59, 116, 145
New Synthesis, 29, 79–81, 99, 108
    challenges to, 38–39, 45–46, 100–1
    impact on biology careers, 30
    See also Dobzhansky, Theodosius;
        Mayr, Ernst
New York Academy of Sciences, 266
news media
    on Alabama school board, 235
    on California trial, 230–31
    on Darwin Centennial, 229–30
    on Galileo, 236
    and news frames, 233, 236
    on pope and evolution, 236
    on science-religion, 241
    on Scopes trial, 227–34, 237–39, 241
    on Supreme Court ruling, 233–34
    on Tennessee law, 237
    on Vista school board, 234
    See also Arkansas; Kansas
Newsweek, 19, 47, 100, 187, 195, 230, 231,
    240
Newton, Isaac, 79, 124, 199, 202, 265
Nichol, Francis D., 223
Niebuhr, Reinhold, 31
Noah's Ark, 103, 116, 173, 194, 224
Noll, Mark, 55–56, 142
nonreductionist physicalism, 251
Nord, Warren, 147–48, 150, 262
North American Paleontological Conven-
    tion, 43
Notre Dame University, 36, 71, 215, 221, 224
    science taught at, 177
NOVA, 123, 126, 130, 143
Novik, Jack, 85
Nowers, William, 145, 156
Numbers, Ronald, 4, 49, 54, 113, 202, 213,
    263
    Creationists, The, 113
Nye, Bill, 7, 238

O'Connor, John, 192
Olsen, Roger, 220
Osborn, Henry Fairfield, 26, 183
    on evolution, 30–31, 188, 244
    and William Jennings Bryan, 26, 259
Overton, William R., 61–62, 85, 232

Padian, Kevin, 63–64, 222
Palau, Louis, 195
Paleobiology, 168
Paley, William, 65, 76, 121, 131, 201
    Natural Theology, 131, 201
papacy
    John Paul II, 36, 49, 91, 192, 236
    Pius XII, 177, 192
    Urban VIII, 9, 236
Papineau, David, 4
Patterson, Colin, 66, 96
Peacocke, Arthur, 40, 48, 211, 214
Pennock, Robert T., 8
    Tower of Babel, 8
People for the American Way, 147, 261
    polls on evolution, 54, 55, 149, 241
Peters, Karl, 253
physics
    and biology, 29, 46, 82–83, 87, 163,
        220
    century of, 29
    and God, 51, 194, 199, 210
    and number of physicists, 29
    in science-religion, 28, 36, 50, 47, 35,
        48
Pigliucci, Massimo, 71–72
Plantinga, Alvin, 71, 221
plate tectonics, 11, 77, 96, 106, 112
Polanyi, Michael, 220
Polanyi Center, 223, 261
politics and evolution, 133–46
Polkinghorne, John, 47–48, 225
    Scientists as Theologians, 48
Pollard, William, 28
Popper, Karl, 34–35
Porter, Duncan, 12–14, 167, 172
    Portable Darwin, The, 13
postmodernism, 9, 53, 55
Price, George McCready, 32
Price, Derek J. De Solla, 207
Primack, Joel, 205
process philosophy, 31–32, 35, 195–97